Production Metrology

by
Prof. Dr.-Ing. Tilo Pfeifer
RWTH Aachen

Oldenbourg Verlag München Wien

Die Deutsche Bibliothek - CIP-Einheitsaufnahme

Pfeifer, Tilo:
Production Metrology / von Tilo Pfeifer. – München ; Wien : Oldenbourg, 2002
  Dt. Ausg. u.d.T.: Pfeifer, Tilo: Fertigungsmesstechnik
  ISBN 3-486-25885-0

© 2002 Oldenbourg Wissenschaftsverlag GmbH
Rosenheimer Straße 145, D-81671 München
Telefon: (089) 45051-0
www.oldenbourg-verlag.de

Lektorat/Editor: Sabine Ohlms
Herstellung/Producer: Rainer Hartl
Gedruckt auf säure- und chlorfreiem Papier
Druck/Printing: R. Oldenbourg Graphische Betriebe Druckerei GmbH

Preface

On the basis of a quality understanding, which increasingly places the quality and economy of all processes of the production chain at the forefront, the use of production metrology and the consequential utilisation of quality data is continually gaining in importance.

The book is based upon the second revised edition of the german book "Fertigungsmesstechnik". It is identical in its contents and the established strusture of the book has been maintained.

This book is the background for new master studies in relation to mechanical engineering at German universities, specially directed to guide the masters degree in production engineering at the RWTH in Aachen, taught in English. Furthermore, the book is of great help for the engineers already working in international companies related to mechanical engineering, specially in the English-speaking countries.

I would like to take this opportunity to thank my colleagues, which have made possible the translation of this book. I would like to name individually:

Ubaldo Aleriano Sanchez, Mafred Benz, Sascha Driessen, Gerd Dussler, Olga Garcia, Michael Glombitza, Frank Mönning, Andreas Napierala and Domonic Sack.

I also want to thank Dorothea Mc Kay, who was of great importance for the translation of the book.

My special thanks to Robert Münnich, who has been the co-ordinator of the whole translation.

Aachen   Prof. Dr.-Ing. Dr. h.c. Prof. h.c. T. Pfeifer

Preface to the second german edition

On the basis of a quality understanding, which increasingly places the quality and economy of all processes of the production chain at the forefront, the use of production metrology and the consequential utilisation of quality data is continually gaining in importance.

The great interest in this topic of industry, research and science is reflected in the brisk demand of this book, which is now, only three years after its first publication, published in a revised second edition. The established structure of the book has been maintained and supplemented by current topics. Particularly optical measuring and inspection technology has increasingly moved into practical application over this short period of time. Therefore, relevant application examples have been included in the respective chapters.

I would like to take this opportunity to thank everybody for their helpful comments, which have been included in this second edition. Furthermore, I would like to thank my colleagues who have been working on this second edition. I would like to name individually:

Alexander Bai, Manfred Benz, Frank Bitte, Benno Bröcher, Gerd Dussler, Sascha Driessen, Dirk Effenkammer, Jörg Feldhoff, Michael Glombitza, Ingo Krohne, Frank Lesmeister, Andreas Napierala, Dominik Sack, Karsten Schneefuß and Michael Zacher.

Lastly, my special thanks go to Mr Feldhoff, who again supervised the coordination of the publication.

Aachen, May 2001   Prof. Dr.-Ing. Dr. h.c. Prof. h.c. T. Pfeifer

Preface to the first german edition

"To learn from one's mistakes" is not only a core requirement of modern quality management but also the maxim of every company, which is interested in economic success and securing its competitiveness.

To see errors, or deviations, of product or process critical characteristics as an opportunity and as a basis for improvements, requires that the respective dimensions are reliable, i.e. taken with a high rate of precision, dynamics and robustness at the correct place and at the correct point in time. Therefore, suitable measuring and testing methods and process-capable sensors, as well as regulators and actuators must be available and integrated into the complex scenario.

In order to realise the strategy of error avoidance by feeding back the corrective dimensions derived from measurement signals, it is, however, also mandatory that unambiguous correlations are developed between the quality features of a product and the parameters of the processes involved which originated them. Ultimately, powerful algorithms and regulatory or control mechanisms are required, which stabilise the quality features of the relevant parameters within the permitted tolerances, using the measured deviations of monitored products and process parameters by single or multi-sized control devices. On the whole, it is a complex task which requires a profound knowledge of production metrology, signal processing, control and communication technology.

We say today that the knowledge society is a decisive condition for our future development and for coping with the tasks that lie ahead. We also speak in this context of "knowledge management". However, before we can manage our knowledge, it must to be imparted. The aim of this manual is to contribute to this by focusing on the area of production metrology as a key technology for controlling efficient production systems. It is the result of my lecture at the RWTH Aachen and is the joint effort of a dedicated team of young scientists and industry experts, who have contributed their basic knowledge as well as their experience from numerous research and development projects and a multitude of industrial applications to the book.

The history of origin also provides an understanding of the manual's target group, which includes students at universities and technical colleges studying mechanical engineering with the main focus on metrology and automation technology, as well as practicing technicians and engineers within industry.

I would like to thank all of the experts from science, research and industry whose helpful and useful contributions on current developments and applications for production metrology were incorporated in the contextual structure of the book. My thanks go to my colleagues, who have made this book possible through their personal commitment. I would like to name the following:

Frank Bitte, Benno Bröcher, Dirk Effenkammer, Jörg Feldhoff, Christian Glöckner, Michael Glombitza, Jürgen Großer, Dietrich Imkamp, Stefan Koch, Frank Lesmeister, Stefan Meyer, Horst Mischo, Peter Scharsich, Peter Sowa, Dietmar Steins, Harald Thrum and Lorenz Wiegers.

Mr Feldhoff has again earned my special thanks for coordinating the overall publishing of the book.

Aachen, April 1998   Prof. Dr.-Ing. Dr. h.c. Prof. h.c. T. Pfeifer

Content

1 Introduction

Production metrology is the generic term for all activities connected with measurement and testing functions to be provided in the industrial development process of a product. This global definition results from changed production conditions, a high level of automation, short product life span, a reduced vertical range of manufacture and increased demands on product quality. These aspects form the functions and objectives of production metrology. It has developed from being a pure checking procedure to an important component of quality management.

1.1 Functions and Objectives of Production Metrology

The central function of production metrology is the recording of an object's quality criteria measurements. An object of measurement is often a workpiece, however it can also be a tool, a machine or even a measuring device within the scope of test equipment monitoring. .

The concept of production metrology is closely connected with the concept of product testing (Section 2.1.2), where it is established whether a characteristic of an object meets set requirements. Testing can be aided by measurements, which comprise the fundamental purpose of production metrology. In industrial production the concept of product testing is often brought to the forefront, which is expressed, for example, by the introduction of terms such as "test planning" and "test data recording". Even it if is specifically pointed out that the functions of production metrology go far beyond testing, the common terms are used here in connection with the concept of "testing".

The measurable quality criteria of a product have various parameters. Fundamentally, a differentiation is made between criteria that relate to material attributes, the geometry and the function of a product (**Figure 1.1-1**).

The objective of *material testing* is the ascertainment of material parameters such as hardness or the elasticity module, as well as the evaluation of macro structures such as cracks or textures.

Material Testing	Operational Testing	Geometric Testing
Fissure	Force	Shape
Texture	Moment	Dimension
Hardness	Rotation Speed	Location
E-, G- Module	Noise	Roughness

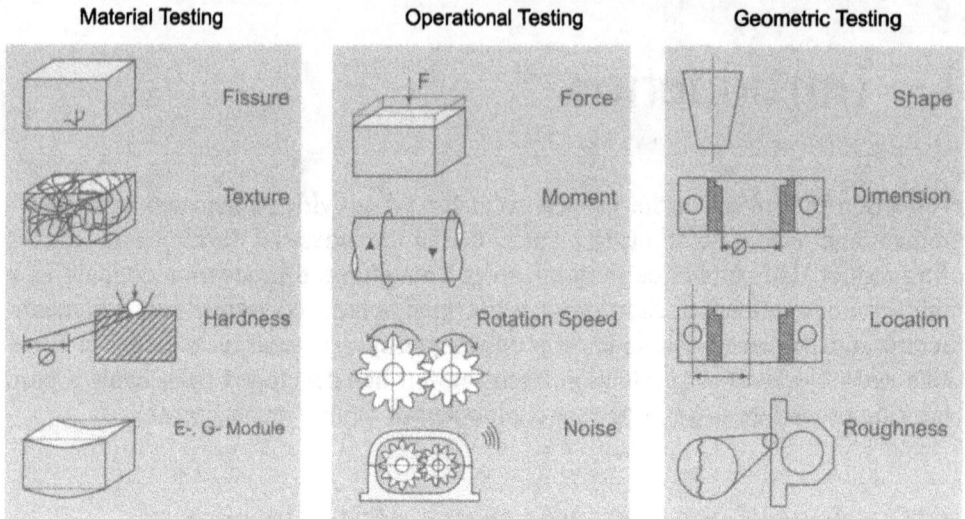

Figure 1.1-1:   Testing and measurement functions within production metrology

This type of metrology application primarily applies during the initial inspection or after treatment of materials such as tempering [Blu 87], [Ste 88].

As a rule, the *operational test* involves testing the functionality of the entire product and is therefore predominantly found at the end of a process chain, for example, in the final testing stage. The spectrum of functions range from manual visual inspection by a human tester to automatic electronic systems which, for example, allow a conclusion to be drawn about the quality of gears through the measurement of noise from gear wheels [Pf 96b].

The most common test within production metrology, with a proportion of nearly 90%, is the *geometric testing* . of workpiece properties [Dut 96]. In addition to the measurement of shape, mass or location of geometric elements, surface texture is often also an important workpiece component for the operational efficiency of a future product.

Due to the great importance of "geometric testing", the main focus of this book is dedicated to this field of application. This is where dimension, location, shape and position parameters are fundamentally established, such as characteristic values for surface texture.

The field of production metrology is, however, not constrained to measurement and testing procedures. Production metrology faces new challenges, which are marked by changed production conditions on the one hand and increased quality demands on the other.

The production environment is characterised by a high level of automation, short product life span and a declining vertical range of manufacture. The high degree of automation has also led to the automation of many measurement functions and the accompanying reduction in cycle times has made it necessary to accelerate measurement procedures. The shorter product life span requires a continual adjustment of measuring devices to the changed test criteria. The proportion of supplied parts from sub-contractors increases through the declining vertical range of manufacture. In order to minimise expenditure by the customer in testing these parts, it is necessary that the sub-contractor's measurements be reliable. A fundamental condition for this is the unrestricted comparability of measurement results. It would only take one incident of the customer's measurement results differing from that of the sub-contractor to seriously disturb mutual trust.

These technical measures alone, however, are not sufficient to fulfil the demands on production metrology, especially under the influence of higher quality expectations by the market. The higher expectations of quality are expressed, on the one hand, in terms of shrinking tolerance levels and on the other hand, in terms of lower error rates. Nowadays the accepted error rates lie in the ppm (parts per million) region.

Increased demand on the quality of workpieces

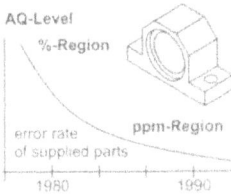

AQ-Level
%-Region

error rate
of supplied parts   ppm-Region

1980   1990

Not fulfillable with post-process measurement technology !

Quality improvement through controls accompanying the process
Minimisation of error rate within the process
Preventative sensors near to the process
Process monitoring through post-process measurement technology

Increased demands on the integration of dependable quality data

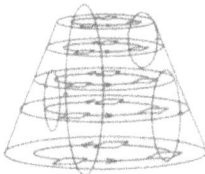

Archiving of measurement results does not create improvement !

Construction of small and large quality control loops
Intervention in negative process trends and processes
Use of suitable machines for the processes
Process orientated product construction

Figure 1.1-2: Demands on production metrology

In order to meet these demands, production metrology is confronted with the task of contributing to the continual improvement of the product development process.

Controls which accompany a process are constructed by feeding back production measurement data to those who are responsible for production. Information is sent back to planning areas in the form of large control loops spanning various levels. In this way, production metrology supports the early phases of product development by defining product criteria relevant to quality and by planning the acquisition of their measurements. The rules of thumb, in the form of measurement data archives and practical experiences of production metrology, aid in the choice and layout of production processes, such as process orientated construction (**Figure 1.1-2**).

Through the targeted recirculation of measurement data, which is obtained during and after the production process, production metrology provides the production area with information about deviations from target conditions. This becomes the starting point for all quality improvement measures and forms a substantial part of the quality loop [Pf 96]. It accompanies all phases of product development, from product planning to sales. (**Figure 1.1-3**).

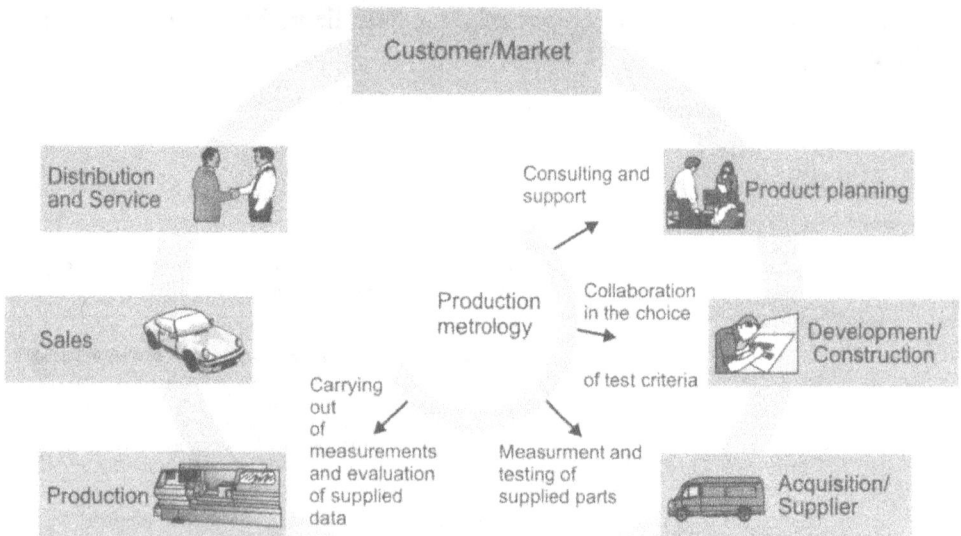

Figure 1.1-3 Production metrology as part of the quality loop

This justifies the definition, which describes production metrology as a generic term for all activities connected with measurement functions to be provided in the industrial development process of a product. This definition goes far beyond the previous understanding of production metrology, which describes it as measurement technology in industrial production [Dut 96].

1.2   Historical Development of Production Metrology

Long before the Age of Industrialisation, it was already necessary to compare geometric sizes of different goods. As a direct comparison was not always possible, geometric dimensions needed to be defined and suitable dimensional representations produced. In Egypt in 4000 B.C., the dimensions of a person, for example the pharaoh, were used as a base factor. This type of definition was used right into the Middle Ages and even beyond. Typical measurements were finger widths, hand widths, feet, cubits and steps [Dut 96]. The definition of these various terms was by no means always clear cut.

In the 16th and 17th centuries, several proposals were made to unify the definitions for units of length. During the French Revolution, the proposal to derive a unit of length from the circumference of the earth was adopted. Based on the underlying geodetic measurements, a first primal meter was produced [Dut 96], [Lot 88]. This was the starting point for the future development of meter definition (Section 2.1.1).

One of the oldest surviving measuring instruments is a Chinese calliper, which according to an inscription, was produced during the 9th century. Measuring instruments were already being produced in Mediterranean Europe before the birth of Christ. They played an important role in the making of astronomical instruments and navigation devices. The demands of astronomy were instrumental in promoting the construction of precise measurement instruments. In the early 17th century, callipers and micrometers were used on astronomic instruments. They were initially constructed of wood and later of brass, already employing the vernier to ease the determination of interim values [Häu 95].

Nineteenth century industrialisation is characterised by the emergence of interchangeable manufacturing and mass production. In contrast to the previously conventional pairing of matched parts, where the parts to be combined are produced together and the matching function was achieved by direct comparison, interchangeable manufacturing requires the production of several parts in connection with a particular unit, such as the length unit meter and small tolerances. The idea of interchangeable manufacturing can be traced back to the 18th century. However, the lack of appropriate measurement instruments could be a reason why it was not put into practice until later.

For reasons already mentioned, industrialisation was associated with the development and production of measurement devices. Instruments such as the vernier calliper (developed in 1790) and the micrometer (developed in 1848) (**Figure 1.2-1**), are still in use today. In order to further improve measurement precision, bench micrometers were constructed. From this the first mechanical measurement machines were developed. These were already at that time employed within the framework of test device monitoring, to calibrate gauge blocks. Shortly before the turn of the century, Johansson began to produce entire gauge block boxes in Sweden. An examination of these gauge block boxes established that they already then adhered to tolerances of 2μm [Häu 95].

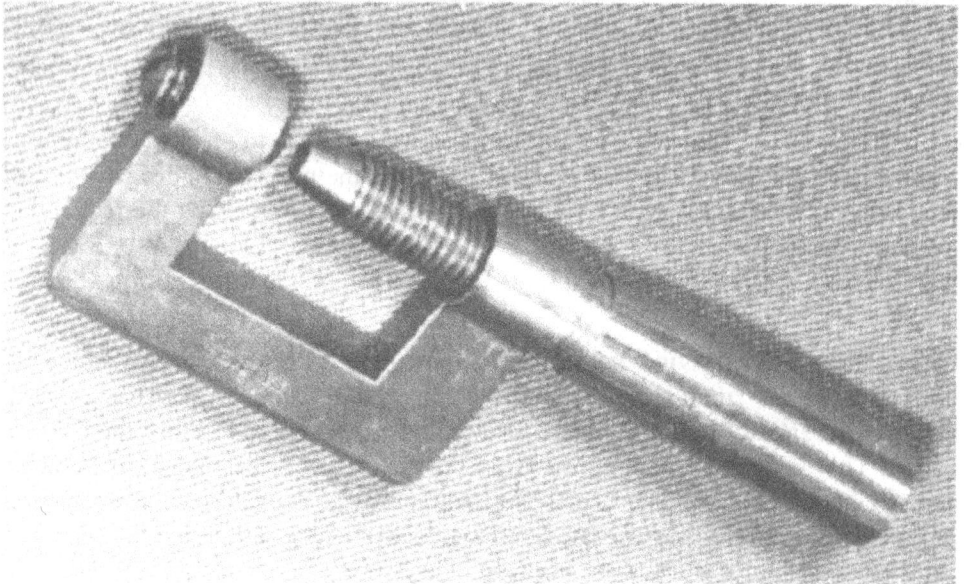

Figure 1.2-1: Micrometer by Palmer, France 1848 [Häu 95]

At the same time, optical methods were primarily employed to enlarge the measurement reading. The development of the interferometer by Michelson used light properties, such as the wavelength of light, directly for measurement. The first optical devices suitable for workshops were profile projectors developed in 1920 [Häu 95].

The first pneumatic measurement instruments emerged in 1930 for the testing of cross-sectional dimensions [Häu 95].

Around 1970 electronics entered geometric metrology. Today electronics even make their mark on simple measuring devices, such as a calliper with a digital display. A highlight in the use of electronic components is certainly their application in coordinate measuring devices and image processing systems, which would not be feasible without the employment of modern computers. Today IT also plays a vital role in the evaluation and transfer of measurement data.

In addition to technical development, metrology's role and scope of functions has changed within the framework of quality assurance (**Figure 1.2-2**).

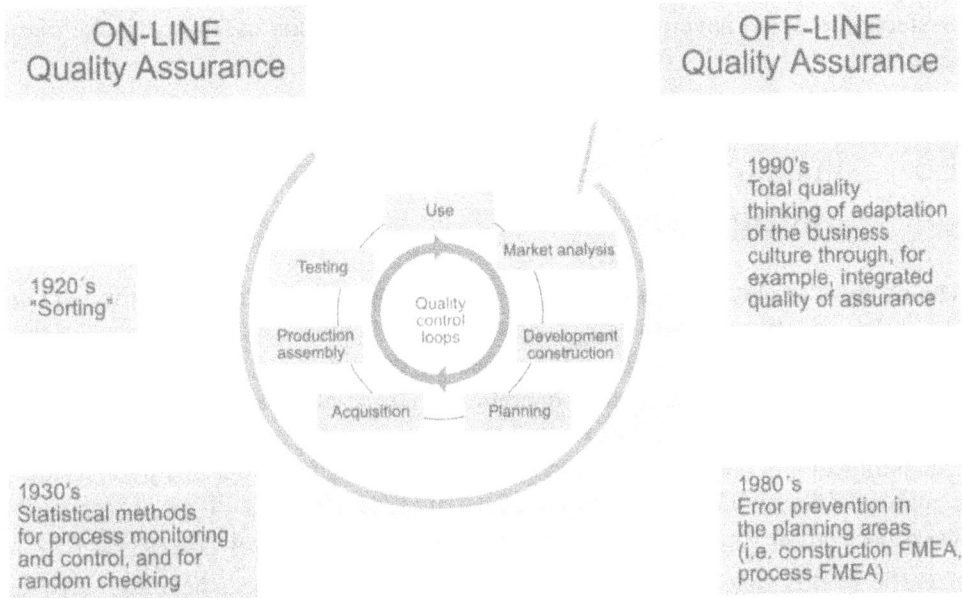

Figure 1.2-2 Development of quality assurance

The onus of checking the results of the production process fell to the 1920's. The sorting of good and bad parts took place based on measurement results. Following the saying that "quality cannot be tested, but must be produced", rudimentary safeguarding of production processes through statistical methods for process monitoring and control could already be found in the 1930's. Here, the product was no longer placed solely in the foreground; in fact metrology provides the actual data for the optimisation of production processes and production machinery [Lei 54]. With the increasing complexity of products and the higher demands on production precision, techniques were developed in the 1980's, which were applied in the planning areas and followed a product and process structure, which was orientated towards the interests of production. These preventative methods contribute to "production friendly" construction, minimisation of possible sources of error for subsequent production and the restriction of demanding production tolerances to vital parts. The fundamentally new aspects of the 1990's are seen in the integral approach to quality thinking. This means that quality can no longer be limited to the production areas of an organisation, but must be borne by the entire organisation. This is conditional upon a suitable business culture and the integration of quality assurance in all business areas in the sense of "Total Quality Management" [Pf 96]. In this way, production metrology's scope of functions has expanded ever further away from the end of the product formation process towards the early phases of the process chain. Online quality assurance functions in the product and production process have been augmented by offline quality assurance in the planning areas.

1.3   Production Metrology as a Component of Quality Management

The changes in general conditions (see above) intensified product liability and an increasing "quality competition". These are some of the reasons, which ensure that organisations are increasingly made aware of their duty to implement a quality management system within their operations. The goal of a quality management system is to ensure the quality of tangible and intangible products, while observing technological and economic boundary conditions. In order to put this into practice, a number of activities are required which include all areas of an organisation. Taken as a whole, these functions form the quality management system [Pf 96].

Specific quality management systems in different organisations have the same or similar elements of company organisational structure and procedural structure, albeit with individual selection and specification. This has led to the definition and documentation of general requirements for quality management systems [Pf 96]. The defined requirements also serve to make the system open to scrutiny by customers and third parties. In this context, the set of norms DIN EN ISO 9000ff. is undoubtedly the most well known and widely used. In the set of norms DIN EN ISO 4500ff., the requirements of quality management systems are defined within the field of accreditation and certification, particularly with regard to the considerations, which must be made for the certification of testing and calibration laboratories.

Production metrology plays a primary role in the quality management system. The main focus of its functions often lies in the testing of products. This becomes evident when looking at the quality management elements of "tests", "test device monitoring", "test status" and "management of quality records" within DIN EN ISO 9001, which are directly linked with production metrology. (**Figure 1.3-1**) [Pf 97a].

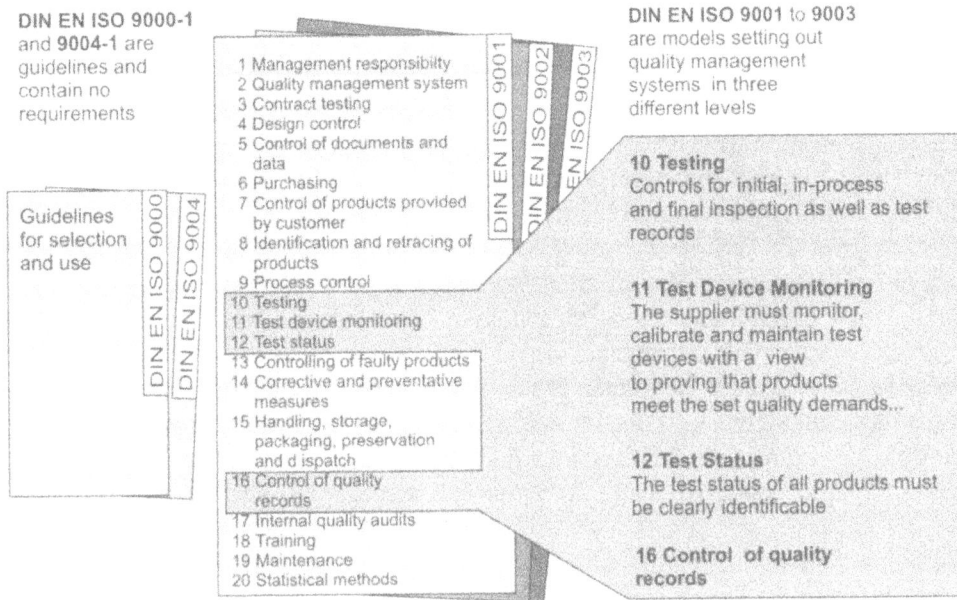

DIN EN ISO 9000-1
and 9004-1 are
guidelines and
contain no
requirements

Guidelines
for selection
and use

DIN EN ISO 9000

DIN EN ISO 9004

1 Management responsibilty
2 Quality management system
3 Contract testing
4 Design control
5 Control of documents and
   data
6 Purchasing
7 Control of products provided
   by customer
8 Identification and retracing of
   products
9 Process control
10 Testing
11 Test device monitoring
12 Test status
13 Controlling of faulty products
14 Corrective and preventative
   measures
15 Handling, storage,
   packaging, preservation
   and d ispatch
16 Control of quality
   records
17 Internal quality audits
18 Training
19 Maintenance
20 Statistical methods

DIN EN ISO 9001
DIN EN ISO 9002
DIN EN ISO 9003

DIN EN ISO 9001 to 9003
are models setting out
quality management
systems in three
different levels

10 Testing
Controls for initial, in-process
and final inspection as well as test
records

11 Test Device Monitoring
The supplier must monitor,
calibrate and maintain test
devices with a view
to proving that products
meet the set quality demands...

12 Test Status
The test status of all products must
be clearly identificable

16 Control of quality
records

Figure 1.3-1:   Production metrology and DIN EN ISO 9001

Within element no. 10 "tests", a quality management system is required to establish controls for initial, in-process and final inspection. The *initial test* protects the
organisation from purchasing parts lacking in quality. As long as the supplier can
prove the quality of his products through his own testing procedures, or demonstrate the capability of productions processes within the framework of a certified
quality management system; the initial inspection can largely be waived. *In-
process testing* provides information about the quality situation within the production process. In order to do this, various data evaluation and process assessment
techniques (Section 5) can be used. The *final inspection* ensures that no products
reach the customer without having fulfilled quality demands. Even the final
inspection often uses only random sampling or is completely omitted if product
quality is ensured through appropriate measures, such as process capability studies
(Section 5).

Test device monitoring involves the monitoring, calibration and maintenance of all
devices used for testing. This follows the goals of ensuring an accurate test result
(Section 6).

According to element 12, "test status", the test status of a product, which shows
conformity or non-conformity in respect of the quality test carried out, must be
recognisable through a clear identification. Within the framework of controls to
manage quality records, the documentation of test records is set out, amongst other

things. The activities connected with these elements are often directly linked with production metrology.

The concentration of production metrology on testing devices within the scope of standardised quality management systems, is attributed to the classic role of production metrology as a control authority. In this role, however, it can not meet increased demands. Within the organisational strategy of total quality management, which requires integral quality thinking with a simultaneous assimilation of business culture [Pf 96], production metrology is to a much greater extend called upon to make contact with all organisational areas which require its information. The purpose is not only to determine the deviation from a target state, but also to reduce these deviations with the co-operation of other operational areas. In order to do this, the information that has been determined is fed back to the responsible areas through control loop structures, where a distinction is made

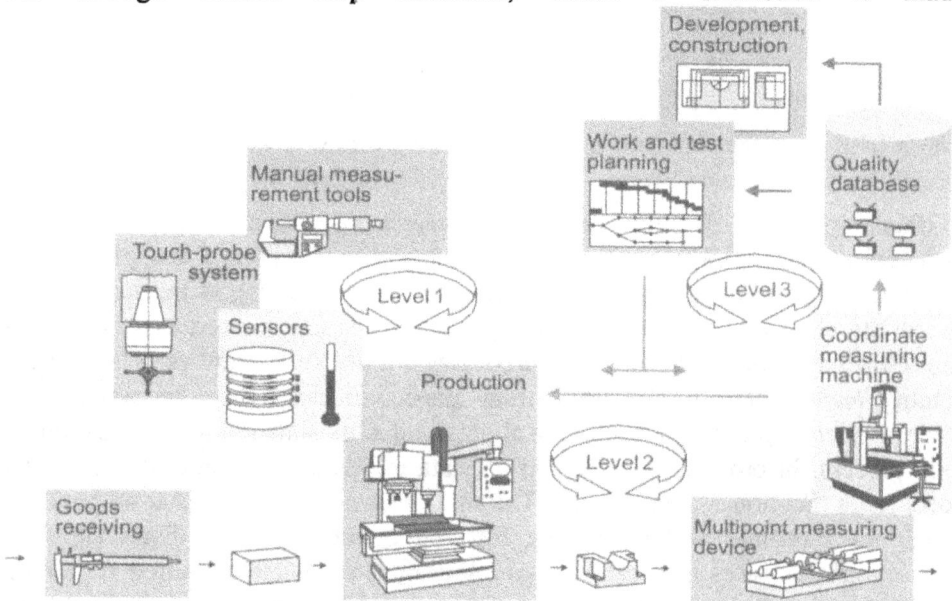

Figure 1.3-2: Quality control loop structures within the production process

between three levels (**Figure 1.3-2**). In level 1, a process is directly measured, either on a continuous or intermittent basis. The information thereby gained is used for process control. The classic post-process metrology (level 2) is often used in the form of statistical process control (SPC) to monitor processes which are, in principle, more capable (Section 5). In order to be used for the long term, measurement data must be consolidated and made available in networked data structures in order to support the indirect production areas such as "development and construction" and "work planning" in large quality control loops (level 3). Levels 1 and 2 are illustrated as control loops operating within levels, while level 3 spans various levels. (Section 1.1).

The structure of quality control loops is explained using the following two examples. **Figure 1.3-3** shows a small control loop (level 1) using the example of process intermittent control of a drilling process.

The drilling order is given through the production order. After a hole has been drilled, the testing order states that the diameter produced on the machine is measured by a substituted measuring scanner. The evaluation of the measurement and the comparison with the target value enables the calculation of a correction factor with which the adjustable drilling tool is regulated and the quality of the process can be improved. This method is conceivable, for example in order to monitor and regulate the process capability of a drilling machine, within the framework of statistical process control (SPC). In order to facilitate this, the testing order provides a sample size which is measured on the machine.

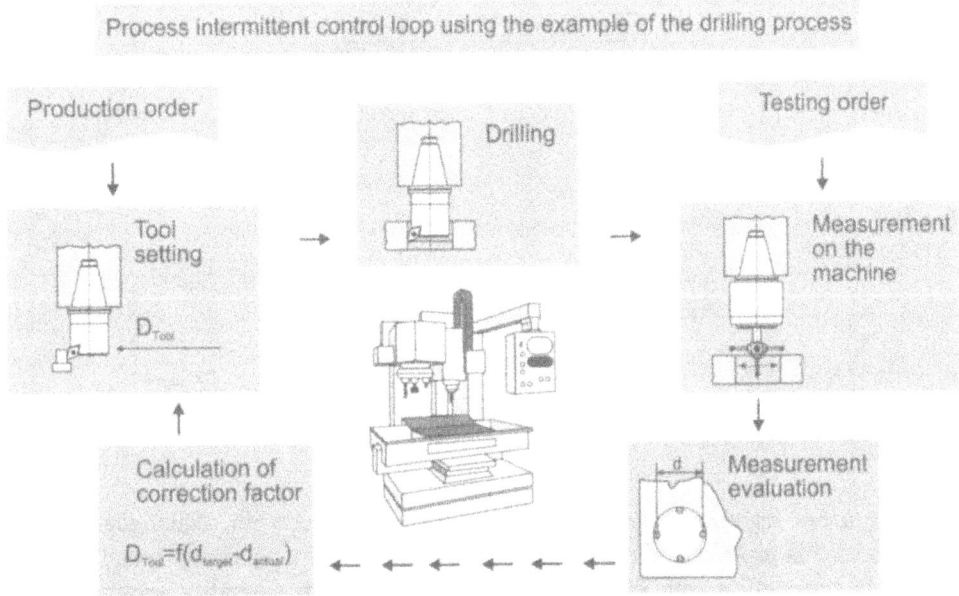

Process intermittent control loop using the example of the drilling process

Production order

Drilling

Testing order

Tool setting

D_{Tool}

Measurement on the machine

Calculation of correction factor

Measurement evaluation

$D_{Tool}=f(d_{target}, d_{actual})$

Figure 1.3-3:   Example of a process intermittent control loop

Figure 1.3-4 shows control loops spanning various levels (level 3). The test data from the production area is processed in test data evaluations and consolidated to characteristic values (e.g. c_p- and c_{pk}-values) in a quality database. Planning areas use this database for support.

In this way, the production manager can obtain an overview of the current quality standards of the various characteristics, for example "inner-diameter" with a target value of "35mm" and can query the feasibility of achieving the respective production characteristic. Hence, this test data analysis is called feature-oriented. The analysis can produce values oriented towards specific machinery (e.g. c_p-value of a machine) or values covering a wider range of machinery (e.g. medium c_p-values of all relevant machines). It provides a positive or negative feature assessment for a characteristic and its tolerances, based upon which the production manager will decide whether or not to allow production to go ahead.

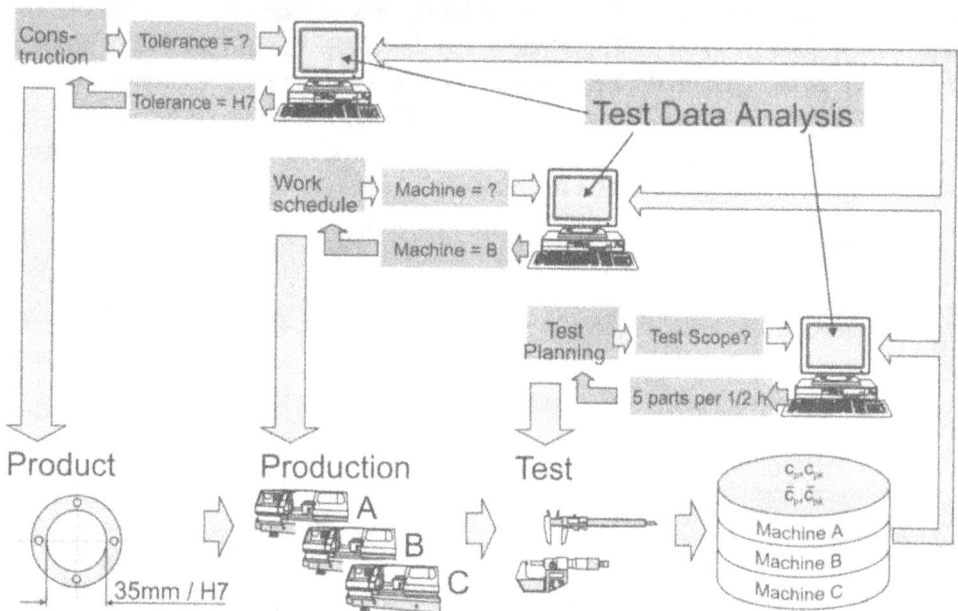

Figure 1.3-4: Quality control loops spanning various levels

Planning areas regulate issues such as at what time and with which machine a characteristic is produced. Criteria such as availability of a particular machine and its hourly production rate play an important part. From a quality improvement point of view, choosing the right machine can be aided by the analysis of the c_p-values within the quality database.

The aim of test planning is to ensure that the test type and scope of the test is sufficient to ensure adherence to tolerances. For surety reasons the test scope should be as wide as possible for critical characteristics. However, from the viewpoint of cost and time used it should be at the minimum still acceptable under quality control standards. The analysis of the test data of already produced parts gives the planner the possibility to accelerate the tests. Parts, which have been produced to a good quality standard in the past undergo less testing than parts, which have falling capability values.

1.4   Overview of Production Metrology

Production metrology, as it is understood today, is more than just the technology used to record quality characteristics in the production areas. It is of much more importance to master the methods of production metrology in all variations. It is the only possible way to choose the optimal systems and processes for the various measurements and testing tasks so that processes, which add value can be judged appropriately without wasting time. The following section enables the reader to gain an overview of the content of this book as the relevant elements of production metrology are summarised.

The methods of production metrology can be divided into "classic" quality control and preventative processes of quality assurance. **(Figure 1.4-1)** [Pf 96].

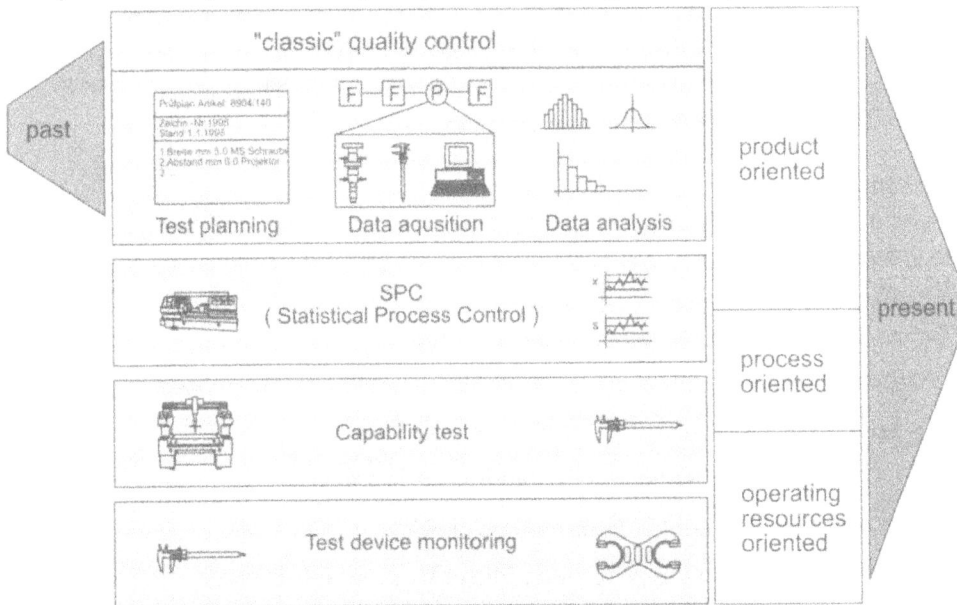

Figure 1.4-1: Production metrology methods

The aim of classic quality control methods is to ensure perfect product quality through quality testing during or immediately after production. Traditionally, this test includes all aspects of *test planning*, *data collection* and *data analysis* .

In accordance with the saying that quality cannot be tested, but must be produced, quality assurance has changed over the past few years to achieve a high quality standard by adopting the right processes from the start. It is therefore targeted at discovering changes in product quality as early as possible and to intervene in the production process with corrective measures. The objectives of production metrology, as already described, have changed from being purely product orientated to include processes and materials.

Statistic Process Control (SPC) is the statistical evaluation of product characteristics, allowing for a process-near control loop and fast correction of the production process before faulty parts are produced. By testing the *capability* of a process or a machine, a stable and secure production can be assured. Hence, the planning basis can be introduced into construction and product planning. Both methods are part of test data analysis in the widest sense (Chapter 5).

Finally, quality assurance processes only work if the test devices supplying the actual situation are monitored during production and therefore produce reliable data. Therefore, *test device monitoring* is an essential part of production metrology.

The following paragraphs will give a short overview of the above mentioned methods of production metrology. A more detailed description will follow in the respective chapters of the book.

Inspection Planning

The term test planning is still closely linked with the testing of workpieces. Test planning in the traditional sense ensures that the production areas are given complete and clear instructions for carrying out their tests (Chapter 3) .

The areas include, in a traditional understanding:

- What to test, i.e. which characteristics of the workpiece are chosen
- When to test, i.e. at which point in time during production
- How to test, i.e. measurement or comparison to a gauge
- How much to test, i.e. 100% or random checks
- Where to test, i.e. in the production line or in a measuring room
- What to test with, i.e. choice of testing devices
- Which analysis to use, i.e. how are the test results implemented

Planning therefore includes workpieces and test equipment. Depending on test documents, such as drawings, work schedules and test regulations, the test plan is drawn up on a form. Computer-aided systems make test planning much easier. A close link between production and test planning makes sense, as both are comparable in their approach and fields of planning, such as the timing of tests during production, can overlap.

Test Data Acquisition

As described in paragraph 1.1, acquisition of test data is the central task of production metrology. To fulfill this task, a wide range of measuring aids are available which differ in precision, measuring speed and level of automation.

The traditional test data acquisition uses measuring aids to measure after each production step and after production has been completed. The object of this metrology is the workpiece or the finished product.

The most commonly used form of post-production-metrology is the geometric test of a simple workpiece close to the production machinery in the form of a self-test by the operator, e.g. with the aid of manual measuring aids. Test result collection is done in test plans and has, in recent times, been increasingly aided by computer systems. In contrast, for realising SPC control loops during large series production, specific measuring aids are useful which allow quick measuring of relevant characteristics.

Using mobile measuring sites allows a flexible usage of the available measuring aids and is especially useful for cost-intensive standard devices. Laptops are used for the collection of test results. Mobile measuring sites are used in statistical quality control, such as random checks in a smaller production series.

To be able to monitor complex geometric aspects of workpieces in a short period of production time, multi-point measuring systems have been established. The negative side of these is that they are tailored to a specific measuring task. It therefore takes some effort to make them useful for any other task. Investing this amount of time and money is generally only justifiable for large production series.

In addition to the multi-point measuring devices such as inductive measuring systems, which are used in various production sections, optical and opto-electric measuring aids offer the possibility to collect test data quickly during production. This way the systems cover a large variety of measuring tasks in the area of production metrology. As these systems are usually connected with modern computer systems, they offer the ideal pre-condition for automatic test data collection and evaluation and feeding the data into the process control loops. As for the development of the computer technology, industrial image processing has become a key technology in many production areas. The production metrology widely uses these systems for collecting object dimensions or monitoring the process status optically.

Coordinate measuring machines have emerged as a flexible tool to measure complex workpiece dimensions. Tactile and optical measuring tools are also being used. When a high level of accuracy is required, coordinate measuring machines are used in measuring rooms (Section 2.4). To realise quick process control loops in the production process, the use of coordinate measuring machines is needed which compensate disturbing influences effectively, as they arise during production. Usually form and surface characteristics will be tested by special measuring tools. These are based on various tactile and optical measuring principles.

Test Data Evaluation

The task of evaluating test data is to evaluate and consolidate the test results into test statements. This is the fundamental basis on which to build quality control loops, which span individual, as well as various, levels (Chapter 5).

The lowest level of evaluation consolidates the initial results for test devices into statistical characteristic values. The interim data evaluation is divided into measurements such as mean value $\bar{x}$, mean variation s or range R within a production series. These are defined and used for lot release or the steering of rejects and repairs. This way the manufacturer can meet his documentation regulations.

The long term monitoring of statistical values of many lots is of interest and its combination of production quality with process and product data. The aim is finding weak points within the production process. A quality database is built to describe the product, process and machine data on the basis of statistical characteristic values on various abstraction levels.

This quality database will help the planning areas to make decisions on future planning steps. Especially within test planning the test scope can be adapted to the necessary demand and costs of testing can be optimised. Quality steering uses the data for introducing methods to improve product quality.

Statistic Process Control (SPC)

Statistic Process Control (SPC) is one of the most commonly used quality technologies used to monitor and regulate the process capability. Quality control cards describe the process situation. These show process interruptions and allow the set-up of control loops close to production. This system is only useful for the regulation of capable processes. Hence, capability has to be proven before the loops are installed (Section 5.3).

After the forerun the intervention limits for the controls are set up and recorded on the control card. After a set of random tests, the operations manager registers the dimensions which need regulation on the card. Should these limits be exceeded, the process is corrected and if the tolerance span is overstepped, a system error has occurred and the process must be examined.

Capability Investigations

There are three different capability investigations: process capability, machine capability and test equipment capability.

Test equipment capability shows the capability of the chosen test equipment to judge a process in accordance with appropriate parameters. The test equipment has to be monitored in regular intervals to ensure its capability (Chapter 6).

While process capability shows whether or not a process is appropriate to fulfill the demands put upon it in relation to a specific characteristic, machine capability describes the performance of the machine under ideal conditions. The machine capability test is therefore often carried out during an inspection. As ideal conditions are usually not given during production, the main control instrument is the process capability.

The calculation of process and machine capability is given by the statistic evaluation of measurements of produced items. The method for both capability tests is identical (chapter 5).

Inspection Equipment Monitoring

Continual monitoring ensures that parts are produced with the correct measurements. This monitoring process is aided by sensors installed closely to the production processes. The quality of these parts is ensured by post-production measurement equipment. To be able to use these control mechanisms over a long period of time, reliability of the means of production has to be ensured and the engaged metrology must deliver meaningful results. Monitoring devices, therefore, have to be installed, which test reliability and quality capability of all machinery, sensors, manual and automated measuring devices, which form a part of the production line (Chapter 6).

Depending on the level of automation, the monitoring of the production means and test devices must be done offline, i.e. outside the production line or by online measurements of suitably calibrated standard parts and be integrated into the production line.

For monitoring machines close to production, these sensors either need to be dissembled or tested offline within defined boundary conditions. Alternatively and usually administered in a practical situation, these are exchanged against tested samples.

Monitoring the capability of manual test devices and measuring instruments can be done offline in the measuring rooms or online via a recycling process on calibrated gauge blocks. Whilst here especially the monitoring in the measuring room is used for universal calibration, monitoring the test device in the production line can be used as a comparator; the measuring instrument is only calibrated for special, and not overall, measuring tasks.

For the offline-monitoring of all universal coordinate measuring machines conventional test devices are used. In contrast the faster, online monitoring, e.g. calibrated ball plates in different sizes have been established. Due to the growing complexity of testing tasks on free formatted surfaces, such as gears, task specific test elements have been developed which assist in the recording and evaluation of the coordinate measuring machines' capability to perform special measuring tasks.

Further Reading

[Blu 87]   Blumenauer, H. (Hrsg.): Werkstoffprüfung. Leipzig: VEB Verlag für Grundstoff-
             industrie 1987

[Dut 96]   Dutschke, W.: Fertigungsmeßtechnik. Stuttgart: Teubner Verlag 1996

[Häu 95]   Häuser, K.: Die Meßschraube, Technikgeschichte Modelle und Rekonstruktion.
             München: Deutsches Museum 1995

[Lei 54]   Leinweber, P. (Hrsg.): Taschenbuch der Längenmeßtechnik. Berlin, Göttingen,
             Heidelberg: Springer Verlag 1954

[Lot 88]   Lotze, W.: Die Entwicklung der industriellen Fertigungsmeßtechnik. Technische
             Rundschau. (1988) Nr. 41 S. 38-45

[Pf 96a]   Pfeifer, T.: Qualitätsmanagement: Strategien, Methoden, Techniken. Wien, Mün-
             chen: Carl Hanser Verlag 1996

[Pf 96b]   Pfeifer, T., Imkamp, D., et. al.: Optimierungspotential Fertigungsmeßtechnik.
             Wettbewerbsfaktor Produktionstechnik, Aachener Perspektiven. Aachener Werk-
             zeugmaschinenkolloquium (Hrsg.). Düsseldorf: VDI-Verlag 1996

[Pf 97a]   Pfeifer, T., Imkamp, D.: Auswirkungen der DIN EN ISO 9000 ff. auf die Ferti-
             gungsmeßtechnik. GMA-Jahrbuch. Düsseldorf: VDI-Verlag 1997

[Pf 97b]   Pfeifer, T.: Ohne Fertigungsmeßtechnik geht nichts. Qualität und Zuverlässigkeit
             QZ 42 (1997) 9 S. 942-943

[Ste 88]   Steeb, S., Basler, G., Deutsch, V., Gauss, G., Griese, A.: Zerstörungsfreie Werk-
             stück- und Werkstoffprüfung. Ehningnen: Expert Verlag 1988

Norms and Regulations

DIN EN ISO 9000ff.   DIN EN ISO 9000ff. Normen zum Qualitätsmanagement und zur
                            Qualitätssicherung/QM-Darlegung. 1992 bis 1994

DIN EN 45001ff.   DIN EN 45001 ff. Normen zum Betreiben, Beurteilen und Akkreditie-
                            ren von Prüflaboratorien. 1990 bis 1995

2 Fundamentals of Production Metrology

2.1 Basic Concepts

Today it is a matter of course, that independently manufactured components, e.g. screws and nuts, fit together. This principle, known as exchangeability, is a fundamental prerequisite for modern manufacturing. In order to guarantee exchangeability, it must be determined whether certain features of a workpiece move within given tolerances. This is established by investigating the relevant feature. In addition to a uniform system of dimensions, knowledge about the applied measurement strategy is also a basis for establishing the comparability of measurement results.

2.1.1 Introduction to the SI-System of Units

A basic condition for measurement is that the quantity being measured must be uniquely defined. .

This prerequisite is always fulfilled with physical dimensions. In general production metrology, however, the conditions must first be created by special agreements for certain measurement functions. With important technological quantities, the testing methods are determined only by standardisation (e.g. hardness testing).

The second prerequisite is that a unique reference standard must be determined for the dimension.

All base quantities, which can serve as a measurement reference and the units, derived from them form a system of units. In the course of technical development, the most diversified quantities were defined as base units and accordingly introduced as extremely differing units. This variety of quantities constitutes a source of probable errors and misunderstandings and, in particular, obstructs the worldwide exchange of goods, services and, not least, researches results.

In view of the variety of systems of units developed over the course of history, a uniform system of units, the SI ("Systeme International d'Unités") was proposed in 1948 [Tra 97]. With the acceptance of this system in 1960 and its subsequent modifications, an internationally recognised system of units was thereby created and introduced for the first time. This system is based on seven physical quantities with their units and symbols (**Table 2.1-1**).

Table 2.1-1:   Base units of the International System of Units (SI)

Quantity	SI Unit	Unit Symbol
length	meter	m
mass	kilogram	kg
time	second	s
thermodynamic temperature	kelvin	K
electric current	ampere	A
amount of substance	mole	mol
luminous intensity	candela	cd

The SI units are derived as products of powers of the base units, with no numerical factor other than 1. This is referred to as a coherent system. The derived units are subject to the same algebraic relationships as the corresponding base quantities. Various derivations have received their own name and special unit. An example of this is the principle equation of mechanics with the quantities (and units):

$$Force = Mass \cdot Acceleration; \; (1N = 1\frac{kg \cdot m}{s^2}) \tag{2.1-1}$$

With the exclusive use of coherent units, unmanageably sized (large and small) numerical values become necessary to indicate quantities. In order to maintain a reasonable size of numerical values, the units can be multiplied by decimal factors (example: 1 km instead of 1000 m). These factors are called SI prefixes (**Table 2.1-2**).

In addition to the SI units, further units are used in technology and commerce. The use of these units is legally regulated. Likewise, the representation, supply and distribution of the units are assigned to organisations, which have been designated with these tasks by the legislator (Federal Republic of Germany: The Federal Standards Laboratory (PTB) in Braunschweig and Berlin). Additionally, the units must either be permanently realised (e.g. international kilogram prototype as a definition, realised by the base unit mass) or be reproducible (caesium atomic clock as time standard).

Table 2.1-2:   SI prefixes and prefix symbols

SI Prefix	Prefix Symbol	Factor	Name
Exa	E	10^{18}	trillion
Peta	P	10^{15}	quadrillion
Tera	T	10^{12}	billion
Giga	G	10^{9}	milliard
Mega	M	10^{6}	million
Kilo	k	10^{3}	thousand
Hekto	h	10^{2}	hundred
Deka	da	10^{1}	ten
Dezi	d	10^{-1}	tenth
Zenti	c	10^{-2}	hundredth
Milli	m	10^{-3}	thousandth
Mikro	μ	10^{-6}	millionth
Nano	n	10^{-9}	milliardth
Piko	p	10^{-12}	billionth
Femto	f	10^{-15}	quadrillionth
Atto	a	10^{-18}	trillionth

The definition of a base unit expresses nothing about its representability. Therefore, the valid definition of the meter since 1983 links the base unit with a natural constant:

Definition: The meter is the length of the path travelled by light in vacuum during a time interval of (1 / 299 792 458) s [PTB 94].

In this run time definition, the natural constant c is connected with length and time through the relationship:

$$c = \lambda \cdot f \qquad\qquad\qquad\qquad (2.1-2)$$

As the speed of light is constant c, only one of the two quantities of length λ or frequency f can be freely defined. As the second unit can presently be represented more accurately than the meter unit, the unit of length was linked to time via the speed of light c.

Three realisation possibilities are currently available. The most important one for production metrology is based on a frequency measurement. For interferential length measurement technology, a frequency-stabilised laser is attached to the caesium atomic clock time standard, thus determining the frequency of an emitted beam with a very high level of accuracy. As with every other realisation, this one is afflicted with uncertainties, while a definition is always accurate.

The necessity for increasingly reducing uncertainties of realisation for base units becomes particularly clear with the length quantity, which is important in production metrology. On the basis of a "natural dimension", a 10,000,000th of the meridian quadrant running through Paris in 1799 (first original measure), further original measures were first produced as measure embodiments with an uncertainty of ± 0.2 µm. The first (1927) and second (1960) wavelength definitions based on a certain spectral line of special lamps, had a realisation uncertainty $\pm$ 4 mm [War 84]. Today, the uncertainty of realisation with iodine or calcium-stabilised lasers is down to 10^{-11} m. Current PTB research work is already achieving uncertainties of 10^{-12} m.

These more accurate realisations became necessary, not least, due to progress in production technology, as the uncertainty of certain measurements is usually situated in an order of magnitude above the uncertainty of realisation. The living space of a human is limited to a ball with a radius of $6.4 \cdot 10^6$ m (the earth). A human's direct sensory perceptions are developed, such that they procure information about an area from several 10^3 m (sight, hearing) to some 10^{-5} m (touch). Thus, the smallest organisms already escape perception. Today, admissible deviations of precision workpieces already lie two or more orders of magnitude below this perception threshold (**Figure 2.1-1**). The length measurement of the units which must be mastered by modern production metrology, therefore range from several meters (vehicle bodies) down to several micrometers (micromechanical components). At the same time, the workpiece tolerances are becoming ever narrower. Therefore, production metrology finds itself in a constant process of development, in order to stay abreast of increasing demands.

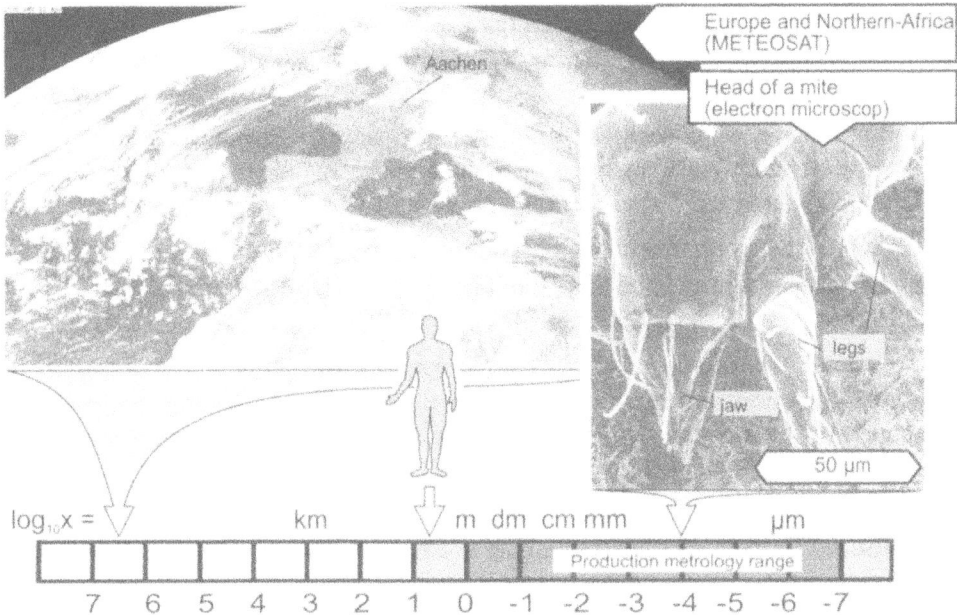

Figure 2.1-1: The "length" quantity

2.1.2 Definitions

The following fundamental metrological definitions are laid down in DIN 1319 Part 1. They are also defined in a similar form in the International Dictionary of Metrology [DIN 94].

Definition: The **measurand** is the physical quantity subject to measurement (e.g. length, density, temperature).

A measurand is generally dependent on several physical dimensions; in particular, it can be dependent on time or location.

Definition: Measurement is the execution of planned operations to quantitatively compare the measurand with a unit.

The term "measurement" includes the evaluation of measurements, as far as obtaining the measurement result, while the further application and utilisation of the measurement results is not part of measurement.

Definition: Counting is the determination of the value of the measurand "number of items in a quantity".

Counting is done using sensory perception or counting mechanisms. This always involves determining the number of homogeneous items separately from those, which differ (numerical value).

Metrology often avails itself of counting to determine measured values. One example is measuring frequency by counting the periods of a signal within a given time interval.

This example shows that counting can be used for the purpose of measurement. However, measurement can also be used for the purpose of counting. For example, the number of items in the same mass can be determined by weighing the all items as a whole (counting balance).

Definition: Testing means establishing to what extent a unit fulfils a demand.

Testing always involves a comparison with a demand, which can be specified or agreed upon. The inspected object can be a specimen, a sample or even inspection equipment. Particular given conditions are margins of error and tolerances. Testing can refer to measurable or countable features (**Figure 2.1-2**).

A test of qualitative features often takes place with a non-dimensional test using the sensory perceptions of an inspector. In particular, predications can already be made about the condition of workpieces, e.g.

- "The workpiece is very hot."
- "The surface of the workpiece is very rough."
- "The workpiece shows cracks and ruptures."
- "The surface of the workpiece is uneven."
- "The gear is too loud."

In contrast, dimensional testing with the aid of inspection equipment leads to an objective predication about whether an item being inspected fulfils the required conditions.

- "The length of the shaft is 149.97 mm."
- "The diameter is situated within the required tolerance."

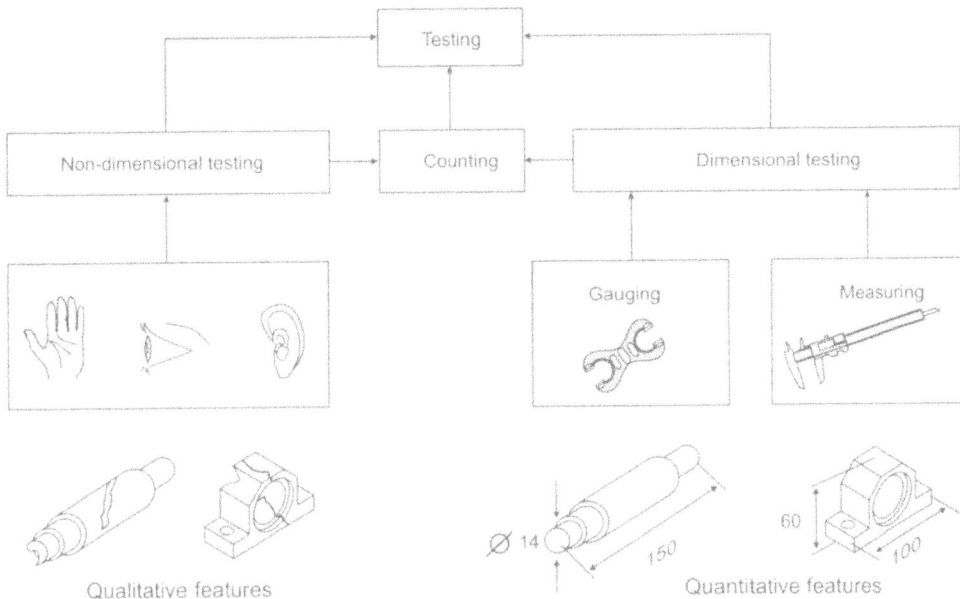

Figure 2.1-2: Definitions: testing, gauges, measurement

A characteristic of testing is that it is always connected with a decision. The decision refers

- directly to the inspected object, e.g. the further processing or rejection of the workpiece,
- the status of the production process, e.g. its capability,
- or to the supplier regarding the adherence to assured characteristics of the supplied product.

Definition: **The measurement result** refers to the estimated value of the true value of a measurand, which is gained from measurements.

The foundations for estimating the true value are measured values and known systematic measurement deviations. Known physical relationships and other knowledge and experiences are also included. A complete specification of the measurement result includes information about the measurement uncertainty [DIN 94] (Section 2.3).

Definition: **A measuring device** is an instrument, which is intended to measure a measurand, either on its own or in connection with other mechanisms.

A device is still considered to be a measuring device when its output is transformed, processed or stored, but is not suitable to be directly received by the ob-
server. Further terms regarding the implementation of measuring devices can be
found in DIN 1319 Part 2.

Definition: **Calibration** determines the relationship between the measured or
expectation value of the output quantity and the appropriate true or
correct value of the existing measurand available as the output quantity, for a measuring device under given conditions.

During calibration there is no intervention, which changes the measuring device. In
addition to the term "calibration", there is also the term "official calibration",
which should only be used when a calibration authority – the national weights and
measures office – inspects a standard or measuring device according to legal regulations (calibration statutes). [Dut 96].

Definition: **Adjustment** is the positioning or alignment of a measuring device, in
order to eliminate systematic measurement deviations as much as possible, to the extent that it is necessary for the intended measurement.

In contrast to calibration, adjustment involves intervention, which permanently
changes the measuring device.

Definition: The **measurement principle** forms the physical basis of the measurement .

An example of a measurement principle is the utilisation of the thermoelectric effect as a basis for temperature measurement. The measurement principle enables
the measurement of another quantity in place of the measurand, in order to use its
value to uniquely determine that of the measurand.

Definition: The **measurement method** is a special type of procedure for measurement, which is independent from the measurement principle.

Examples of this are the comparison measuring method, the difference measuring
method or the zero comparison measuring method.

Definition: A **measuring procedure** defines the practical application of a measurement principle and a measuring method.

Measuring procedures are sometimes differentiated and named according to the measurement principles upon which they are based (e.g. interferential length measurement). An example of a measuring procedure is the amperage measurement with a moving coil measuring device (magnetic induction) using the deflection measurement method.

2.1.3 Measuring Methods

Measurement basically distinguishes between direct and indirect measuring methods.

With the direct measuring method (**Figure 2.1-3**), the quantity being measured is directly compared with a standard, which has the same physical dimensions. An example of this method is comparing the length of a workpiece with that of a parallel gauge block.

a) Comparison with a known material measure

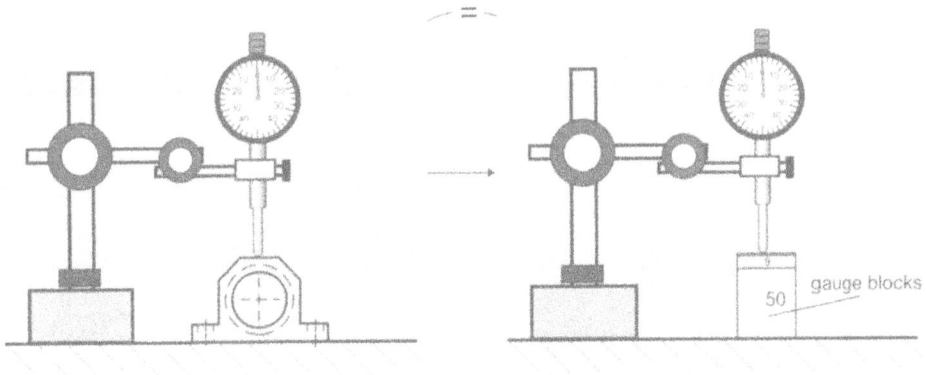

b) Direct display of the value of the measurand

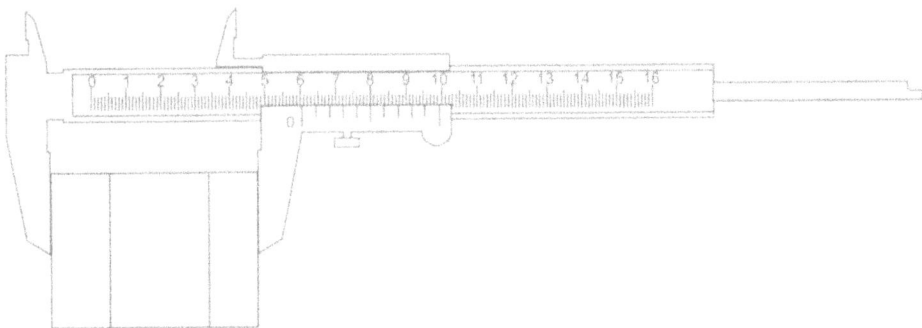

Figure 2.1-3: Direct measuring methods

First, a measuring device, such as a dial gauge, is adjusted with a stand, such that it displays a certain value – preferably the value "zero" – when touching the workpiece. In a second step, a combination of gauge blocks are measured in place of the workpiece, which are adjusted until the previously set value is displayed (**Figure 2.1-3,a**). The combination of gauge blocks then exactly corresponds to the desired length of the workpiece. This methodology is also called the zero comparison measuring method. The advantage of this methodology is that both measurements are carried out under exactly the same conditions, thereby minimising possible error influences caused by differing measurement conditions.

A typical measurement using the deflection measuring method is the measurement of a length using a calliper gauge (**Figure 2.1-3, b**). Here, the measurement surfaces of the calliper gauge are brought into direct contact with the workpiece and length is directly read off of the incremental scale, which serves as the measure embodiment.

With the indirect measuring method, it is not the quantity to be measured, which is directly recorded, but an ancillary quantity, which has a known and describable relationship with the measurand. The measurement of a drilled hole with a pneumatic plug gauge is an example of an indirect measuring method (**Figure 2.1-4**).

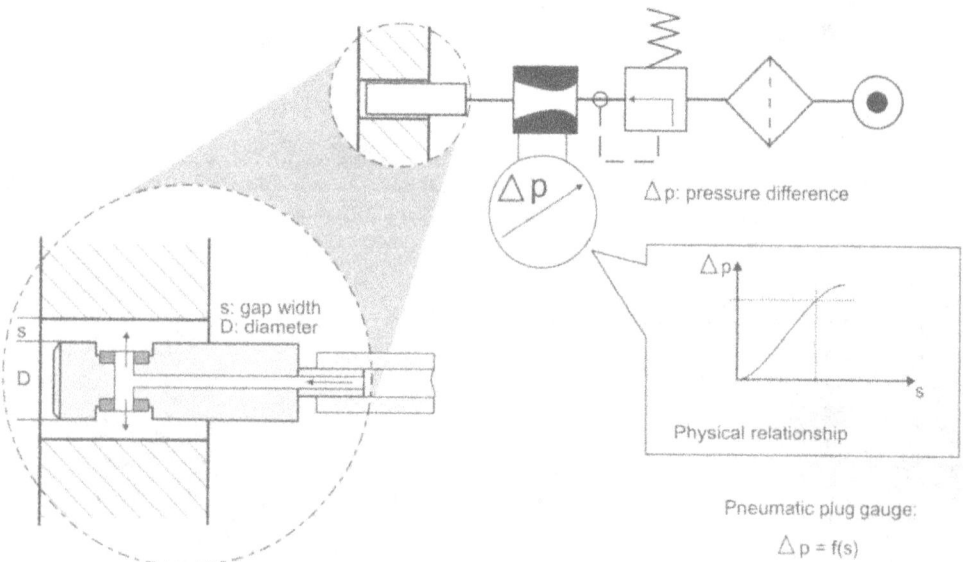

Figure 2.1-4: Indirect measuring method

Pneumatic length measuring devices rest upon the principle that the length to be measured influences the narrowest diameter of a flow channel, through which air flows. The flow rate is determined by the narrowest diameter of the flow channel, which is formed between the outlet of the plug gauge and the drilling wall. Thus, the flow rate represents the desired measurand, which is the width of the gap. If a plug gauge is subject to constant pressure, a pressure differential, which corresponds to the flow volume and thus the width of the gap, can be registered using a nozzle.

The differentiation between direct and indirect measuring methods is independent from the measurement principle. The measurement principle determines how the quantitative comparison takes place between the measurand and the unit of physical quantity. It is based on known physical effects, which indicate a known dependency between the measurand and other physical quantities (e.g. the change in resistance of an electrical wire when expanded). Measuring devices can rely upon different physical effects to determine the same quantity. Apart from that, the recording and representation of the input and output variables (analogue and digital) are device-specific (Section 4.2).

2.1.4 **Measurement Strategies**

Measuring procedures and methods must be selected by keeping in mind minimisation of error influences, flexibility, time and cost efficiency. This is approached, for example, by the zero comparison measuring method already introduced, where the unknown length of a workpiece is compared with a known exact length. However, as this concerns a comparative measurement, where the measurement deviation is aligned to zero, error influences do not affect the result of the measurement, due to conditions being identical during both parts of the measurement.

The selection of the probing strategy has a crucial influence on the result of the measurement. Basically, one differentiates between 1, 2 and 3 point probing. The term is designated according to the number of probing points used for the length measurement with contact or non-contact probing elements (**Figure 2.1-5**).

- 1 point probing supplies a measurement value by probing one surface of an measured object. A prerequisite is that a common reference has been established between the measured object and the measuring device. This can take place when the measure embodiment and the measured object have a common reference surface. This is the case, for example with the depth measurement of a drill hole, where the measuring aid is put on the drilled surface. The reference can also be given by the defined positioning of the measuring aid with respect to a featured axis. An example of this is form testing using the rotation axis of a turntable.

- With 2 point probing, the measured object is probed on two points, which are given by the intersections of a vertical line piercing the probed surfaces. Examples are the measurement of diameters or thickness of small parts with the aid of micrometer gauges or callipers.

- 3 point probing determines external and internal diameters. This is based on the characteristic that three points accurately define a circle. The measurement is carried out using either a probe element and a v-shaped prism as a two point base or with three radially positioned probe elements. Examples are the specification of diameters or drillings.

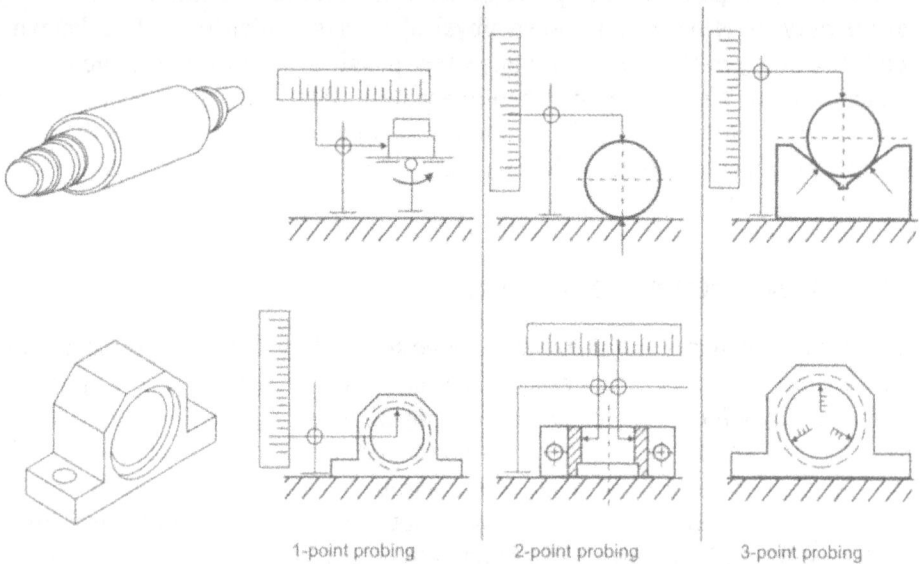

1-point probing   2-point probing   3-point probing

Symbols of length inspection technology according to DIN 2258

Figure 2.1-5: Probing strategies

2.2   Measure Embodiments

As this book primarily deals with methods for measuring geometrical features, the measure embodiments considered below are limited to those, which represent geometrical dimensions. Geometrical measure embodiments can be divided into material and immaterial measures, according to their physical structure. While material measure embodiments represent the dimension through their geometrical shape, immaterial measure embodiments represent the dimension by their own feature. Examples of this are the wavelength of light or the distance which a light or sound signal covers within a defined period of time.

Definition: **A measure embodiment** is generally a tangible object or even a natural phenomenon which represents the dimension to be embodied by a certain constant feature. Measure embodiments have no parts that move during measurement.

A further definition is given in [DIN 94]. Here a measure embodiment is defined as a device "intended to reproduce or supply, in a permanent manner during its use, one or more known values of a given quantity".

2.2.1 Gauge Blocks

Gauge blocks belong to the group of material measures. They are a common basis for industrial measuring and testing. Parallel gauge blocks embody a length through the distance of two parallel surfaces, which are the measuring surfaces. Parallel gauge blocks are usually cube shaped bodies with lengths varying from 0.5 mm to 3000 mm and are made of a hard wearing, stable material, such as hard metal or ceramic. The measuring surfaces must be free from defects.

Various length dimensions can be embodied by the so-called "wringing" of gauge blocks of different lengths. Wringing is the complete contact between the measurement surfaces of two gauge blocks, where attraction between the surfaces holds them together. In this way, many different dimensions can be formed with a small number of meaningfully stepped gauge blocks (**Figure 2.2-1**). Parallel gauge blocks are therefore supplied in sets of different compositions, whereby the compositions of the sets determine the measuring range and the smallest possible graduation of measurement values.

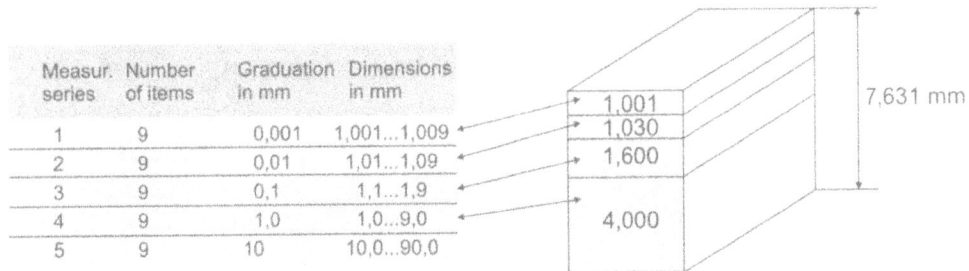

Measur. series	Number of items	Graduation in mm	Dimensions in mm
1	9	0,001	1,001...1,009
2	9	0,01	1,01...1,09
3	9	0,1	1,1...1,9
4	9	1,0	1,0...9,0
5	9	10	10,0...90,0

1,001
1,030
1,600

4,000

7,631 mm

Figure 2.2-1: Wrung parallel gauge blocks

Table 2.2-1:   Appropriate use of parallel gauge blocks in different precision classes

precision class	appropriate use
00:	Monitoring of measuring instruments in precision inspection rooms and highly precise measures
K:	Inspection of gauge blocks situated in lower precision levels (Precision class K is also called the degree of calibration)
0:	Exact length measurements and checking of production gauge blocks
I:	Checking of gauges and setting of measuring devices
II:	Measuring and testing jig construction and in mechanical engineering

According to the definitions in DIN 861, gauge blocks are divided into precision classes 00, K, 0, I and II, which define the parallelism of the measuring surfaces, the allowable deviation from the nominal size as well as the trueness tolerance between the measurement surfaces and side surfaces [DIN 861].

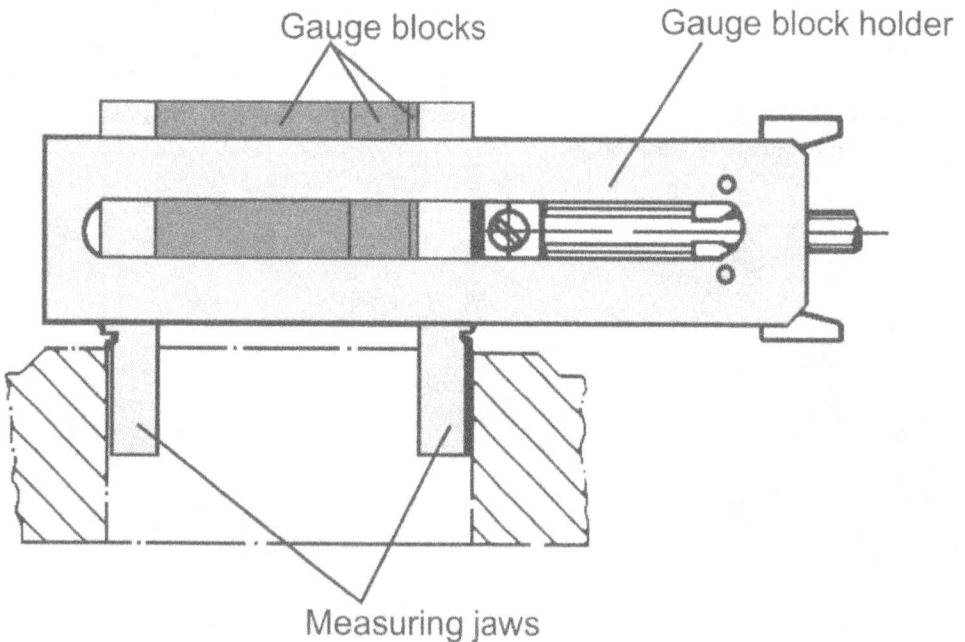

Figure 2.2-2:   Measuring tool with gauge blocks for testing groove widths

The parallel gauge blocks are marked by their nominal size and degree of precision, whereby the degree of precision should be selected according to the intended purpose (**Table 2.2-1**). For example, it does not make sense to use highly precise gauge blocks, such as class 00, in the production environment.

By using special accessories, dimensions can be transferred directly from gauge blocks to workpieces.

In this way, for example, different tools can be assembled for measuring and testing, in conjunction with measurement arms or scriber points (**Figure 2.2-2**). Manufacturers therefore offer complete accessory kits, in addition to the gauge blocks themselves.

An important unidimensional inspection device for the monitoring of coordinate measuring machines is the bi-directional stepped gauge block (**Figure 2.2-3**). This carries out several types of measurements (inside, outside, front and rear stepped measures, centring point distance from block and gap) and embodies graduated interval sizes for recording short and long periodic errors.

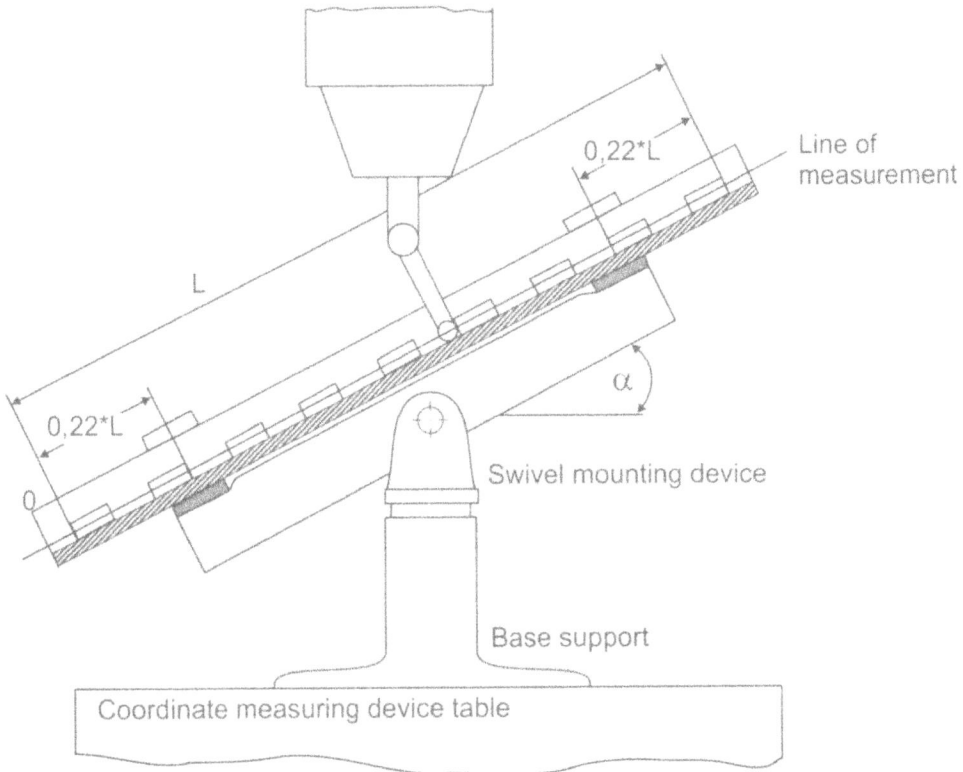

Figure 2.2-3: Bi-directional stepped gauge block

Bi-directional stepped gauge blocks have cylindrical gauge blocks embedded in the neutral (i.e. unexpanded) fibre of a supporting structure. In this way, the distances between the gauge blocks remain approximately constant, even when the supporting structure is deflected. Nevertheless, in order to minimise the deflection of the supporting structure due to its dead weight, the supports are placed at appropriate points. From strength theory, it can be deduced that a beam with two support points experiences the least deflection precisely when the supports are placed at (0.22 x 1). These points are also called Bessel points, as they can be determined using a Bessel function.

The stepped gauge block can be mounted in various spatial positions with the aid of a swivel mounting device and probed along a measurement line. In this way, it is possible to investigate the length measurement uncertainty of a coordinate measuring machine in several spatial directions.

Angle gauge blocks embody an angle through two measuring surfaces. In a similar way to the parallel gauge blocks, different angles can be represented by wringing the different angle gauge blocks. By positioning individual angle gauge blocks, it is possible to add and subtract angular pitches, so that many different angles can be embodied by a few angle gauge blocks. For example, a set of only 14 angle gauge blocks can form angles from 0° 0' 0" to 90° 0' 0" with an interval of 10".

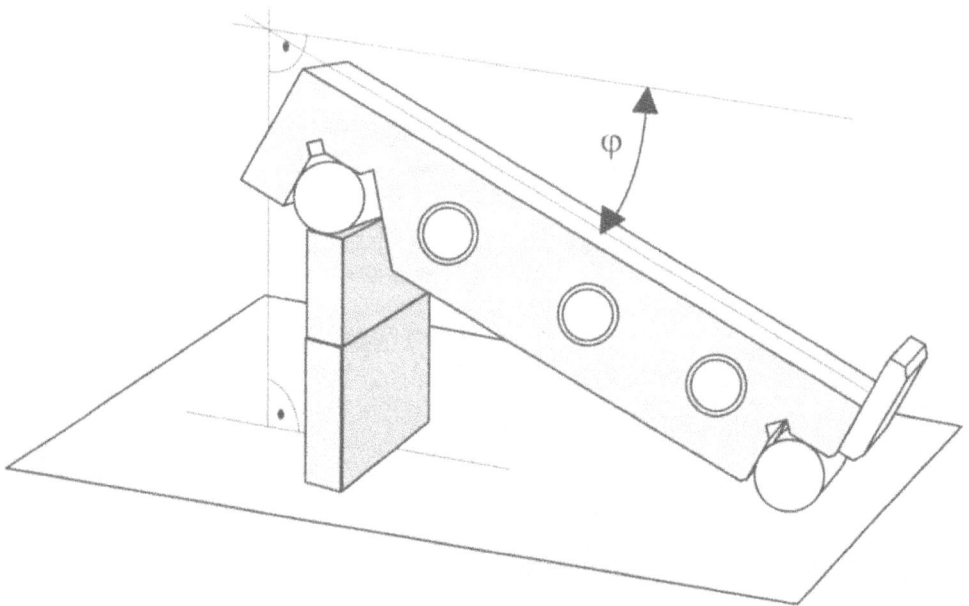

Figure 2.2-4:   Sine ruler

Angle gauge blocks are, however, not commonly used, as they can be replaced by positioning a sine ruler in combination with parallel gauge blocks. The sine ruler has a fixed length and lies on two parallel gauge blocks of differing height combinations. Thus, it forms the hypotenuse of a right-angled triangle. A side of the triangle is formed by the known height difference between the combinations of parallel gauge blocks. With these two dimensions, a defined angle can be formed through the sine relationship (**Figure 2.2-4**).

Gauges represent a special form of measure embodiment. They embody the entire dimension or form of the object to be inspected. Unlike those measure embodiments considered so far, there are neither units nor sections of units. For this reason, gauges cannot be used to determine a quantitative result in the form of a measurement value. The inspected object is compared qualitatively with the gauge in order to obtain a result in the form of "good", "re-work" or "reject". Gauges can be positioned for various types of inspections. Plug gauges or limit roller gauges are typically used for checking internal or external diameters. (**Figure 2.2-5**).

Section 4.6 provides a more detailed description of using gauges for inspection.

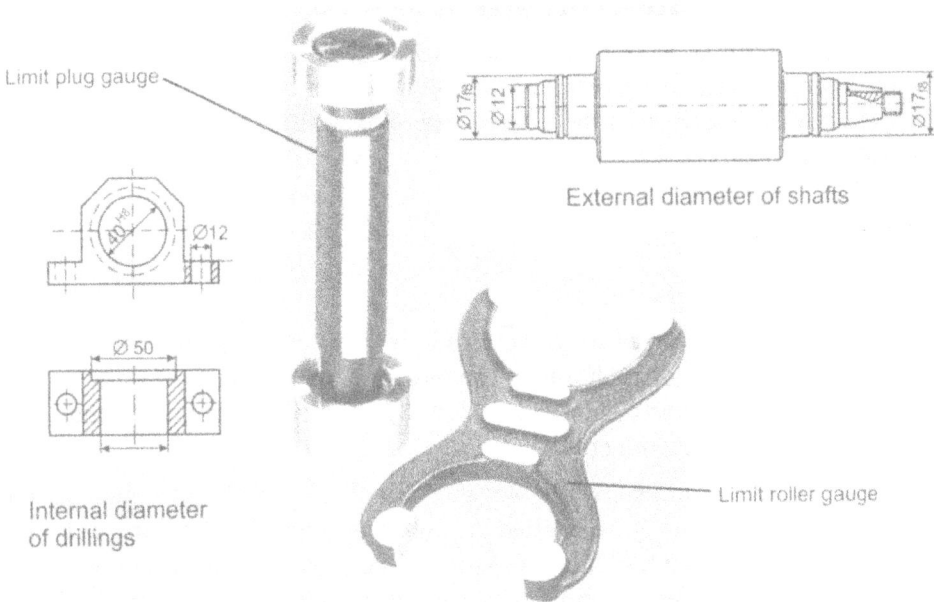

Figure 2.2-5: Examples of gauges

2.2.2 Incremental Measure Embodiments

When using incremental scales, the number of steps are recorded in measuring a shift. This number is multiplied by the increment and provides the total shift. However, with incremental scales, the direction of the shift can only be determined using additional devices (e.g. a directional signal), or by using two standards which are offset against one another.

Furthermore, it is important, especially when applied to NC machines, that a reference mark is initialised on start-up, in order to determine the absolute position. However, in the worst case, this requires scanning of large parts of the measuring range. Therefore, modern measurement systems are provided with distance encoded reference marks, where distances have defined, varying intervals attached to them. The absolute position can then be determined after scanning only two neighbouring reference marks.

The crucial advantage of incremental scales is that they only require a single path for encoding the length, which makes them relatively economical (Section 2.2.3) and that is why they are used in numerous industrial applications.

Examples of common incremental scale embodiments are incremental scales, polygon mirrors, interferometers and various mechanical and electrical scales which are briefly described in the next section. For further details, please refer to Sections 4.2 and 4.3.3.

2.2.2.1 Incremental Scales

Glass scales
With glass scales, two plates of glass, to which strips of chromium are attached equidistantly, lie on top of each other. The measure embodiment is formed by the widths of the strips and the distances between them. Both dimensions are the same amount and denoted as increment τ.

One of the two plates is fixed, while the other is movable. When the scanning disk is shifted in relation to the scale around the splitting period $T = 2\tau$, precisely one light impulse is registered through the change in the effective translucent or reflective surface. Through the number of light impulses, which are registered photoelectrically, the path of the shift can be determined with a known splitting period (Section 4.2.6).

For the splitting period T, values of 8, 10, 20, 40 or 200 μm are common, whereby a dissolution of up to 0,1 μm can be achieved through interpolation. To optimise the accuracy of scanning, the strips must be particularly high in contrast. Therefore the grids are produced using either evaporation or photochemical processes.

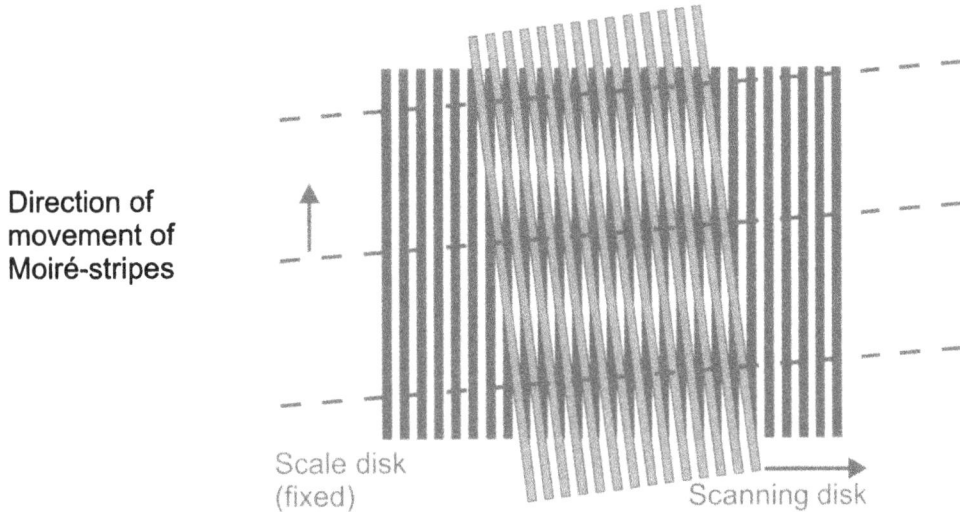

Direction of movement of Moiré-stripes

Scale disk (fixed)

Scanning disk

Figure 2.2-6: Moiré effect

The lines of the scanning disk and the scale disk are typically positioned parallel to one another. However, with the Moiré procedure, the lines of the scanning disk are rotated at a fixed angle (e.g. 5°-10°) against the scale (**Figure 2.2-6**). In this case, there is no longer a complete overlay of the strips of the scanning disk and the scale disk. So-called Moiré strips develop, which move in a vertical direction when the scanning disk is shifted horizontally. While the Moiré procedure requires more extensive evaluation, it permits a finer measurement resolution than with the strip width. Further details on the Moiré procedure can be found in Section 4.3.2.5.

Incremental measure embodiments are used in geometrical metrology for both length and angular measurement. The angle dimension is embodied by division markings, which are located on a circular course (**Figure 2.2-7**). Alternatively, the angle measurement is carried out according to the same principle as the length measurement.

Electrical Incremental Scales
The gridding of incremental scales is not always done optically. It is often done electrically, whereby the principle of induction or capacity modification is essentially utilised. The respective measure embodiment takes place through appropriate intervals and scopes of conductive strips, dialectrics, magnets or capacitor disks. Electrical scales generally require a relatively large amount of adjustment, especially for large length measurements, but they are durable, wear-free and com-

paratively insensitive to contamination such as dust or oil mist. **Figure 2.2-8** provides an overview of common electrical scales.

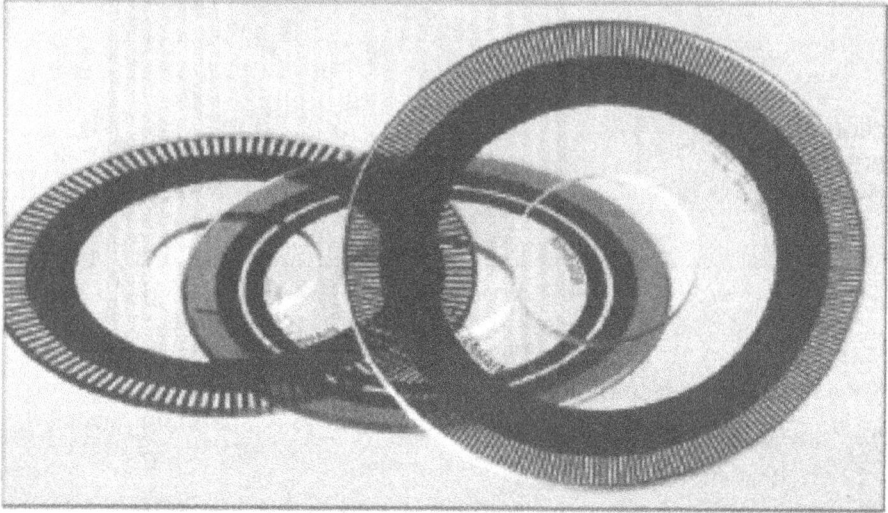

Source: Heidenhain

Figure 2.2-7:   Glass graduation substrates for incremental angle measurement

With the inductosyn scale (**Figure 2.2-8, a**), an alternating voltage is set on a
winding conductive strip (ruler) which functions as the primary coil of a transformer. Through the movement of a winding secondary coil (rider) over the ruler,
the magnetic coupling moves periodically between the primary and secondary coil
and with that, the level of the secondary induced voltage. The shift direction can
be determined if two offset spools are used in the rider.

The accupine scale (**Figure 2.2-8, b**) changes the magnetic coupling between the
two coils with magnetic cores, through adjustable ferromagnetic pivots. If an alternating voltage is set on the primary coil, the induced voltage, which is dependent
on the position of the pivots, can be measured and analysed at the secondary coil.

With the magnetic scale (**Figure 2.2-8, c**), a series of coils moves over series of
permanent magnets and thereby, through a magnetic field. According to the law of
induction, voltage impulses are thereby induced in the coils. These are counted and
their number is proportional to the shift which has taken place.

The capacitive scale (**Figure 2.2-8, d**) consists of thin metal strips on the scale and
scanning disks, which together form capacitors. When the scanning disk is moved

over the scale disk, the effective disk surface of the capacitors, and thereby, the capacitor voltage, which is linked through a resistance, changes. The shift is proportional to the maximum or minimum voltage which have occurred in the capacitor.

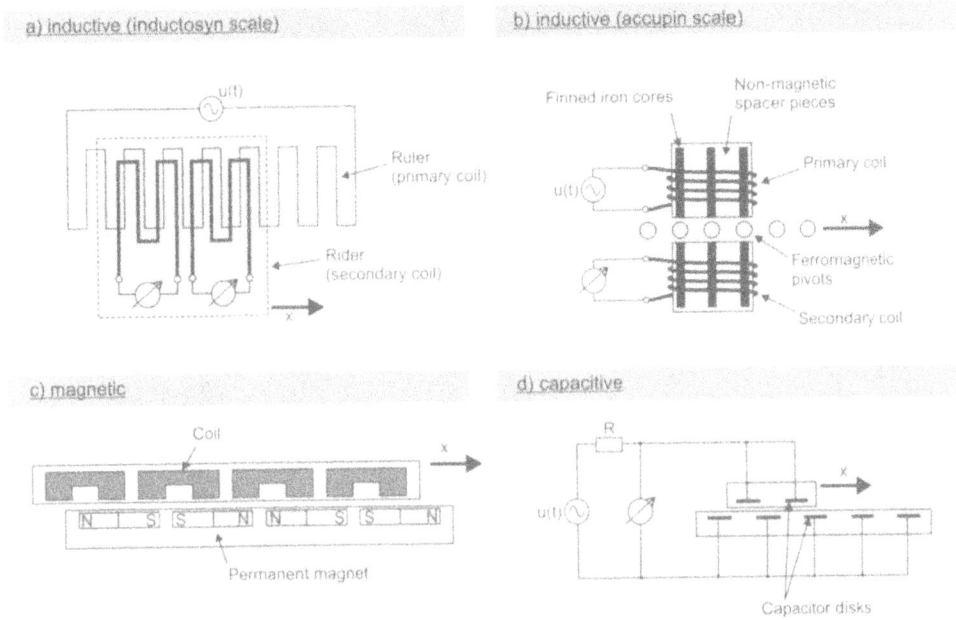

Figure 2.2-8: Overview of electrical scales

Further details about optical or electrical scanning of the described incremental scales can be found under Section 4.2.

2.2.2.2 Polygon mirrors

Polygon mirrors, which usually have 4, 8, 12, 36 or 72 surfaces, represent a relatively roughly gridded angle embodiment. Mirror polygons are usually made of a stress free block (ceramic(s), Zerodur or steel) or of individually positioned mirrors. The scanning of the mirror polygon can be done with an autocollimator ("AKF"), which is dealt with under Section 4.3.4.2. With every rotation of the polygon around the angular pitch α, the beam is reflected back into itself (**Figure 2.2-9**).

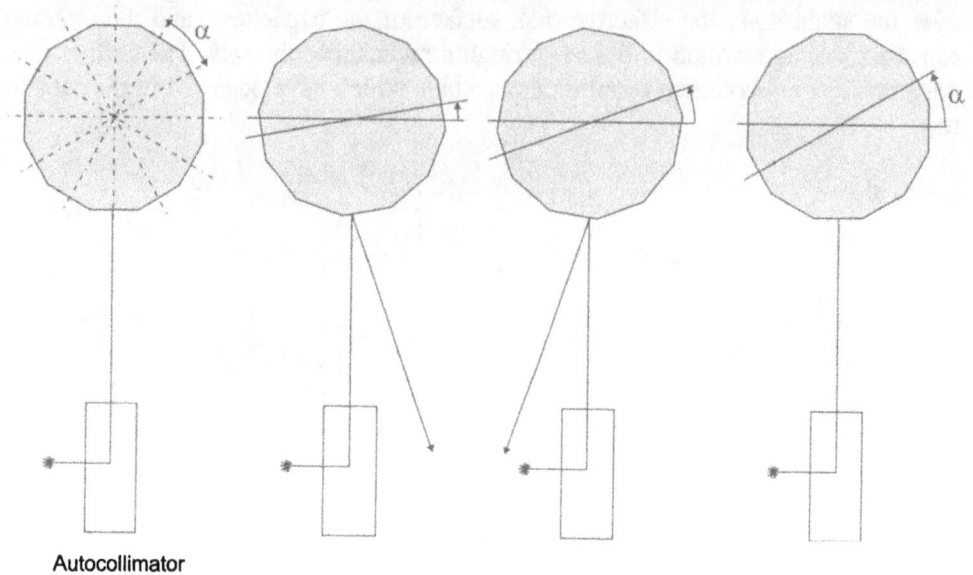

Autocollimator

Figure 2.2-9:   Mirror polygon in conjunction with an autocollimator

2.2.2.3 Light Wave Length as a Scale (Interferometer)

In contrast to the previously described measure embodiments, the wave length of light is an immaterial measure embodiment.

In order to use light wave length as a scale, a so-called interferometer is used, which was first introduced by Michelson in 1882. With the Michelson interferometer, . a laser beam is split into a reference beam and a measurement beam, each travelling different paths. Both parts of the beam are reflected back into themselves and recombined, after travelling through the reference and measurement paths. Depending on the difference in paths, a phase variance appears, which results in a wave with modified intensity. If the length of the measurement path changes, the minimum and maximum intensity are collected by a photodetector, with number of them being proportional to the shift which has taken place. Even with the Michelson interferometer, the direction of movement can only be detected using additional measures. For details on interferometry, please refer to Section 4.3.3.6.

2.2.2.4 Mechanical Scales

Length and angular dimension can be simply embodied through the pitch of pinions or spindles. Such mechanical measure embodiments have the advantage that they embody the size to be measured, while also enabling a transmission of force, for example, in positioning machine components. .

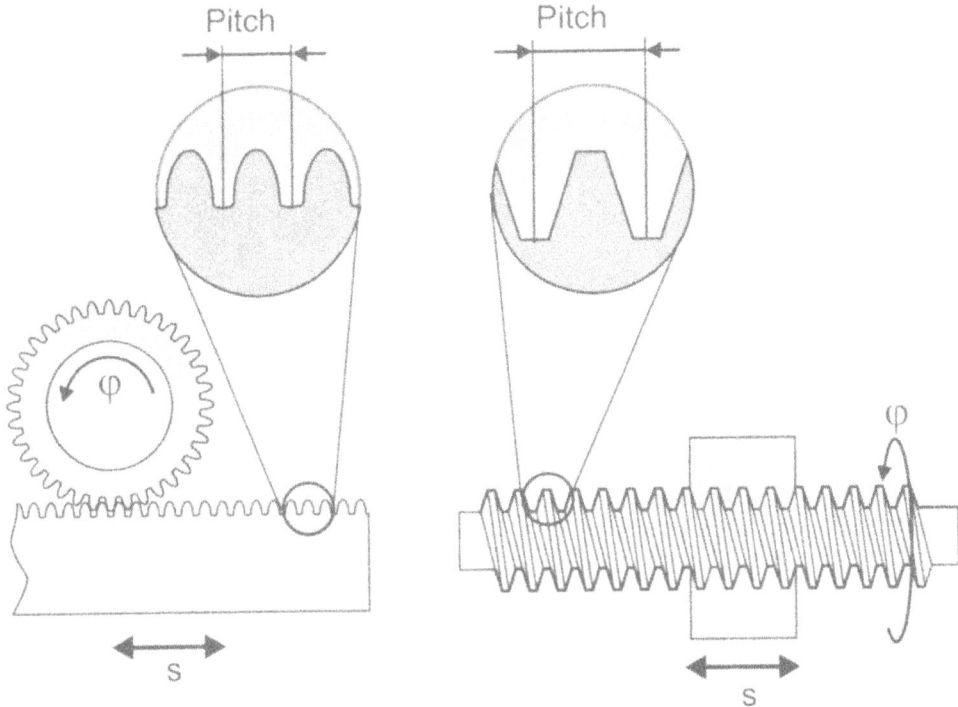

Figure 2.2-10: Measure embodiment through tooth pitch

For a gear rack and pinion arrangement, the rack is scanned by the winding of the pinion on the rack (**Figure 2.2-10**). The relative displacement between the rack and pinion is determined by measuring the angle of rotation.

The arrangement of a spindle and nut is based on a similar principle. The shift of the nut is determined by measuring the angle of rotation of the spindle.

2.2.3 Absolute encoded measure embodiments

With absolute encoded scales, . there is a specific allocation between position and display. Like incremental scales, they can also be used to embody lengths as well as angles. In order to encode the respective position, the binary code or Gray code is used. However, with the binary code, slight positional deviations of code markings can already lead to false results. The Gray code has the advantage that precisely one code place changes between the two positions, thereby reducing errors. **Figure 2.2-11** clarifies the differences between the scale codings.

Incremental scale

Absolute encoded scales

a) binary code

b) Gray Code

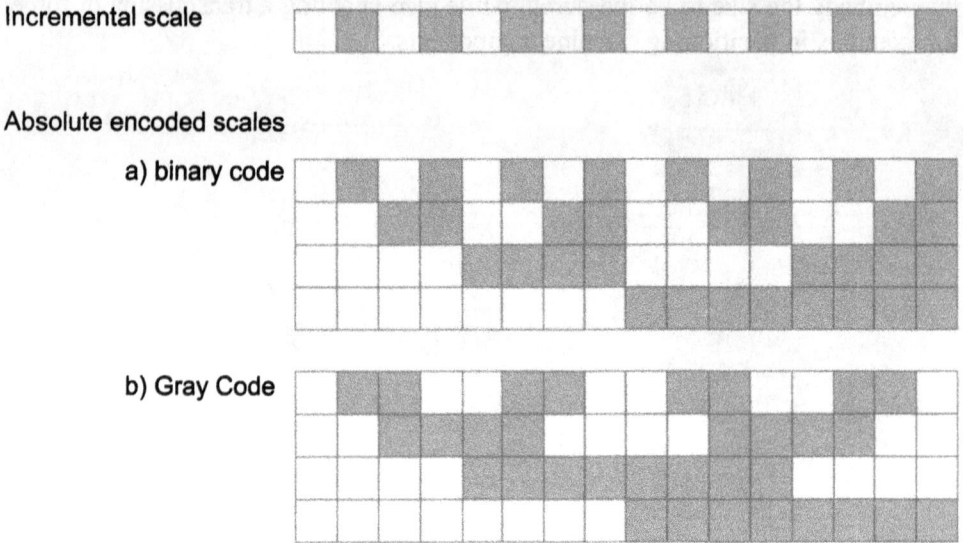

Figure 2.2-11: A comparison of different length scales

For the embodiment of angles, division markings are positioned appropriately on a circular disk.

In contrast to incremental scales, absolute encoded scales require a multitude of tracks and detectors. Generally, the number N of required tracks when using a binary code results from the length L of the measuring range and the desired resolution a, rounded up to the next whole number:

$$N = \log_2\left(\frac{L}{a}\right). \tag{2.2-1}$$

For example, with a measuring range of 300 mm and a comparatively small resolution of 10 µm, 15 tracks and detectors are required.

The relatively high number of tracks make absolute encoded measuring systems relatively expensive, which is why they have scarcely caught on in practice [Dut 96]. Furthermore, interpolation is not possible, so that the resolution is solely determined by the number of tracks used.

The advantages of absolute encoded scales are primarily their insensitivity to malfunctions which cause counting errors, as well as their ability to detect the direction of movement without additional devices. Furthermore, encoded scales do not the require the time-consuming start-up of reference marks to determine position.

2.3 Measurement Uncertainty and Measurement Error

A measurement result is useless without indicating measurement uncertainty. Although this is actually a trivial point, it is nevertheless still a regular occurrence, that measurement results are indicated without a measurement uncertainty, although a measurement result is always made up of the result and an appropriate measurement uncertainty [DIN 1319-1, DIN 1319-3, DIN 95, DIN EN ISO 14253]. As long as the measurement uncertainty is seen to be "small" compared with the inspected tolerance, this often still remains disregarded today (Section 2.3.4). However, due to the increasing demands on production and the accompanying decrease in production tolerances, measurement uncertainty is becoming an influence factor which must be considered when deciding whether a feature lies within the tolerance (Section 2.5). Defined rules ensure that the decisions made can be understood by all parties involved (Section 2.5.4.1).

2.3.1 Definitions and Terms

The goal of measuring a measurand is to determine its true value. However, due to influences affecting the measurement, unavoidable measurement errors occur. There is a differentiation between systematic influences and coincidental influences (Section 2.3.2). Due to influences on the measurement result, it is not possible to find the exact true value, so an agreed correct value is normally used.

Definition: A **correct value** is recognised by agreement and assigned to a particular quantity with an uncertainty appropriate for the respective purpose [DIN 94, DIN 95].

The measurement values and other information about the measurement serve only to determine the measurement result, as an estimate of the true value of the measurand, and the measurement uncertainty [DIN 1319-3].

Definition: Measurement uncertainty is a parameter associated with the result of a measurement that characterises the dispersion of values which could reasonably be attributed to the measurand [DIN 94, DIN 95].

Measurement uncertainty is often determined by taking into consideration knowledge about the measurement errors. It does not, however, correspond to the measurement error, but indicates an area assumed to be greater than or equal to the actual measurement error [DIN 1319-1].

Definition: Measurement error is the deviation from the true value of a value gained from measurements and assigned to the measurand [DIN 1319-1], or the measurement result minus the true value of the measurand [DIN 94].

As, in practice, the true value cannot be determined, a correct value is used in its place.

Measurement uncertainty and measurement error refer to a specific measurement. In this context, the term margin of error is used to identify parameters of a measuring device.

Definition: The **limit of error** is the maximum amount of measurement deviation of a measuring device [DIN 1319-1].

The limit of error is an amount and is therefore stated without signs [DIN 1319-1]. The extreme values of the measurement error of a measuring device, which naturally have a sign, are defined as the limit variances. This term is also used to indicate extreme values of parameter deviations (**Table 2.3-1**).

The term *measurement accuracy* is a qualitative term [DIN 94], which should not be used in connection with quantitatively determining the measurement uncertainty of measurements [DIN 55350].

The first step of measuring is the specification of the measurand in the form of a description (**Figure 2.3-1**). For example, when measuring embodied lengths, the specification of a temperature at which the measurement is to be carried out is an important component of the description. In principle, however, this description is never complete without an infinite quantity of information. That is why it contributes to the uncertainty of the measurement result to the extent that there is room for interpretation [DIN 95]. The true value itself is thus afflicted with uncertainty.For simplicity purposes, however, a unique true value is assumed here. Measurement is carried out on the realised size, which is ideally the specification of the measurand. The influence of coincidental measurement errors can be reduced by measurement repetition (Section 2.3.3). In this case, the average value of all measurements is determined. The unadjusted observed values are corrected on the basis of information about recognised systematic influences. Correction can take place before or after forming the average value. The measurement result determined after correction represents an estimate of the measurand. By determining the measurement uncertainty, the complete measurement result is subsequently determined, which consists of the measurement result and measurement uncertainty. The remaining residual error which, cannot be determined, just like the true value, forms a component of measurement

Value	Proceeding	Example

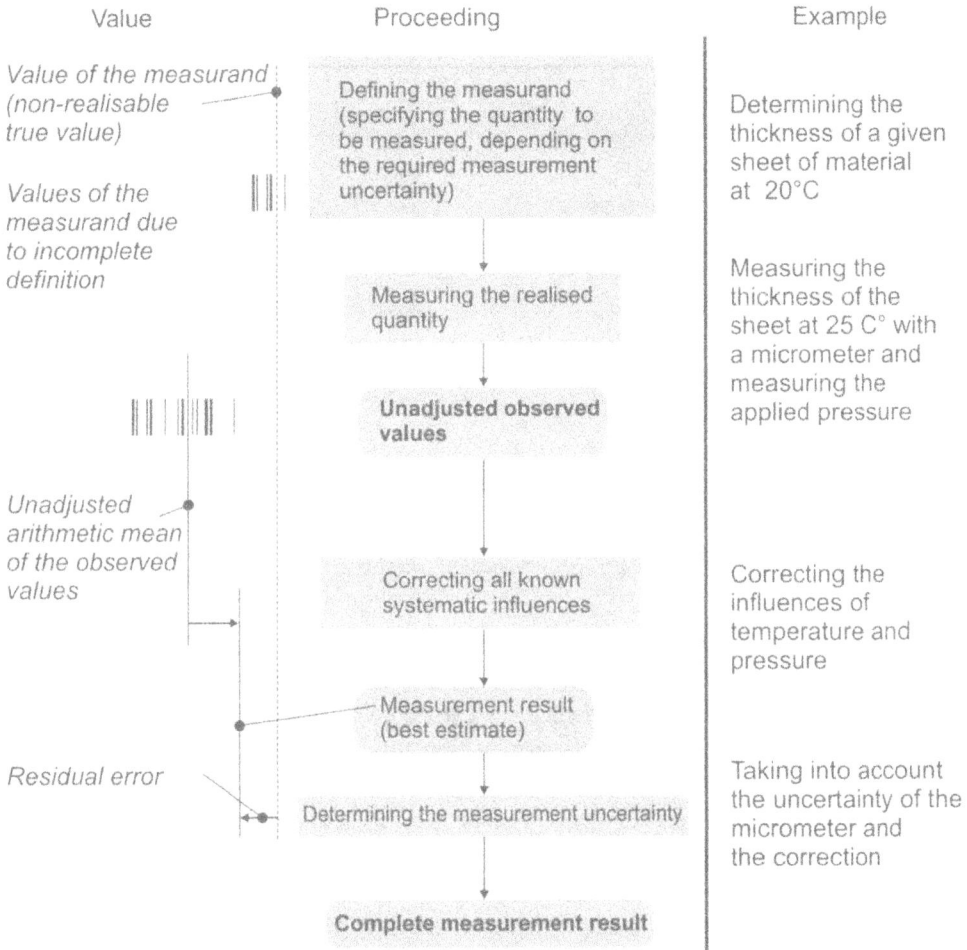

Value of the measurand (non-realisable true value)

Defining the measurand (specifying the quantity to be measured, depending on the required measurement uncertainty)

Determining the thickness of a given sheet of material at 20°C

Values of the measurand due to incomplete definition

Measuring the realised quantity

Measuring the thickness of the sheet at 25 C° with a micrometer and measuring the applied pressure

Unadjusted observed values

Unadjusted arithmetic mean of the observed values

Correcting all known systematic influences

Correcting the influences of temperature and pressure

Measurement result (best estimate)

Residual error

Determining the measurement uncertainty

Taking into account the uncertainty of the micrometer and the correction

Complete measurement result

Figure 2.3-1:   Measurement factors (example [DIN 95])

uncertainty. It must be considered that even the residual error cannot be specifically determined due to the incomplete definition.

2.3.2 Factors Influencing Measurement Error

Due to influences affecting a measurement, inevitable measurement errors occur, which are gauged upwards due to the measurement uncertainty. The influences and errors resulting from them can be classified, on the one hand, according to their cause and, on the other hand, according to their type.

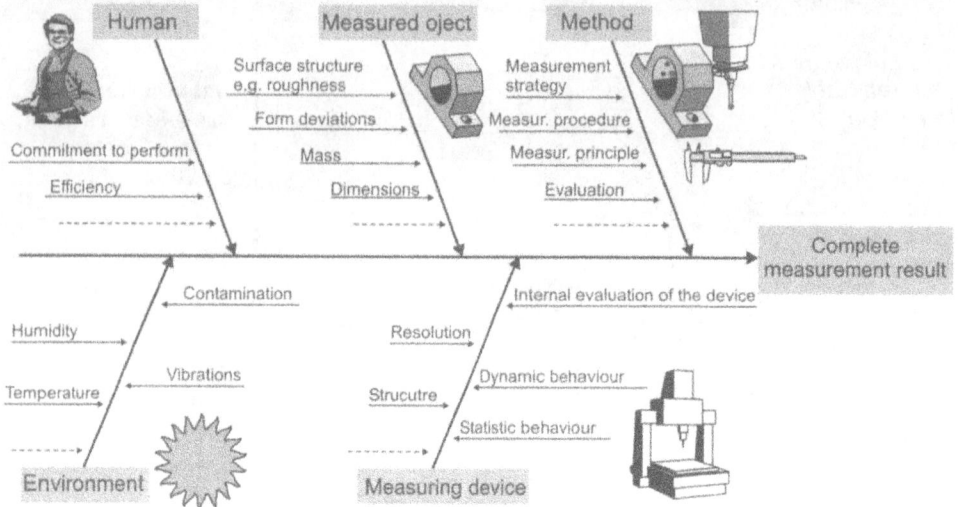

Figure 2.3-2: Cause and effect diagram of production metrology

In order to represent the causes of measurement errors, a cause and effect diagram (Ishikawa diagram) [Pf 96] can be used for the analysis of production processes and also in coordinate measurement technology [Wec 96], for example. Using the main factors influencing measurement, which are similar to the 6 elements of a production process (human, material, method, environment, machine, measurement), a cause and effect diagram can be set up for the measurement process within production metrology (**Figure 2.3-2**). As the measurement itself is omitted as an influence factor, production metrology is left with 5 factors. The list of individual influences does not pretend to be complete. Depending on the individual measurement, it must be checked which influences are so great that they must be taken into consideration, in the form of a correction or measurement uncertainty factor when determining the complete measurement result (Section 2.3.3).

Types of influence factors or measurement errors are basically differentiated between systematic and coincidental influences (**Figure 2.3-3**). The errors caused by known systematic influence factors are used to correct the measurement result. The residual error resulting from the uncertainty of the correction, the unknown systematic error and the coincidental errors must be gauged upwards by determining the measurement uncertainty. The following sections briefly describe the effect of substantial influence factors on measurement error, postioned according to the main influence factors from **Figure 2.3-2**. Depending on the measuring device being used, there are additionally device-specific influences, which are represented in connection with the individual devices in Section 4.

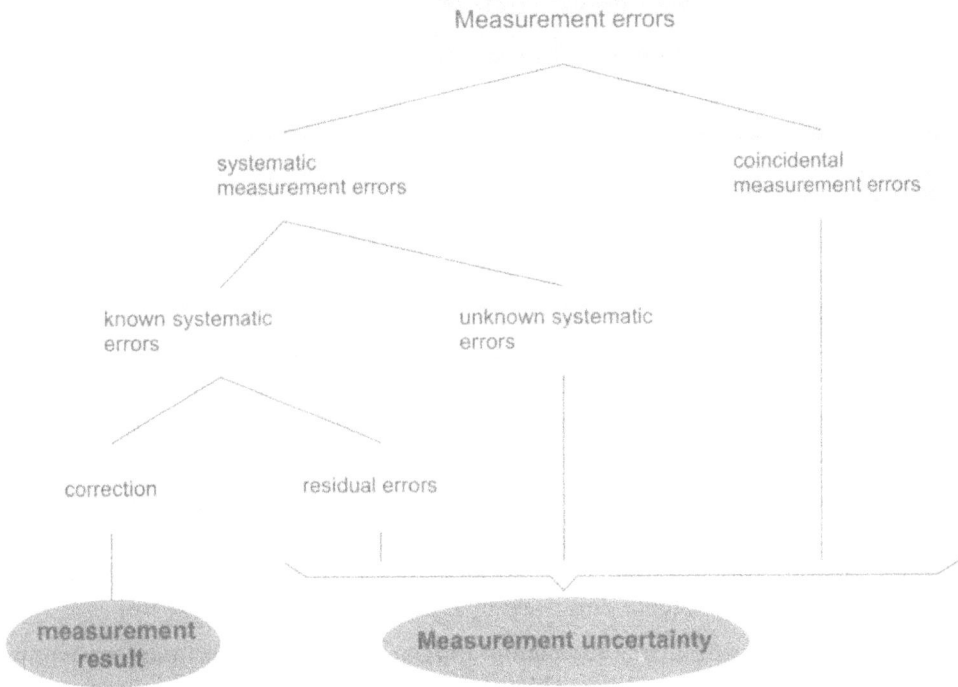

Figure 2.3-3: Types of measurement errors [Her 96]

2.3.2.1 Human Factor

The human, or worker, who carries out the measurement must be suitable in terms of efficiency as well as commitment to perform the measurement. Apart from physical suitability (e.g. visual acuity for the visual inspection), efficiency mainly refers to appropriate education and training.

Commitment to perform is influenced by personal motivation. An ergonomically arranged measuring station also increases commitment. This refers to measuring equipment and measuring aids as well as lighting, noise level, etc.

Nevertheless, error in taking readings and handling, as well as transfer and calculation errors cannot be ruled out during evaluation. Particularly readings taken off of a scale (parallax) from a diagonal viewpoint can influence the result substantially, as it is an error of the first order [Dut 96].

The human influence can be reduced by automating the measurement, however, it can never be completely ruled out, as even with fully automatic measuring equipment, a human is responsible for building the equipment and interpreting the results.

2.3.2.2 Environment: Vibrations, Contamination, Climate, Temperature

The substantial measurement errors in this area result from vibrations, contamination, humidity and temperature influences.

Floor *vibrations* particularly play a part with measuring devices which are installed directly on the floor, such as coordinate measuring devices [Pre 97]. Measuring device parts, such as drives, can also cause vibrations which influence the measurement result. Furthermore, the influence of vibrating electromagnetic fields and sound pressure are worth mentioning in this regard.

Contamination of both the measuring device and the measured object can influence the measurement result by simulating dimensional, form and surface defects. Causes of contamination are oil, dust and coarse dirt particles, for example in the form of filings. Particularly in areas close to production, coolant and lubricant mist are to be expected.

Humidity causes corrosion in both the measuring device and the measured object, which disturbs the measurement and possibly even the functionality of measuring devices. Granite, which is often used for measuring plates and other measuring device components, can become deformed by humidity penetration [Pre 97].

Temperature has a particularly substantial influence on measuring functions within the area of production metrology. This especially refers to length measurement, so it is fair to say: "Temperature measurement is not everything in length metrology, bit it cannot be done without [Pre 97]". A temperature influence is caused by heat transfer, which can take place via thermal conduction, convection or heat radiation.

There is a differentiation between three types of temperature influences [Pre 97]:

- Deviation of the temperature level from the reference temperature,

- temporal temperature fluctuations (temperature gradients), which can be long periodic (summer and winter, day and night) as well as short periodic (per hour) and spatial temperature fluctuations (temperature gradients).

Through the influence of temperature, most – particularly metal – materials expand reversibly with increasing temperature. This linear expansion characteristic is described by the equation

$$\Delta L = L \cdot \alpha \cdot \Delta t \tag{2.3-1}$$

ΔL defines the length variation, L the nominal length, α the linear expansion coefficient (**Table 2.3-1**) and Δt the change in temperature.

Table 2.3-1: Coefficient of thermal expansion for solid bodies [Her 97, Pre 97].

Material	Coefficient of linear expansion α [10^{-6}/K]	Limit variance [10^{-6}/K] u_α
Aluminium alloy	23..24	0,5..2
Glass	8..10	0,5
Grey cast iron	9,5..10	0,5
Steel	10..12	0,5..1,5
ZERODUR (glass ceramic)	0..0,05	0,05

For precise measurements, the heat radiating from the examiner is already suffi-cient to crucially change a measurement. According to equation 2.3-1, a 100 mm long steel ruler will already stretch by more than 1 μm with a temperature differ-ence of 1°K.

When measuring, the change in the length of scale must be considered as well as the change in length of the measured object. Then

$$\Delta L = L \cdot (\alpha_{Work} \cdot \Delta t_{Work} - \alpha_M \cdot \Delta t_M)$$ (2.3-2)

applies. The index *Work* indicates the workpiece or measured object and the index *M*, the scale for instance measuring device. The reference temperature for length metrology is 20°C [DIN 102, ISO 1]. The temperature change or deviation Δt re-fers to the deviation from this reference temperature. Equation 2.3-1 can only be used for a measurement when both the scale and the measured object have the same coefficient of expansion. Δt then represents the temperature difference be-tween the scale and the measured object.

The effects of the change in length are nil or negligible under the following condi-tions [Dut 96]:

- linear expansion coefficients are very small ($\alpha \approx 0$)
- linear expansion coefficients and temperatures of workpiece and measuring device are equal ($\alpha_{Work} = \alpha_M$ and $t_{Work} = t_M$)
- workpiece and measuring instrument both have reference temperature

The first case is only achieved with measured objects and measure embodiments made of glass or Zerodur. The second case occurs with measurements of steel workpieces and the utilisation of scales made of the same material. The third case is realised by carrying out the measurement in an air-conditioned measuring room (Section 2.4).

The effects described by equations 2.3-1 and 2.3-2 are limited to a temperature influence, which arises when there is a deviation from the reference temperature or a temperature gradient between the measuring device and the measured object. Spatial temperature gradients within the measuring device or measured object and temporal temperature gradients cannot be taken into consideration in this way. Their mathematical description is substantially more extensive. In order to compensate for temperature influences within the measuring devices, complex corrective procedures are implemented [Bre 93, Tra 89]. The temperature gradient within the workpiece can only be equalised through sufficiently long storage of the workpiece at a constant temperature, which can be shortened, for example with an air shower. The influence of temporal temperature gradients in measuring devices is reduced by using components with high heat conductivity such as aluminium. By using air conditioning, for example in a metrology laboratory or measuring chamber, the temperature gradients can be specifically limited.

The quantities determined with specific equations can be used to correct the measurement results by subtracting the calculated deviation from the measured length. To the extent that a correction is waived, it can be used to estimate the systematic measurement error caused by the influence of temperature [Her 97].

The calculable influence of temperature contains an uncertainty which results from the uncertainty of the temperature measurement and the uncertainty of the coefficients of expansion (**Table 2.3-1**). This uncertainty influence must also be taken into consideration. This is computed by partially deriving the equations 2.3-1 or 2.3-2. For equation 2.3-2, the following expression results from the squared addition of the individual parts (Section 2.3.3), where u_α represents the uncertainty of the coefficient of expansion (**Table 2.3-1**) and $u_{\Delta t}$ the uncertainty of the temperature measurement:

$$u = L \cdot \sqrt{\left(u_{\alpha Work} \cdot \Delta t_{Work}\right)^2 + \left(u_{\Delta t Work} \cdot \alpha_{oerk}\right)^2 + \left(u_{\alpha M} \cdot \Delta t_M\right)^2 + \left(u_{\Delta t M} \cdot \alpha_M\right)^2}$$

$$(2.3-3)$$

The uncertainty of the temperature measurement must be exactly determined for each individual case [VDI/VDE 3511]. The limit variations for temperature measurements are usually situated between 0.2 and 1 Kelvin, depending on measurement conditions [Her 97]. The uncertainty for equation 2.3-2 can be determined in a similar manner. It should be observed that all uncertainty quantities have the same level of confidence. Limit variations of uniformly or evenly distributed quantities, as they are here, should therefore always be converted into a standard uncertainty (Section 2.3.3).

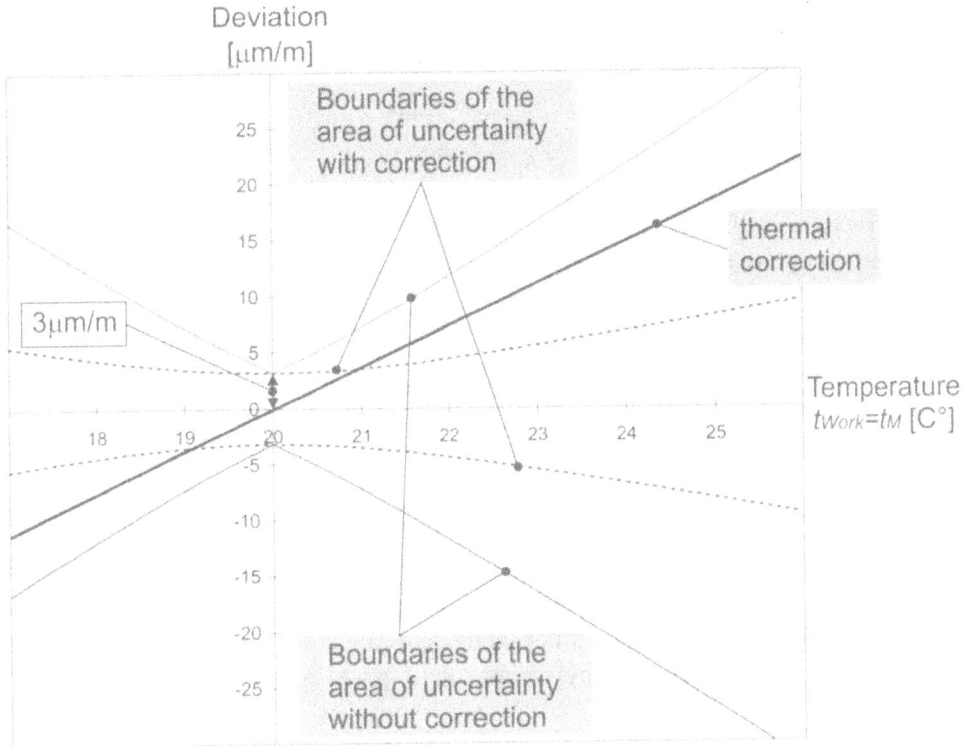

Figure 2.3-4: Length measurement error conditional upon temperature

The influence of temperature, as described by equations 2.3-2 and 2.3-3 is clarified in **Figure 2.3-4** through an example where a steel workpiece ($\alpha_{Work} = 11,5 \cdot 10^{-6}/K$, $u_\alpha = 1,2 \cdot 10^{-6}/K$) is measured with a glass scale ($\alpha_M = 7,8 \cdot 10^{-6}/K$, $u_\alpha = 0,5 \cdot 10^{-6}/K$) under the condition that the workpiece and scale have the same temperature and the uncertainty of the temperature measurement is $u_\vartheta = 0,2$ K. In the case of a measurement which takes into account the thermal correction calculated according to equation 2.3-2, the resulting area of uncertainty is indicated by the dotted line. Its boundaries are described by the equation 2.3-3, where an extended measurement factor was assumed with the extension factor 2. It is also assumed that the specifications regarding the uncertainty of the coefficient of expansion and the determination of temperature are to be regarded as limit variances of a uniform distribution. The result of 2.3-3 is therefore multiplied by the factor $2/\sqrt{3}$ (Section 2.3.3). It is clear to see that even a measurement at 20°C is afflicted with a thermally induced measurement uncertainty 3μm/m.

To the extent that no correction is carried out, the length measurement error described by equation 2.3-2 is a component of measurement uncertainty in the form of a uniformly distributed limit variation. As it has a systematic influence, it is simply added to the component calculated with equation 2.3-3, which was multiplied by $2/\sqrt{3}$.

In both cases, a symmetrical area of uncertainty results, which grows with increasing distance from the reference temperature of 20°C. The growth of the area without correction is clearly stronger. Systematic influence is rapidly becoming a dominant component of measurement uncertainty.

This example makes it clear that, on the one hand, a correction can significantly reduce the measurement uncertainty where there are strong deviations from the reference temperature. On the other hand, a measurement uncertainty remains, despite the correction, which can only be reduced through a more accurate temperature measurement and an improved specification of the coefficient of expansion.

2.3.2.3 Measured Object

Features of the measured object itself can also influence measurement errors. This is especially the case with form deviations (Section 4.5) of the measured object, which result in the probed points not being representative of the shape. The surface structure can also play an important role. Optical features, such as reflection

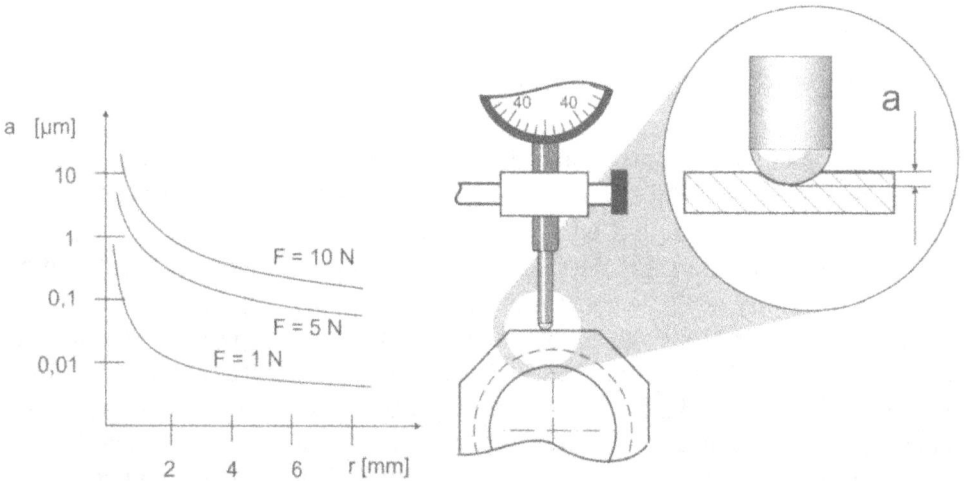

Example: ball-plane (steel)

Flattening (Hertz Pressure) through measurement force

Figure 2.3-5:   Measurement error through deformation [Dut 96]

behaviour influence the measurement when using optical measuring devices (Section 4.3). When using tactile devices, the roughness and waviness can play a role.

$$F = mg \downarrow$$

For the deflection of uniform measured objects (pipes, rulers, plates) under dead weight:

a) Support at the ends: maximum deflection

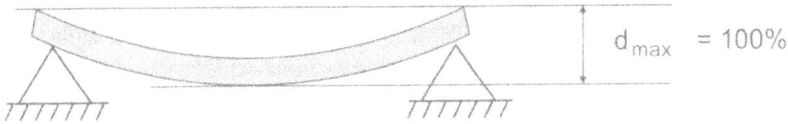

d_{max} = 100%

b) Support at each 0,22 l: minimum deflection

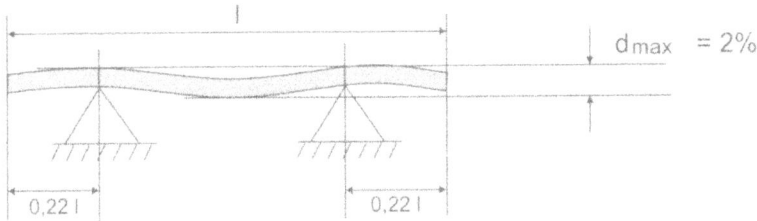

d_{max} = 2%

0,22 l   0,22 l

Figure 2.3-6:   Deflection of slim measured objects

With tactile devices, the shape of the measured object can be deformed through the force applied by a measuring device. This particularly applies to unstable measured objects. Additionally, with steel workpieces which are probed, for example, by a measuring scanner with a steel ball, measurement errors can occur due to a flattening at the sensor tip and the workpiece surface. This flattening through Hertz pressure can be calculated (**Figure 2.3-5**). With a sensor tip radius of more than 1.5 mm and a measurement force of up to 1.5 N, this effect can be neglected [Dut 96]. The influence which is caused by storing and mounting the measured object can be assigned to the measured object influence factor as well as to the measurement method. As storage and mounting are substantially determined by the dimensions and weight of the measured object, the influence is allocated to the measured object. When mounting measured objects, it must be made certain that the workpiece is not deformed. It is preferable to store objects statically with a three point support. It must always be taken into account that measured objects can be deformed through surface application when using magnets.

Measured objects which are slim or have thin walls become deformed under their dead weight. Using pipes, shafts, rulers or plates are their ends for support is the least favourable form of storage (**Figure 2.3-6, a**). Having support at points which

are each 0.22 times the linear distance from the outside edges (Bessel points) reduces deformation to approx. 2% with homogeneous objects (**Figure 2.3-6, b**).

2.3.2.4 Measuring device: Errors of the first and second order, Abbé principle

Even though the device-specific influences are already described in Section 4
when describing the devices themselves, this section nevertheless explains several
fundamental effects and connected terms which occur with many devices.

Measurement errors due to inaccurate guides play an important part in measuring
devices with a built-in measure embodiment. The clearance, which is technically
required in guides for measuring pins, touch probes or eyepieces, causes tilting.
The influence of these on the measurement result are large or small, depending on
how the measure embodiment and the measured object have been positioned.

With a parallel offset of the measuring distance and the reference distance, a small
tilt already causes measurement errors which are no longer negligible. The calliper
gauge is an example of such an arrangement (**Figure 2.3-7**).

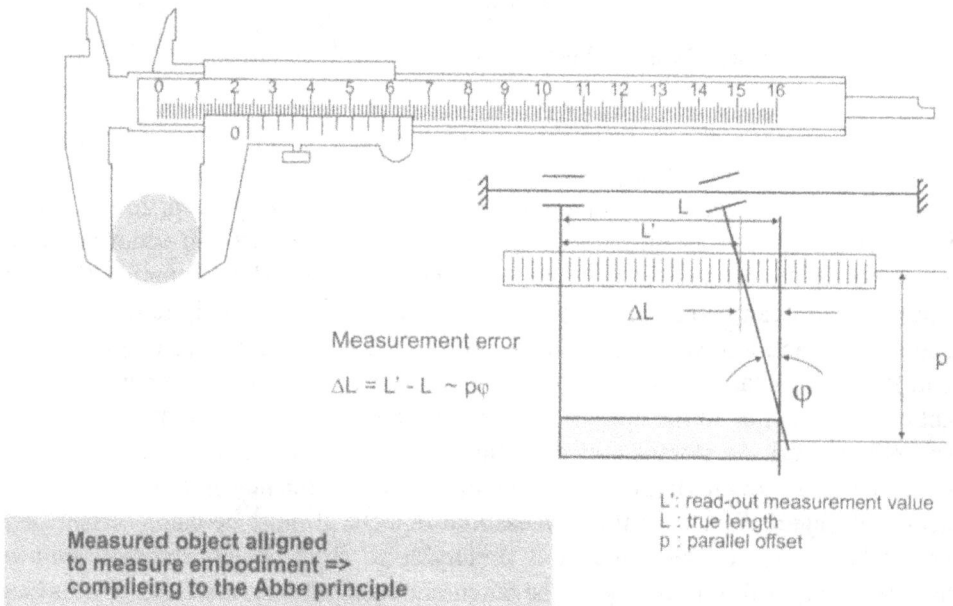

Measurement error

$$\Delta L = L' - L \sim p\varphi$$

Measured object alligned
to measure embodiment =>
complieing to the Abbe principle

L': read-out measurement value
L : true length
p : parallel offset

Figure 2.3-7: Error of the first order

Depending on the point of contact of the inspected workpiece with the measuring
jaws, a parallel offset p occurs between the measured object and the measure embodiment (**Figure 2.3-7**). The measurement error amounts to

$$\Delta L = L - L' = p \cdot \tan(\varphi) \tag{2.3-4}$$

and for small angles ($\varphi \ll 1$) with φ in the radian measure

$$\Delta L = p \cdot \varphi. \tag{2.3-5}$$

Tilts of $\varphi = 0{,}35'$ to $2{,}6'$ (angle minutes) are to be expected [Dut 96]. A tilt of $\varphi = 2'$ and a parallel offset of $p = 30$ mm results in a measurement error of $\Delta L = 17$ μm.

The error arising is called an error of the first order. The term "first order" refers to the order of the influence factor causing the error, which, in this case is the order of the angle φ.

In order to avoid this error of the first order, the scale of the measuring device must be positioned such that the distance to be measured forms a straight-line continuation of the scale. This rule, which was already formulated in 1893, is called the Abbé principle or comparator principle. An example of a measuring device positioned according to the Abbé principle is the micrometer (**Figure 2.3-8**). The measured object, measurement surfaces and screw pillar, with an upward gradient which serves as a measure embodiment are colinear.

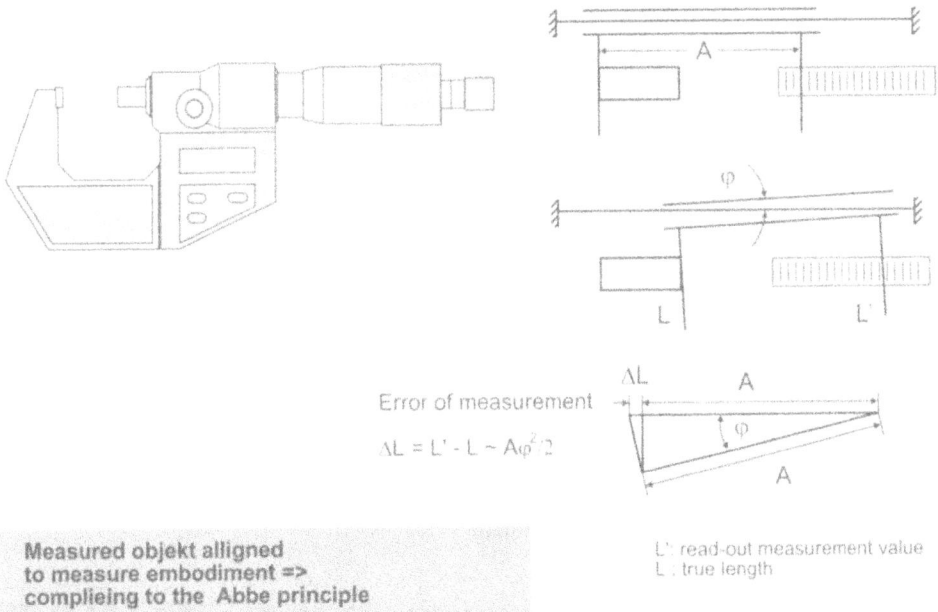

Error of measurement

$\Delta L = L' - L \sim A \varphi^2 2$

Measured objekt alligned
to measure embodiment =>
complieing to the Abbe principle

L': read-out measurement value
L : true length

Figure 2.3-8: Abbé principle, error of the second order

The measurement error thereby amounts to ($\varphi \ll 1$)

$$\Delta L = A\left(\frac{1}{\cos(\varphi)} - 1\right) = A\left(\sqrt{1 + \tan^2 \varphi} - 1\right) \approx A\left(\sqrt{1 + \varphi^2} - 1\right) \qquad (2.3\text{-}6)$$

and is simplified with φ in the radian measure to

$$\Delta L \approx A\left(\frac{\varphi^2}{2}\right). \qquad (2.3\text{-}7)$$

When positioned according to the Abbé principle, an error of the second order occurs, which is proportional to the second power of the tilt angle. As this angle is very small, its square is near to zero, so that the emerging error is usually negligible.

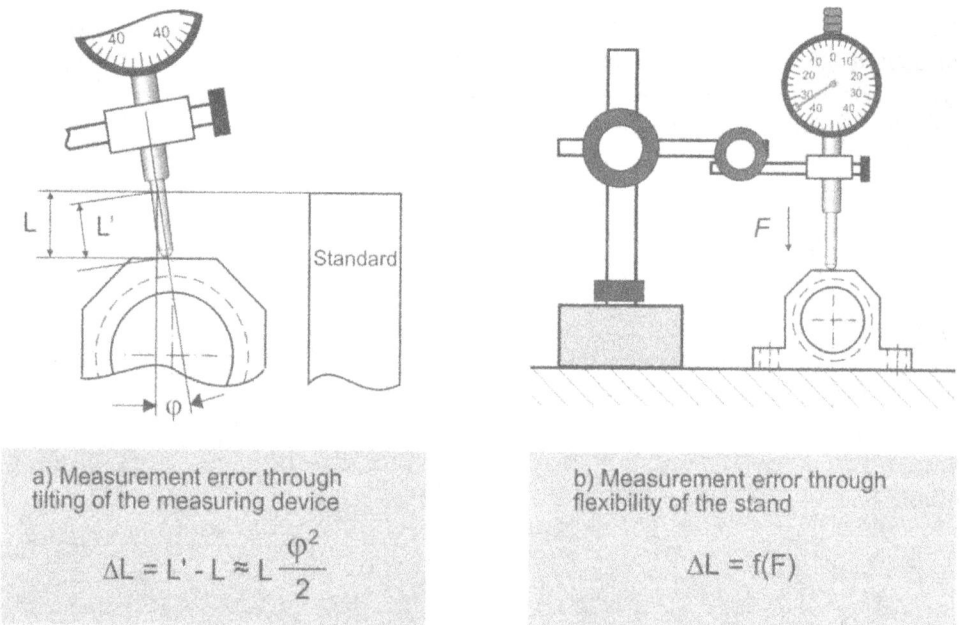

a) Measurement error through tilting of the measuring device

$$\Delta L = L' - L \approx L\frac{\varphi^2}{2}$$

b) Measurement error through flexibility of the stand

$$\Delta L = f(F)$$

Figure 2.3-9: Measurement error through tilting

A tilted measuring device also causes an error of the second (Figure 2.3-9). This can occur, for example, if the previous measurement is set up with a standard which strongly deviates in geometrical dimension from the measured object. Due to the size difference, the deflection of the probe element when measuring the object differs from the deflection when probing the standard gauge. If the measuring instrument is not colinear with the sought-after size, a measurement error results, which is dependent on the tilt angle and the difference in length between the workpiece and the standard (**Figure 2.3-9, a**).

Tilting can also occur through incorrect clamping or non-alignment, but can also be caused by the measuring force applied by the measuring device. The force can cause a deformation of the stand by bending the invidual joints and levers upwards. The tilt angle and thus, the measurement error, are then a function of the measuring force F (**Figure 2.3-9, b**).

In addition to the effects of tilting on guides, friction and clearance can cause reversal span, or hysteresis.

The resolution power of a measuring device can be a very substantial component of measurement uncertainty. The *resolution* defines the smallest difference between two displays which can be uniquely differentiated within a display facility [DIN 94, DIN 1319-1]. Therefore, at any rate, the resolution forms the boundary of the measurement error obtainable by a measuring device.

An additional important cause of measurement errors is incorrect measure embodiment (Section 2.2).

Finally, the entire area of signal transmission and signal processing within a measuring device is a possible cause of measurement errors. Particularly with electrical measuring devices and CNC controlled devices, such as coordinate measuring devices, dynamic effects can also influence measurement error [Pro 97].

2.3.2.5 Method

The selection of the measurement principle is fundamentally important to the influence of measurement error. With a length measurement which is based upon the interference effects of light in the form of a laser (Section 4.3.3.6), smaller measurement errors are realised than with a length measurement where the upward gradient of a screw pillar is used as a measure embodiment, as with a micrometer (Section 4.1).

The measurement procedure, measuring method and practical application of the measurement procedure or measuring method are also crucial to the size of the measurement error. For example, the direct positional measurement of a carriage on a processing unit equipped with a spindle drive using an incremental scale is more accurate than an indirect measurement via a rotary encoder on a spindle. In the latter case, there is also an effect caused by transfer errors between the carriage and the rotary encoder [Wk 95].

As many measurement results only achieve an interpretable form through evaluation, the evaluation itself can also have an influence on measurement error. This plays a substantial role, for example, with coordinate measuring technology (Section 4.4), where only individual points of the object are recorded in the form of coordinates. The actual dimensions are determined through computer-aided evaluation.

2.3.3 Procedures for estimating measurement uncertainty

The measurement errors which are caused by different influences (Section 2.3.2) overlap. They cause a deviation between the measured value and the true value. This means that every measurement is afflicted with uncertainty. In the context of evaluating the measurement, this uncertainty is estimated.

The basis of every procedure for estimating measurement uncertainty is the "Guide to the Expression of Uncertainty in Measurement", also called "GUM" [DIN 95]. In order to apply this extensive manual more easily, a simple procedure is being developed by the international standardisation committees on the basis of GUM to determine measurement uncertainty. With the aid of an iterative method, a cost optimising solution is determined for a particlar measuring task, while taking into account the measurement uncertainty. The method is called the PUMA method" (Procedure for Uncertainty Management) [ISO 14253-2]. In Germany, methods for the evaluation of measurements have been compiled within DIN 1319 which are orientated towards GUM [DIN 1319-3], [DIN 1319-4]. They form the basis of the following representation.

The evaluation of a measurement, including the estimation of measurement uncertainty, can be carried out in four steps [DIN 1319-3], [DIN 1319-4]. (The procedure according to GUM is broken down into 8 steps):

a) Setting up a model which mathematically describes the relationship of the measurands (output variables) to all other quantities involved (input variables).

b) Preparation of the given measurement values and other available data (the average is calculated of multiple quantity measurements and the measurement uncertainty of each individual input variable is determined).

c) Calculation of the measurement result and measurement uncertainty of the measurand from the prepared data.

d) Specification of a complete measurement result and determination of the extended uncertainty.

This methodology is described here in more detail using the example of evaluating an individual measurand as it is usually done in production metrology. A more general representation is given in [DIN 95] and [DIN 1319-4].

The starting point of estimating measurement uncertainty is a mathematical function model which describes the relationship between the measurand y as an output variable and the input variables x_1, $x_2,..,x_n$:

$$y=f(x_1, x_2,..,x_n) \qquad (2.3\text{-}8)$$

As long as the measurand has been measured several times, an arithmetic mean can be calculated from the individual measurement results (Section 5.1). This represents the uncorrected measurement result of one direct individual measurement.

If several measurements are carried out, the standard deviation of the mean u is determined (Abschnitt 5.1). It is calculated from the standard deviation of the observed values s, where n denotes the number of observed values:

$$u = \frac{s}{\sqrt{n}} \qquad (2.3\text{-}9)$$

This captures the influences which have changed between the individual measurements and represents a component of uncertainty. The influences captured through these components are dependent on the number of measurements and the conditions under which they were carried out. All other influences must also be taken into consideration.

The influences on a measurement bring about coincidental and systematic measurement errors. The known systematic error" is used to correct the measurement result. The unknown systematic and coincidental errors" from the measurement uncertainty (**Figure 2.3-3**). Their influence must be estimated, to the extent that it is not taken into account by the repeated measurements. Data from previous measurements, manufacturer data or knowledge about the behavior and characteristics of the measured objects and measuring devices are examples of what can be used.

Determining the proportion of uncertainty through repeated measurements is also defined as method A, while using another approach is defined as method B [DIN 95].

It is crucially important for the subsequent summarising of individual components of measurement uncertainty, that the dispersion parameters used for the individual components are equivalent and thus transferable [DIN 95]. This is achieved by expressing the measurement uncertainty as a standard deviation. This is also ex-pressed to as the standard uncertainty [DIN 95].

Method	Form of distribution	Calculation
A	Normal distribution 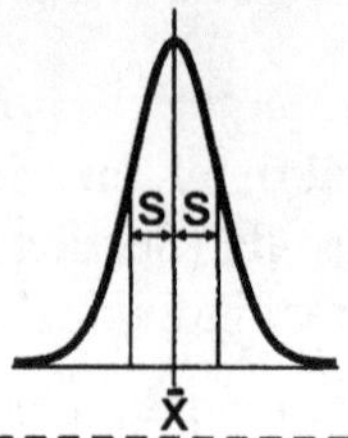	Standard uncertainty of the mean value (s: stand deviation n: number of observed values) $u=\dfrac{s}{\sqrt{n}}$
B	Normal distribution 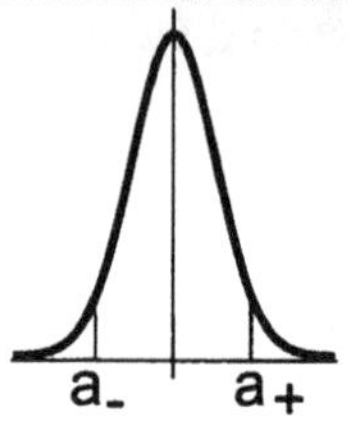	Assumption: the estimated value lies within the boundaries a_+ and a_- with a confidence level of 95% $u=\dfrac{a}{\sqrt{4}}$
B	Uniform distribution 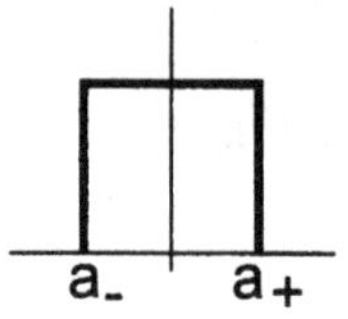	Assumption: the estimated value lies within the boundaries a_+ and a_- with a confidence level of 100% $u=\dfrac{a}{\sqrt{3}}$
B	Triangular distribution 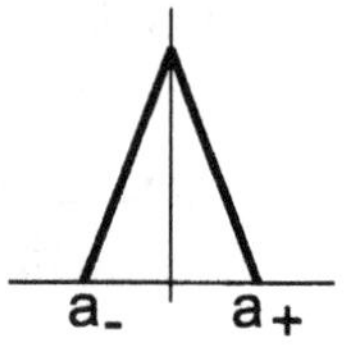	Assumption: the estimated value lies within the boundaries a_+ and a_- with a confidence level of 100% $u=\dfrac{a}{\sqrt{6}}$

Figure 2.3-10: Determining the standard uncertainty u

In order to determine measurement uncertainty using method B, the standard deviation, or standard uncertainty, is determined from the available information. some simple examples are represented in **Figure 2.3-10** [DIN 95].

In order to determine the complete measurement result, the correction, which is the negative value of the measurement error, is added to the uncorrected observed value which is formed by the average value of several measurements. The result represents the best estimated value for the measurand which can be determined through this measurement.

The measurement uncertainty of the measurand, the combined *standard uncertainty* u_c, is determined through squared addition of the individual uncertainty components. As long as all input variables are independent from one another:

$$u_c = \sqrt{\left(\frac{df}{dx_1} u_{x1}\right)^2 + \left(\frac{df}{dx_2} u_{x2}\right)^2 + ... + \left(\frac{df}{dx_n} u_{xn}\right)^2}$$

(2.3-10)

applies. $\frac{df}{dx_i}$ represents the partial derivation of equation 2.3-9.

These *sensitiviy coefficients* quantify the sensitivity of the model with respect to changes in input variables and so determine the contribution, which the individual uncertainty contributions make to the standard measurement uncertainty, to which the output variables are assigned.

If the input variables are not independent of each other, the correlations must also be taken into consideration [DIN 95, DIN 1319-3]. Enlarging the standard uncertainty by the correlation is usually negligible for measurements in production metrology [Dut 97]. If the quantities x_1, x_2 to x_n from equation 2.3-9 are linked together linearly and directly weighted, equation 2.3-10 is simplified to:

$$u_c = \sqrt{u_{x1}^2 + u_{x2}^2 + ... + u_{xn}^2}$$

(2.3-11)

The form of this equation corresponds to that of the error propagation law.

In production metrology, the measurement uncertainty is not given as standard uncertainty but as a range, in which 95% of the measured values are expected [DIN 95, DIN EN ISO 14253-1]. The evaluation methods used usually assume that the measured values are distributed normally [Dut 96]. The method described here according to [DIN 95, DIN 1319-3, DIN 1319-4] is independent of the distribution form.

If normal distribution is assumed for the measured values, the specification of the measurement uncertainty in form of a standard deviation corresponds to a range of confidence of 68.3%. A range of confidence of 95% can be determined by multiplying the standard deviation by two. For a different range of confidence a different multiplier must be used (Section 5.1). If there is no normal distribution the multiplier must also be adjusted. However, the value of 2 is on the safe side.

These factors are called extension factors. Multiplication with the *extension factor* k, shows the *extended measurement uncertainty*:

$$U = k \cdot u_c \qquad\qquad\qquad (2.3\text{-}11)$$

Definition: The extended measurement uncertainty is a characteristic value, which indicates an area around the result of measurement. Coverage of a large proportion of the distribution values can be expected from it and the measured variables are reasonably assigned [DIN 95].

Details about an extended measurement uncertainty must always include the extension factor. It only characterises a range of confidence if the distribution of the measured values is known.

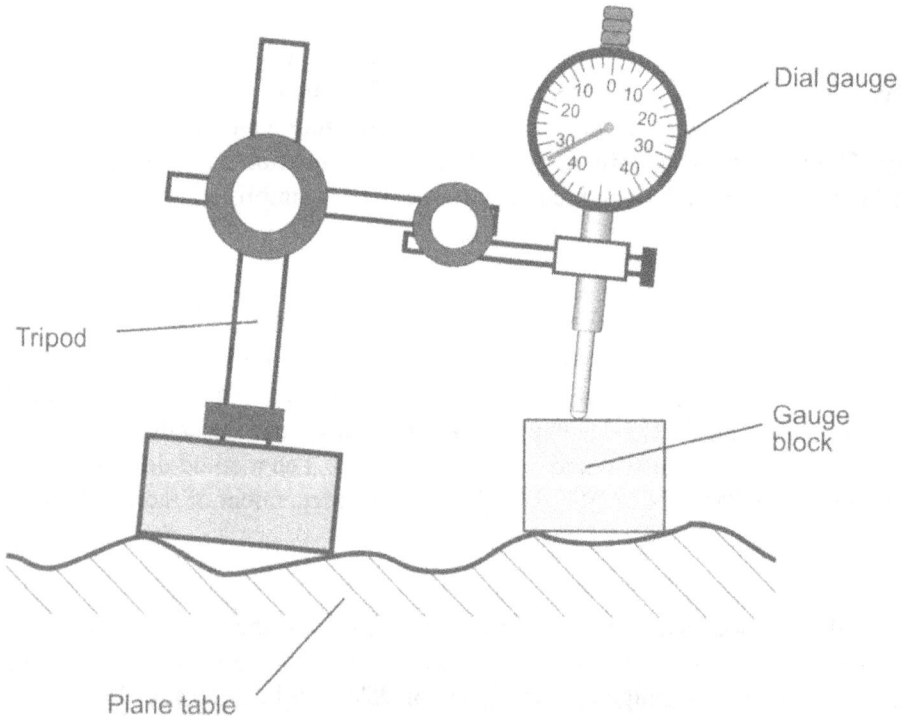

Figure 2.3-11: Example for the definition of measurement uncertainty

The method to define measurement uncertainties can be described with a simple measuring task. Further examples can be found in [DIN 95, Kes 95, DIN 1319-3, DIN 1319-4].

For the measuring task a dial gauge is used to define the height h of a gauge block (**Figure 2.3-11**). The five main disturbances for the measuring result are:

- Human:
  The measurement is carried out by a trained technician, which ensures that no serious errors occur during measurement. The influence, which results from the handling of the measuring device, is not known.

- Surroundings:
  The measurement is carried out at 20°C (error limit of the temperature definition 2 K). The thermal expansion coefficient of the workpiece lies at $12 \cdot 10^{-6}$/K (uncertainty $1 \cdot 10^{-6}$/K).

- Measurement object:
  Form deviations are negligible compared to measurement deviations. Surface characteristics, measurements and values do not lead to significant influences.

- Measuring device:
  Measurement deviations of the dial gauge for a temperature range of 18 to 22°C are, according to the manufacturer, in an area of ±0.02 mm for 95% of the values. It is assumed that the values are distributed normally. A systematic deviation of 0.06 mm is to be taken into consideration. The levelness of the plane table, the support face of the tripod and the formation of the tripod are not known.

- Method:
  If the order is in accordance with Abbé's principle, measuring deviations caused by small angle errors can be neglected. Deviations in the measurements, which are caused by shell distortion, should be neglible by limitation of the probe force and a sufficiently large probe radius (**Figure 2.3-5**).

Twenty measurements are taken from different points of the plane table. The mean value of the observation value is $y = 100.02$ mm. The standard deviation is calculated from the value of $s = 0.09$ mm.

The systematic deviation necessitates a correction of the mean value, so the best estimate can be used as the real value:

$$y = 100,02\ mm - 0,06\ mm = 99,96\ mm \qquad (2.3\text{-}12)$$

By taking repeat measurements at different points on the plane table the influence of the level of the tables, the tripod base and the shape of the whole measurement arrangement are taken into consideration. The standard deviation of the mean value of the repeat measurements is in accordance with equation 2.3-9:

$$u_1 = \frac{0,09mm}{\sqrt{20}} = 0,02mm \qquad\qquad (2.3\text{-}13)$$

The uncertainty of the temperature definition is determined with equation 2.3-3. Since the error limits of the measuring device are for 18 to 22°C, the thermal influences on the measuring device are taken into consideration in this value. There is no deviation from the reference temperature. The error limit for the temperature definition is 2K. When quoting error limits without stating the distribution it is useful to assume a uniform distribution of the values [DIN 95]. The standard uncertainties are calculated in accordance with **Figure 2.3-10**. For the temperature uncertainty influence:

$$u_2 = 99,96\,mm \cdot \sqrt{0 + (2K\!\!\Big/\!\!\sqrt{3} \cdot 12 \cdot 10^{-6}\,1\!\!\Big/\!\!K)^2 + 0 + 0} = 1,4\,\mu m \quad (2.3\text{-}14)$$

follows. Assuming that the measurement deviations for the dial gauge are distributed normally, the statement that 95% of the values are within a range of ±0.02 mm, allows for definition of the standard deviation (**Figure 2.3-10**):

$$u_3 = \frac{0,02mm}{2} = 0,01mm \qquad\qquad (2.3\text{-}15)$$

All components of the measurement uncertainty are independent of one another and equally weighted so that they can be summarised to a combined standard uncertainty in accordance with equation 2.3-11:

$$u_c = \sqrt{u_1^2 + u_2^2 + u_3^2} = 0,022mm \qquad\qquad (2.3\text{-}16)$$

With an extension factor $k = 2$ an extended measurement uncertainty of $U = 0.044$ mm is given in accordance with equation 2.3-11. With the *probability of coverage P*, a complete result can be given as follows:

$$h = 99,96\,mm \pm 0,04\,mm;\ k = 2\ (P = 95\%)$$

2.3.4   Measurement Uncertainty and Tolerance

When choosing a measuring device for testing the tolerance compliance of a feature, the measurement uncertainty plays an important role (section 3.2.7). To ensure that no feature, which has a true value outside the tolerance is within the tolerance due to measurement uncertainties, the tolerance limit for production must be reduced by the amounts of the measurement uncertainty [Neu 85]. Regulations for the conformity test are stated in the standardisation (section 2.5.4.1).

The measurement uncertainty is often neglected if it is much smaller than the tolerance. As a limit the "golden rule of metrology" is used, which says that the measurement uncertainties should not be greater than a tenth or at the very most a fifth of *tolerance T*. In this case the influence of the measurement uncertainty for the measuring dispersion of the production process is small and is summarised by the production process itself and the measurement uncertainties of the measuring device [Ber 68]. For a usual relation of $U/T = 0.2$ this can be the cause of tolerance excesses of 20%, which remain undetected. Their occurrence is dependent on the dispersion of the production process.

2.4 Measurement rooms

The surrounding conditions under which the measurement is carried out can considerably influence the result. These have to be taken into account for the definition of the measured quantity (Section 2.3.1). When looking at the reasons for measurement deviations these are the main disturbances (section 2.3.2) .

At the actual production sites, the surrounding conditions vary greatly. At the same time, the temperature is often very different from the reference temperature of length metrology of 20°C (**Figure 2.4-1**). Therefore, the possibility of accurate measurements is limited. To diminish the measurement uncertainty, it is necessary to decrease the sensitivity of the measuring device with regard to the surrounding conditions or to carry out the measurements where the surrounding conditions can be controlled. Compensation for temperature influences and encapsulation of devices are one way of achieving this [Bre 93, Bet 94]. By carrying out the measurements in special rooms or sections of rooms, which are called measuring rooms and adhere to defined surrounding conditions the influence of these conditions can be controlled. [Zim 94, VDI 2627].

Temperature, humidity and the permissible vibrations define the expectations from the surrounding conditions. These three criteria are the basis when dividing measuring rooms into grades according to the VDI/VDE-regulation 2627 (**table 2.4-1**). Apart from the grades described here, there is also grade 5 for a production measuring point and grade 0 for a measuring room for special tasks.

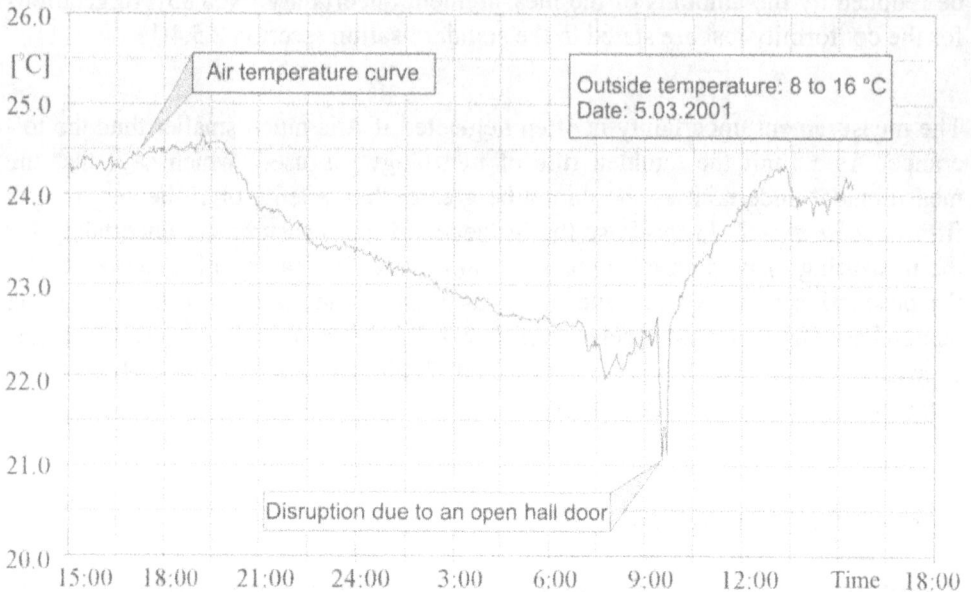

Figure 2.4-1:   Air temperature curve in a production hall

These grades are defined with a view to the measuring tasks, which need to be carried out (task related classification) (**table 2.4-2**) and with a view to the measurable tolerances and therefore the obtainable measurement uncertainty (tolerance related classification) [VDI 2627]. For the tolerance related classification it must be taken into account that the measurement uncertainties for a measurement are not only dependent on surrounding conditions of a measuring room but also on many other influences (section 2.3.1). The explanation for these classifications is not covered in this book.

The requirements for temperature conditions in a measuring room are related to the time and room temperature curve. The time temperature curve is marked by short period deviations from a mean temperature and long period deviation from the basic temperature. To judge a measuring room, large deviations in defined time periods are decisive (**Table 2.4-1**). The room temperature distribution is marked by the deviation from the base temperature at a few spots in the measuring room at the same point in time. When judging a measuring room, the greatest temperature difference is used. In relation to the room distance of the involved sensors, the temperature difference shows the temperature gradient. [VDI 2627].

Table 2.4-1:   Characteristic value for measuring rooms for variables of length metrology [VDI 2627]

Descriprion	Grade class	Basic temperature	Temperature variations in K						Temperatue-gradient in K/m	Humidity variation in %	Nadir acceleration in m/s²	
			during		over					within 30% - 60%	under 10 Hz	over 70 Hz
			15 min	60 min	4 Std	12 Std	24 Std	7 Tag.				
Precision measuring room	1	Reference temperature	0.2	0.2	0.2	0.2	0.4	0.4	0.1	10	0.02	0.2
Fine measuring room	2	According to definition	0.4	0.4	0.6	0.8	0.8	1.0	0.2	20	0.04	0.3
Standard measuring room	3	According to definition	-	1.0	1.5	-	2.0	2.0	0.5	20	0.04	0.3
Measuring room close to production	4	According to definition	-	2.0	3.0	-	3.0	4.0	1.0	30	0.06	0.4

Humidity impacts on the metrological characteristics of the measuring devices (section 2.3.2). To avoid the appearance of corrosion, especially on stainless steel, the relative humidity should not be greater than 60%. To ensure comfort and the probability of static charging of people and electronic devices, humidity should not fall below 30% [VDI 2627].

To meet the requirements for temperature and humidity, climatisation of the measuring room is necessary. The equipment for measuring rooms is not only very expensive to buy but also cost intensive to maintain. Refrigerating capacity is especially costly [Dut 96]. Usually the heat generated in a measuring room is discharged through the air, but now also by water (**Figure 2.4-2**).

The air for the air conditioning unit can be blown directly into the measuring room. In many measuring rooms, the air is introduced through a suspended ceiling from above and removed by suction in the floor area. Guiding the air from the floor to the ceiling would require perforated floor plates and result in dust being whirled up. For people working in the measurement room, the draft resulting from this concept could be very uncomfortable. Drafts do not occur in a room with climatisation which guides the air between the exterior and interior rooms. The solution whereby floor, ceiling and walls are covered with pipes, as for warm water floor heating, which is flooded with water just beneath the measurement room temperature, is also comfortable for humans, as it does not create a draft (**Figure 2.4-2**).

Tabelle 2.4-2: Grading form measuring rooms of the longitudinal measuring technic classification according to measuting tasks [VDI 2627]

Grade	Labelling and samples for tasks
1	Precision measuring room e.g. calibration of standards, dimensioning of measurement devices
2	Special measuring room e.g. calibration of standards, measure individual parts, acceptance of precision elements, tools and equipment
3	Standard measuring room e.g. measuring tasks for process control, measure equipment, tools, test equipments, sample testing for documentation purposes, measurement of parts and samples
4	Measuring room close to the production site e.g. surveillance of production and machines, testing of production aids, tools (depending on the production area)

When planning the air conditioning unit, warmth given out by humans and equipment as well as the need for fresh air have to be taken into consideration just as much as floor area and ceiling height.

Vibrations can be evaluated through amplitude as well as the maximum speed and acceleration in m/s^2. When evaluating measuring rooms, the vibration acceleration is used.

Low frequencies are especially disruptive. They can only be kept away by the foundation of the measurement devices, which is separated by vibration insulation from the building's foundation. Higher frequencies can be filtered by vibration damping on the equipment. For smaller, vibration sensitive measuring devices (e.g. surface measuring device) a heavy stone or steel plate, which has been placed underneath the device may be sufficient. Further ideas for vibration insulation can be found in [VDI 2602].

Further measures to limit other surrounding influences in the measuring rooms, e.g. dirt, have to be taken into account. For this, and for the design and planning of measuring rooms, please see the VDI/VDE regulation 2627 [VDI 2627].

Vertical expel current

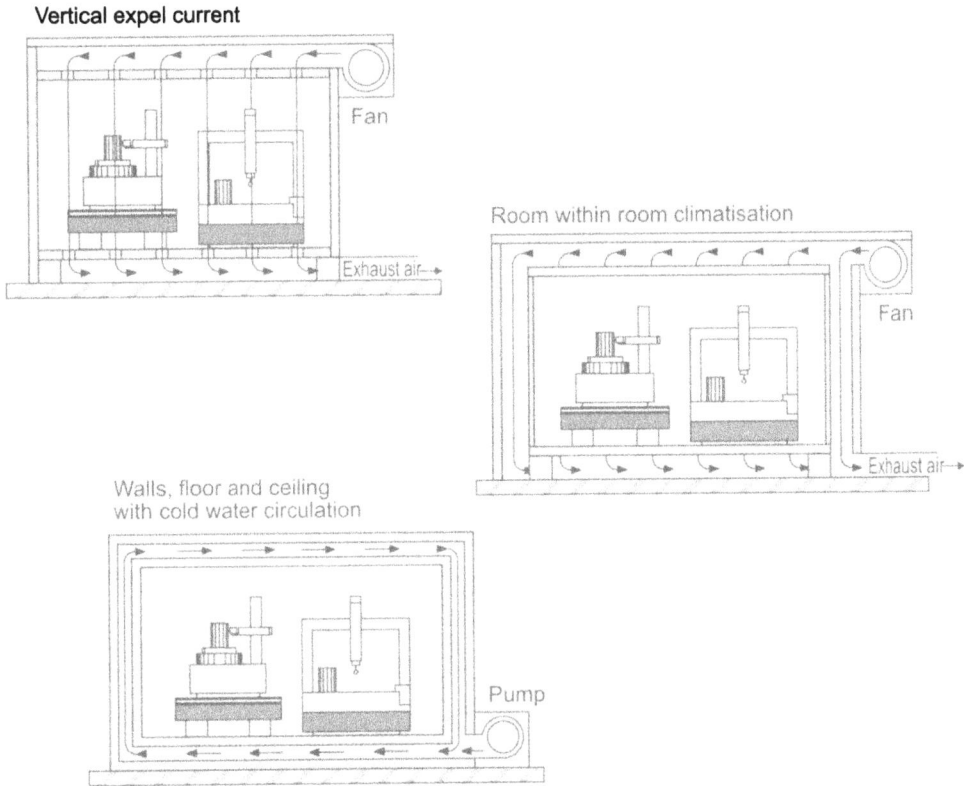

Figure 2.4-2: Climatisation of measuring rooms

2.5   Drawings specifications and tolerances

A work piece is usually designed as a single part of a machine or a device in accordance with its functionality, the load and a cost-efficient production process. The technical drawing shows the three-dimensional workpiece in all the necessary views. By the dimensioning, all geometrical features and dimensions of the component are clearly defined. Apart from the dimensions, the drawing also includes the necessary information on tolerances of dimensions, finish quality, materials etc. needed for production. **Figure 2.5-1** shows the example of a technical drawing of a bearing block.

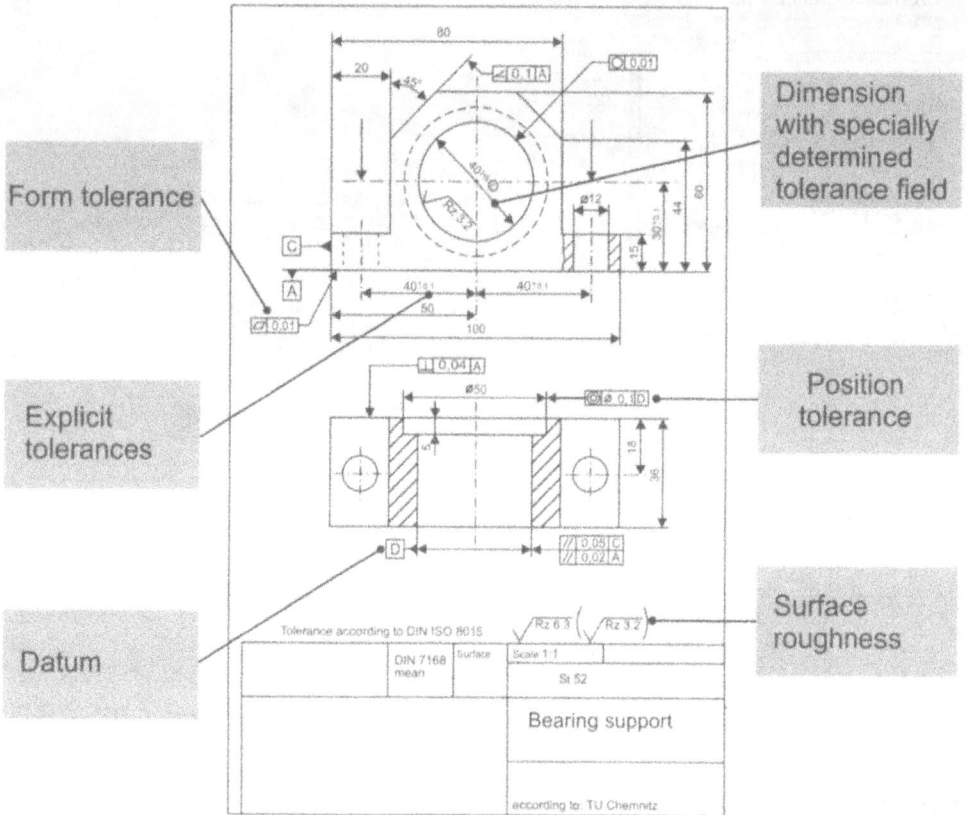

Figure 2.5-1: Measuring tasks and inspection features of a sample workpiece (bearing unit)

Due to the production process, there are always deviations from the shape, as defined in a drawing, which include deviations of dimensions, form and position of geometrical features. A reason is, for example, the workpiece clamping during production, the tool and its fixing as well as the cutting forces [Tru 97]. With respect to the designated function and interchangeable manufacturing, these deviations must not exceed defined limits which are given as tolerances during construction.

With regard to the most economical method of production it makes sense to adjust the accuracy requirements to the feature function, keeping them as low as possible, as production costs will increase with increased requirements. This is especially true for inspection costs as, according to the "golden rule of metrology", the requirements for the measurement uncertainty are higher by factor 10 or at least 5 than for the production tolerance, which is under surveillance [Ber 68]. **Figure 2.5-1** shows the example of a drill-hole for a ball bearing, which has been given a specially small tolerance, while the mounting holes of the bearing unit have been given relatively large tolerances.

2.5.1 Dimensions, dimensional tolerances and fits

2.5.1.1 Dimensions and dimensional tolerances

The dimensioning of technical drawing is defined in DIN 406-10 to 12. These define the representation of the dimensioning elements and the dimensions with and without tolerances. At the moment the standard ISO 129 is being revised on an international level. This includes the international harmonisation of the regulations for dimensions on technical drawings ISO 406. This will lead to a completely new structure of the standard for dimensioning [Bag 97].

Definition: The term **dimension** is used to determine length in production technology. The dimension specified in a drawing is called **nominal size**. The **actual dimension** of a part always differs from the nominal size, due to the production process. The permitted deviation in one direction is called **dimensional deviation**. There is a distinction between **upper** and **lower deviation**, the polarity sign shows its relation to the nominal size. The **minimum size** is given by adding the lower tolerance to the nominal size. Analogous to this, the **maximum** is obtained by adding nominal size and upper deviation. Minimum size and maximum are also called **dimensional limit**. The **dimensional tolerance** is the range between the minimum size and maximum size.

The correlation described above are shown in **Figure 2.5-2**.

Definition: The **maximum material limit** limits the material of a component. For shafts, this equals the maximum, for drill holes, the minimum size. Analogue to this, it defines the **minimum material limit** the lower limit of the material. For drill holes it is the maximum and for shafts the minimum size.

Unless otherwise stated, technical drawings show the parameter with regard to length in millimeters [Hoi 94]. The upper and lower deviations do not have to be of the same amount. They can even have the same arithmetic sign, which would mean that the nominal size itself does not fall within the tolerance range.

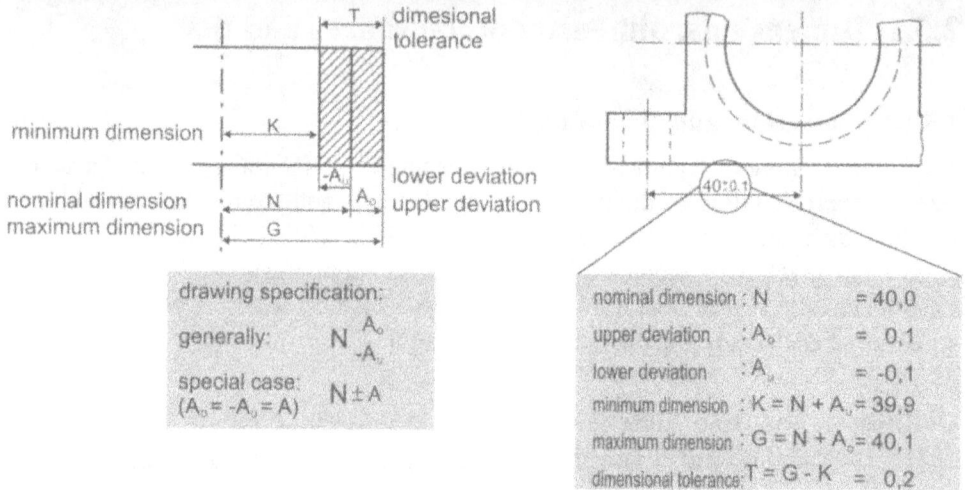

Figure 2.5-2: Dimensional definitions, drawing specifications and explicit tolerances

A tolerance can be explicit or can be defined by specifying the tolerance acronym according to DIN ISO 286-1. To simplify the drawing, all non-explicit tolerances for length and angle sizes general tolerances can be defined in accordance with DIN ISO 2768-1. This contains measurements for various length and angle areas, which are divided into the levels fine (f), medium (m), coarse (c) and very coarse (v). The current tolerance must be noted in the legend of the drawing, e.g.: general tolerance DIN ISO 2768-m. The general tolerances according to DIN 7168 partly match those according to DIN ISO 2768-1, but they cannot be used for new constructions.

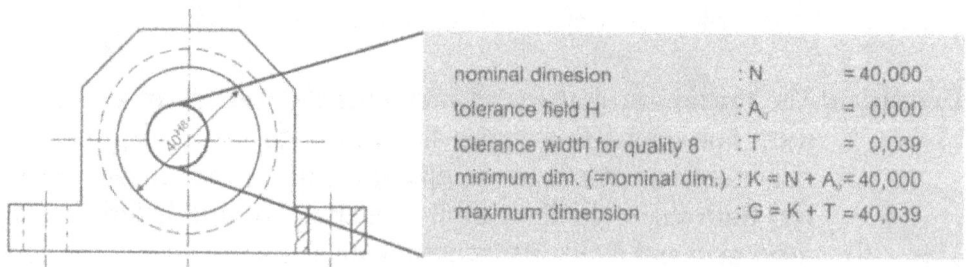

Figure 2.5-3: Drill hole tolerance according to DIN ISO 286-1

Figure 2.5-2 shows an example for an explicit tolerance of a length dimension. From the quoted deviations, the minimum dimensions, the maximum dimension and the dimension tolerance can be calculated directly.

Figure 2.5-3 shows a tolerance according to DIN ISO 286-1, defining the tolerance field with the aid of an acronym. The tolerance field describes the position of the tolerance in relation of the nominal size as well as the size of the tolerance. It is determined by the nominal size and the degree of tolerance. The degree of tolerance is obtained by adding a letter to define the position and a number for the size of the field of tolerance, e.g. 40^{H8}. Fields of tolerance for inner fit surfaces (e.g. drill holes) are shown by a capital letter. For outside fit surfaces, lower case letters are used. Within the ISO-system for dimensional limits, the zero line equals the nominal size (**Figure 2.5-4**).

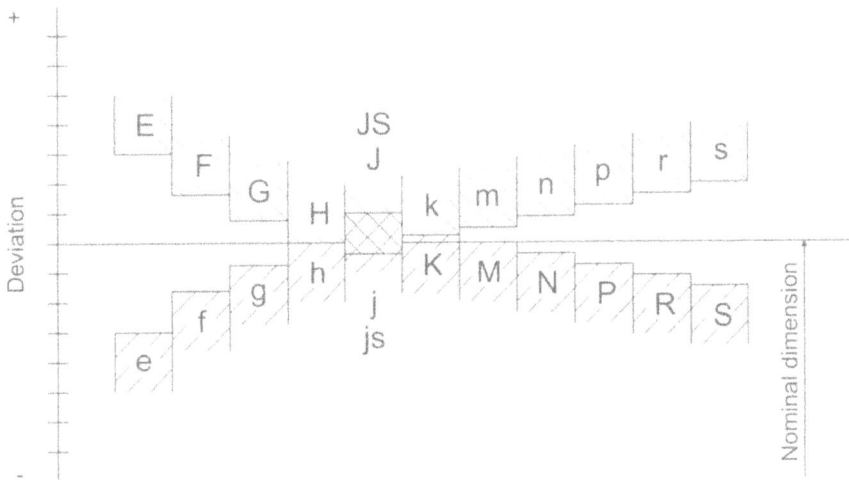

Figure 2.5-4: Position of the ISO-tolerance fields (extract)

2.5.1.2 Fits

A fit results from assembling parts, e.g. shaft and bearing or guide and slide.

Definition: Fit is the relationship resulting from the dimensional difference between two parts before assembly. In this context, a **fitting surface** describes the surface of the assembled parts, which are in contact or could come into contact due to movement.

Principally there are fits with slackness and with excess. Slackness exists when the diameter of the shaft is smaller than the diameter of the drill hole and accordingly an excess exists, when the diameter of the shaft is larger than the drill hole diameter (**Figure 2.5-5**). If two work pieces with excess are connected, the dimensional difference disappears. This is called force fit. Fits for which one cannot predict the fit after pairing off, are called transition fits.

A fit is determined by the nominal size and the tolerance classes of drill hole and shaft, e.g.: 40^{H7}_{r6}. For determining the possibility of joining two parts, the combination of the deviations is relevant for which the maximum or the minimum of a fit is achieved.

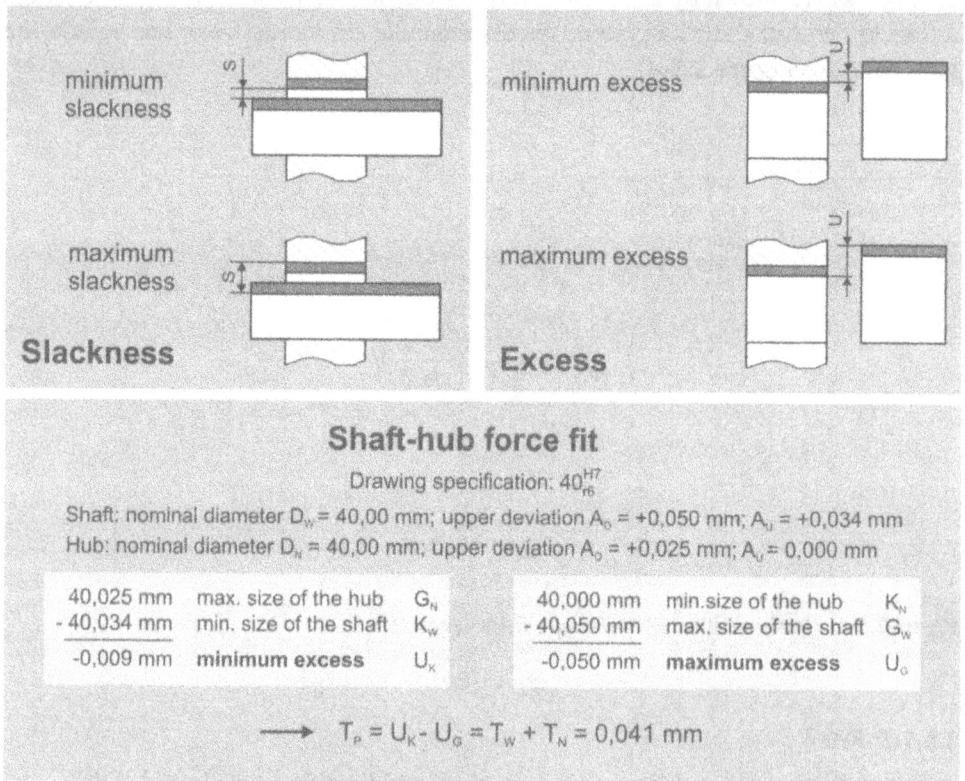

Shaft-hub force fit

Drawing specification: 40^{H7}_{r6}

Shaft: nominal diameter D_w = 40,00 mm; upper deviation A_o = +0,050 mm; A_u = +0,034 mm
Hub: nominal diameter D_N = 40,00 mm; upper deviation A_o = +0,025 mm; A_u = 0,000 mm

40,025 mm	max. size of the hub	G_N		40,000 mm	min.size of the hub	K_N
- 40,034 mm	min. size of the shaft	K_W		- 40,050 mm	max. size of the shaft	G_W
-0,009 mm	**minimum excess**	U_K		-0,050 mm	**maximum excess**	U_G

$$\longrightarrow \quad T_P = U_K - U_G = T_W + T_N = 0,041 \text{ mm}$$

Figure 2.5-5: Fit definition and number sample

Definition: For a slack fit the two possibilities are **minimum slackness** (small drill hole and large shaft) and **maximum slackness** (large drill hole and small shaft). There are also two possibilities for force fits: those with a **minimum excess** or with a **maximum excess**.

To illustrate, we can look at a shaft-hub-squeezed joint, which must not fall below the minimum excess to transmit a required torque (**Figure 2.5-5**).

To describe fit a fit tolerance T_P can also be given. It is derived from the arithmetic sum of the tolerance values of the drill hole and the shaft, which build a connection. It shows the permitted fluctuation of the slackness or the excess. The fit tolerance field defines the position towards the zero line for slackness or excess as well as the tolerance value (**Figure 2.5-6**).

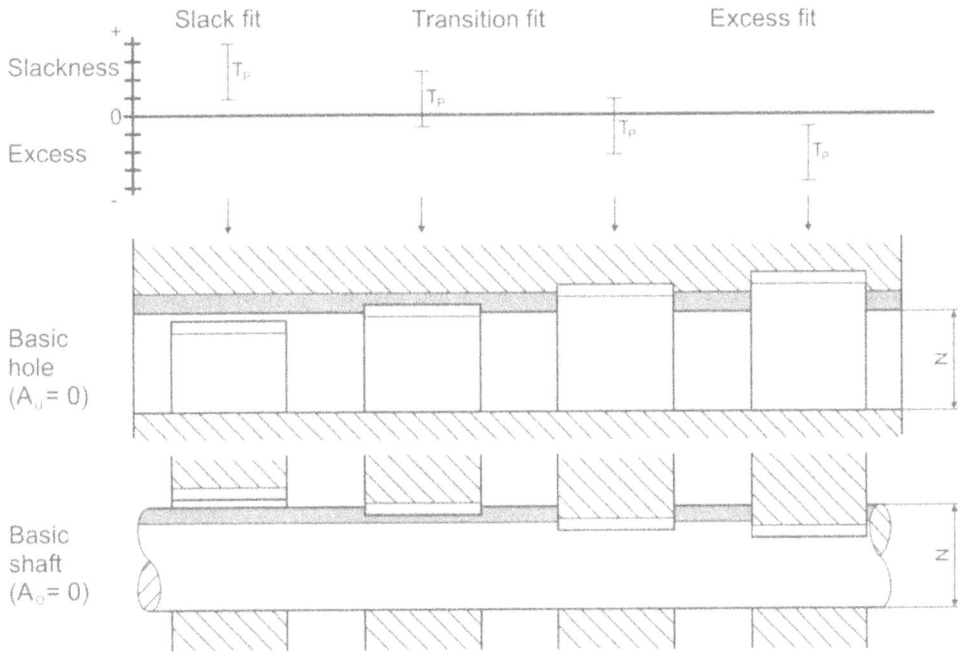

Figure 2.5-6: Fit tolerance fields and ISO fit system for the basic shaft and basic hole

The fit system according to DIN 7150-2 contains a planned series of fittings with various slacknesses and excesses. But as all tolerance fields for drill holes (outer parts) and shafts (inner parts) may be paired off with one another without a strict regulation, multiple possibilities arise. Measuring devices for testing these pairings, especially limit gauges, are very expensive because of their exactness. For cost saving reasons, fits are preferred which are chosen in accordance with the basic hole DIN 7154-1 and 2 or the basic shaft DIN 7155-1 and 2, to reduce the number of necessary gauges [Hoi 94] (**Figure 2.5-6**).

In the system of the basic hole, the nominal size of the drill hole is given the ISO-tolerance field H. The lower dimension of the drill hole A_u is set to nil and equals the nominal size. Depending on the required fit, whether it has slackness or excess, the user can choose an appropriate tolerance field for the shaft. Accordingly, the shaft is given the tolerance field h in this system, which puts the upper dimension A_O to zero. To obtain at various fits, the drilling of the position of the drilling is chosen with respect to the zero-line.

By choosing the fit according to DIN 7157, the spectrum of usable tolerance combinations for the system of the basic shaft and basic hole with view to an economic production is further reduced. The sample fit in **Figure 2.5-5** is taken from this selection.

2.5.2 Tolerance of form and position

The shape of a workpiece is made up of single geometrical form elements such as planes, spheres and cylinders. To describe a workpiece, also geometrical elements can be used, which do not have to be part of the workpiece. Examples are elements of symmetry or centres of circles, which are not directly ascertainable in normal circumstances. In such a case, geometrical aid elements must be determined and with their aid the virtual elements can be calculated with a mathematical equations.

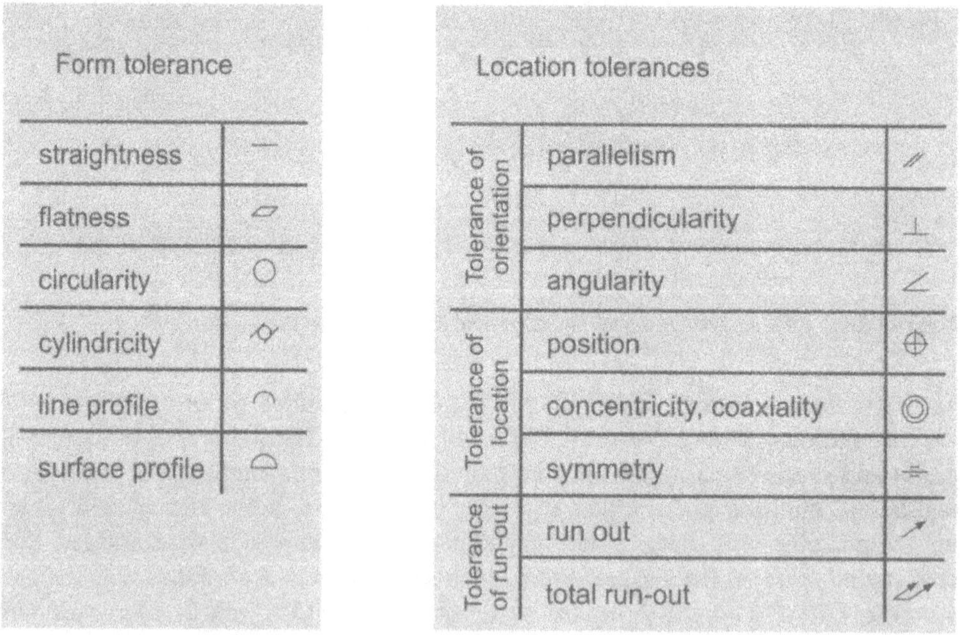

Form tolerance	
straightness	—
flatness	⟋
circularity	○
cylindricity	⌭
line profile	⌒
surface profile	⌓

Location tolerances		
Tolerance of orientation	parallelism	∥
	perpendicularity	⊥
	angularity	∠
Tolerance of location	position	⊕
	concentricity, coaxiality	◎
	symmetry	⌯
Tolerance of run-out	run out	↗
	total run-out	⌰

Figure 2.5-7: Tolerances of form and location

Apart from dimensional deviations shape deviations can occur when producing workpieces. Examples are deviations from the form and position of workpiece features (section 4.5.1). These play an important role in the function of the workpiece and for that reason the permissible deviations are limited by their respective tolerances.

DIN ISO 1101 is used as a basis for determining and specifying form and locational tolerances in drawings. This determines the definitions of the respective tolerance zones and the affiliated symbols for marking the tolerated features (**Figure 2.5-7**). The tolerances of position features are divided into tolerances of orientation, location and run-out.

Form and positional deviations can be limited by DIN ISO 2768-2, like length and angle deviations. For this, the classes H, K and L are differentiated. The entry in the legend is a combination with a general tolerance for length and angle dimensions, e.g. general tolerance DIN ISO 2768-mK. DIN 7168 is also valid but cannot be used for new constructions.

2.5.2.1 Specification of tolerances and references

The so-called tolerance frame, which is connected with the element in question by a reference line with a reference arrow, is used for stating a form or position tolerance in a technical drawing (**Figure 2.5-8**). The frame consists of two or more parts, which have the following entries from left to right:

1. the symbol for the tolerated feature,

2. the tolerance value with the same unit as the dimensions, and

3. if required, capital letters for datum.

If the tolerance zone is in form of a circle or cylinder, the tolerance value is preceded by a diameter symbol. Theoretically exact dimensions, use of the maximum material principle, the minimum material principle, the condition of envelope or the projected tolerance zone are marked with respective symbols (section 2.5.2.4). If necessary, additional information can be included on top of the tolerance frame. If an element should be given various tolerance features, the separate tolerance frames are put on top of each other. If not determined in another way, the tolerance zone covers the whole area of the tolerated element and is valid for the predetermined orientation or vertical to the form of the part.

Definition: A **datum** is a theoretically exact geometrical element, which is the reference for toleranced elements [DIN ISO 5459].

A datum reference is denoted by a capitalised letter within a reference frame, which refers to the element by a reference triangle (**Figure 2.5-8**).

The geometrical element, which is toleranced or marked as datum reference, is dependent on the location of the reference arrows or reference triangle in the technical drawing. Basically, three cases can be differentiated (**Figure 2.5-9**).

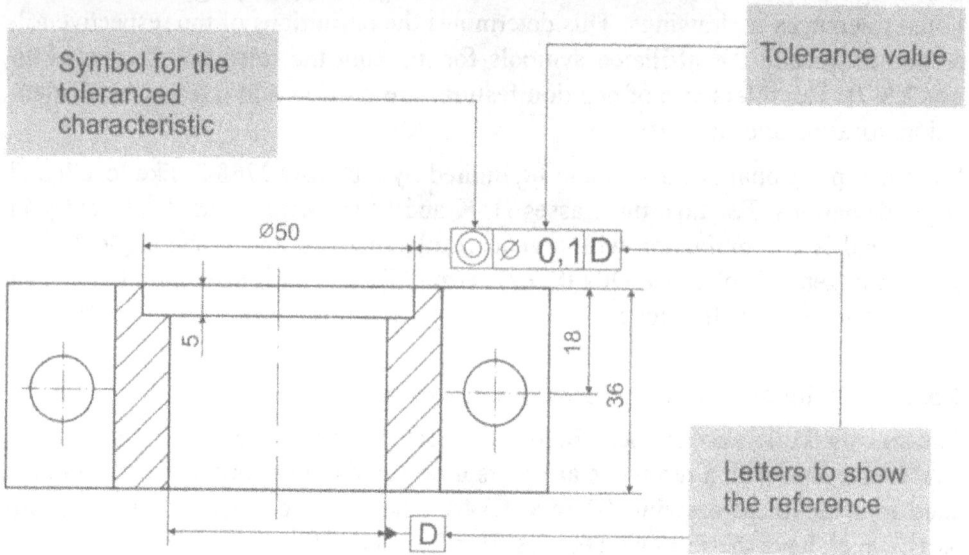

Figure 2.5-8: Elements of a tolerance and datum frame

- The reference arrow/ reference triangle is vertical to the plane or the contour line and is at least 4 mm afar from the dimension lines or edges

⇒ toleranced element/datum reference is the plane or contour line.

- The reference arrow/ reference triangle is in the extension of a dimension of the form element

⇒ toleranced element/datum reference is the axis or the medium plane of the element

- The reference arrow/ reference triangle is vertical to the medium plane or axis, which is shaved by various form elements

⇒ toleranced element/datum reference is the shaved axis or medium plane

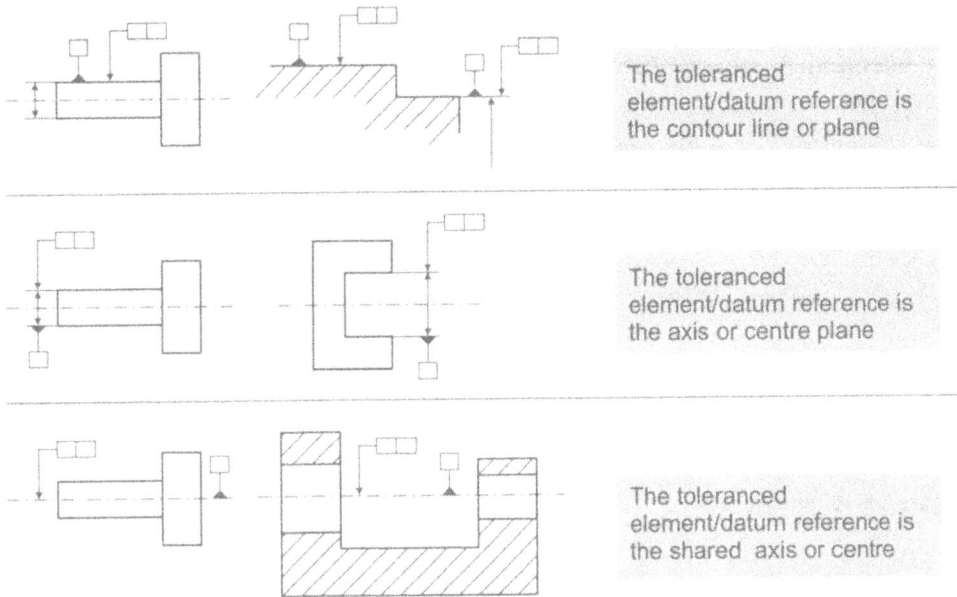

Figure 2.5-9: Relation of tolerances and datum for a geometrical element

2.5.2.2 Tolerance of form

Definition: The **tolerance of form** is the largest possible deviation of form of an element from its geometrical ideal form.

From the definition of the tolerance zones according to DIN ISO 1101, the adjacent element (Tschebyscheff) can be derived as a datum reference for determining form deviations. This element has a geometrical ideal form and touches the real element so that the largest distance to the real element is at a minimum. The deviation is therefore the largest orthogonal distance from the real element. Should other datum references be used [e.g. DIN ISO 4291], larger values are of the form deviation are obtained.

Tolerance of straightness
Depending on the characterisation of the toleranced element, the distinction is made between the tolerance of straightness within a plane and the tolerance of straightness within three dimensions.

For the straightness within a plane, the actual profile of the toleranced plane, for any intersection, must be between two parallel lines, with a distance of the tolerance value.

The straightness of a line-shaped element within three dimensions is divided by tolerance zones which are limited by two parallel planes, a cube or a cylinder.

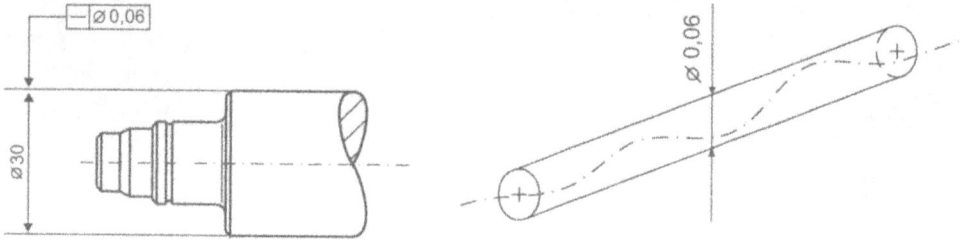

Figure 2.5-10: Straightness

Figure 2.5-10 shows an example for a tolerance of spatial straightness. The reference arrow is shown in the extension of the dimension line of the largest cylinder and therefore the tolerance of straightness is related to the axis of this cylinder. The real axis must be over the whole length within the cylinder with a diameter of the tolerance value, in the example shown 0.06 mm.

Straightness tolerances are necessary if the straightness is not sufficiently tolerated by another form tolerance, e.g. flatness or cylinder form, or by a location tolerance, e.g. parallelism, perpendicularity, angularity, symmetry or position.

Flatness tolerance
For the flatness tolerance the actual surface must lie between two parallel surfaces, which have the distance of the tolerance value. By quoting additional conditions, e.g. "non convex", the tolerance can be limited further.

The bottom of the bearing block has a flatness tolerance of 0.01 mm (**Figure 2.5-11**). It has to fit completely between two parallel planes with a distance of 0.01 mm. If a location tolerance is used at the same time, the explicit tolerancing of the flatness can fall away if it is sufficiently limited by the location tolerance.

Circularity tolerance
The circularity tolerance determines that in each section perpendicular to the axis of a toleranced cone-formed element, the actual profile, or the actual contour line, must lie between two concentric circles within the same plane, which are at the distance of the tolerance value. For sphere-formed elements, this tolerance is valid for each intersection through the centre.

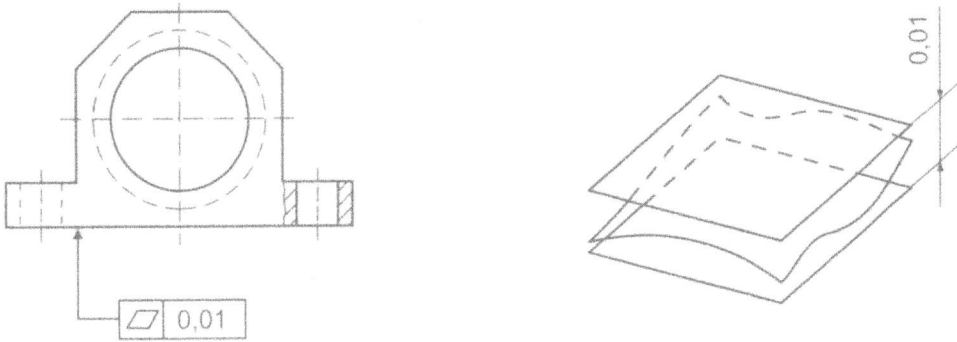

Figure 2.5-11: Flatness

The sample workpiece has a drill hole for the bearing with a circularity tolerance of 0.01 mm (**Figure 2.5-12**). Circularity tolerances can also be used for arcs of circles with less than 360° coverage.

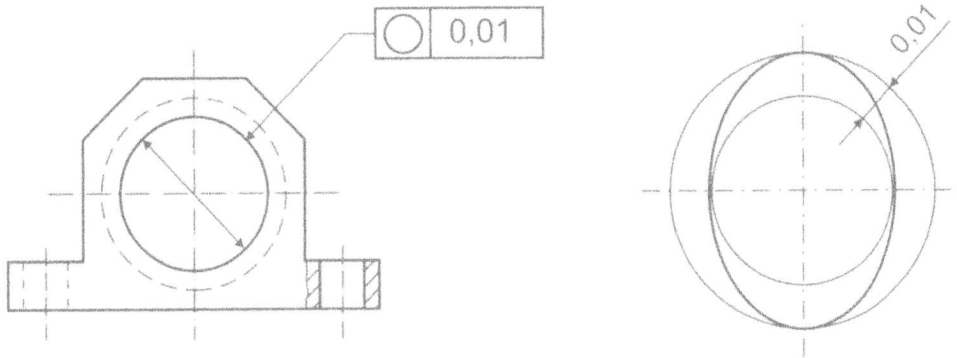

Figure 2.5-12: Circularity

Cylindricity tolerance

For the cylindricity tolerance the actual surface of the toleranced barrel surface must be within two coaxial cylinders, which limit a tolerance zone with the width of the tolerance value. In the example of the shaft, the shoulders for the ball-bearings have a cylindricity tolerance of 0.08 mm (**Figure 2.5-13**).

Cylindricity tolerances always also limit the straightness of the barrel lines, the circularity, the line profile and the parallelism of opposite barrel lines. Therefore, alternatively to a cylindricity form tolerance, the straightness, the circularity and the parallelism can be tolerated. From a metrological point of view it must be said that the cylinder form cannot be ascertained, it can only be approximated by point by point measuring, e.g. by scanning.

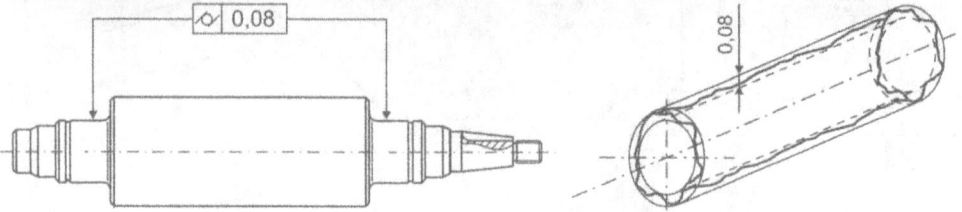

Figure 2.5-13: Cylinder form

Profile tolerance (line profile, surface profile)

For the profile tolerance, a distinction is made between the profile of any line and profile of any plane.

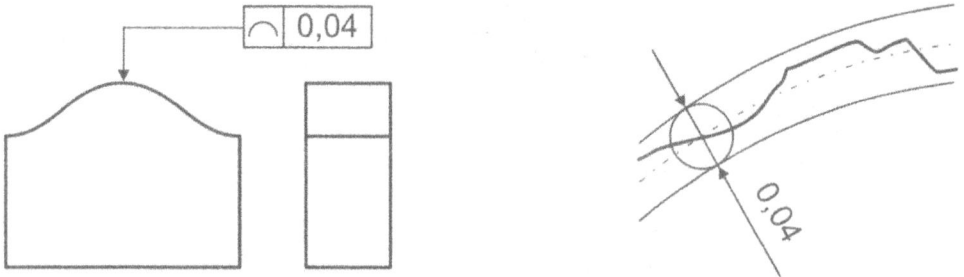

Figure 2.5-14: Surface profile

For the tolerance of profile of a line the tolerance zone is determined by one line on each side of the ideal profile form, which envelopes all circles with a diameter of the tolerance value and centre point on a line of ideal geometrical form (**Figure 2.5-14**). The actual profile resulting from any intersection parallel to the plane of projection must fit into this tolerance zone.

Analogously, the tolerance zone for the profile form of any surface is defined both enveloping surfaces of all circles with a diameter of the tolerance value and centre point on a surface of geometrically ideal form.

By dimensioning the nominal profile with the aid of theoretical dimensions and the datum stated within the frame of tolerances, this tolerance can also define the position of the profile.

2.5.2.3 Location tolerances

Definition: The **location tolerances** limit the permitted deviations of the ideal position of two or more elements with respect to one another. Normally one of the elements is explicitly defined as the datum reference.

When evaluating locational deviations, form deviations of the datum reference elements must be eliminated. For this, the real reference elements are replaced by equivalent elements according to DIN ISO 5459. This is, for example, the adjacent line or surface instead of the real elements or the axis of the cylinder for drill holes or shafts. Instead of the axis and symmetry levels of the real elements, those of the equivalent elements are used [Tru 97]. The precondition is that the real elements have a sufficiently accurate shape.

Should the form deviations of the toleranced element not be taken into account for positional tolerancing, the tolerance needs to be extended by the remark "without form" as this is not intended by DIN ISO 1101.

The American norm ASME Y 14.5 M allows for the elimination of the form deviation also for the toleranced element by marking it with a Ⓣ.

Parallelism tolerance

Parallelism tolerances define the position of lines or planes in relation to other lines or planes. With the example of a bearing block, the parallelism of a line with respect to two planes is shown (**Figure 2.5-15**).

The centre line of the drill hole for the ball bearing is toleranced in relation to the supporting surface of the bearing block, which is marked with the reference letter "A". The position of the axis is therefore limited to the space between two planes, which are parallel to the datum plane and the tolerance value apart, in the example shown at 0.02 mm.

Additionally, the centre line is marked with a parallelism tolerance of 0.05 mm with reference to the "C" marked surface. Therefore, the actual axis of the drill hole must be within a cube-formed area of tolerance, which edge lengths are vertical to the datum planes and of the tolerance values.

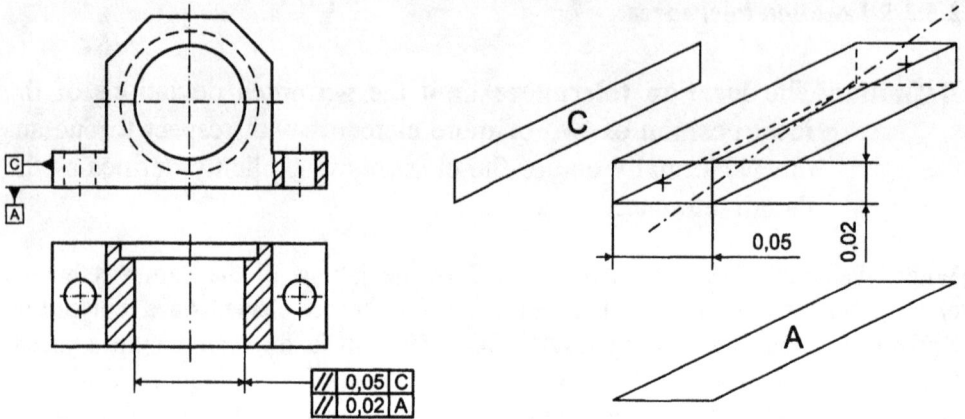

Figure 2.5-15: Parallelism

Perpendicularity tolerance

For the perpendicularity tolerance, which is also a tolerance of direction like the parallelism tolerance, the toleranced elements are lines or planes, which in their turn are related to lines or planes.

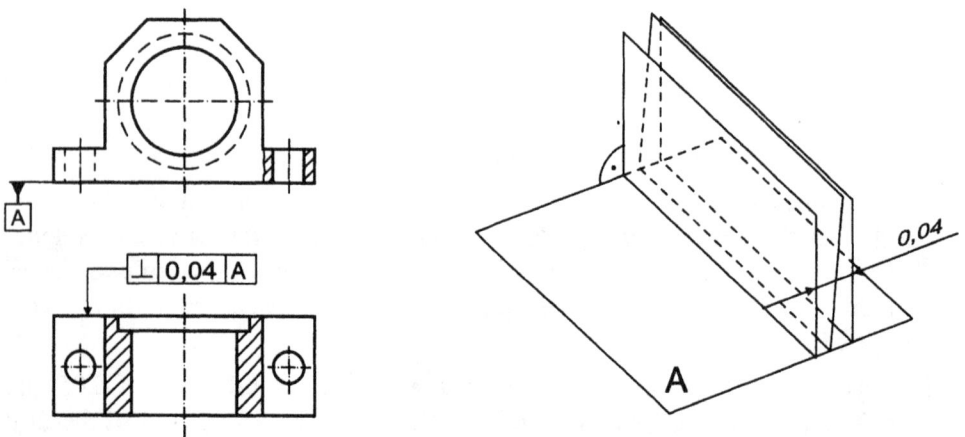

Figure 2.5-16: Perpendiculary

The front face of a bearing block is toleranced with a perpendicularity tolerance, with respect to the surface plane marked with "A" (**Figure 2.5-16**).

The datum plane "A" is the bottom face of the bearing block. By definition of the perpendicularity tolerance, the front surface has to be within two parallel planes, which are vertical to the bottom face of the bearing block and of 0.04 mm apart.

Angularity tolerance

The angularity tolerance is the general form of the earlier discussed tolerances of parallelity and perpendicularity. Angles between 0°and 90° degrees are covered.

In the case of the bearing unit, the bevel of 45 degrees tilting is toleranced (**Figure 2.5-17**). Here also, the datum is represented by the seat, marked as "A". The tolerance value is 0.1 mm. This means the bevel must not exceed the tolerance zone, which is limited by two parallel places of the theoretically exact angle of 45 degrees with reference to the datum plane, i.e. the bottom face of the bearing block, with a distance of 0.1 mm.

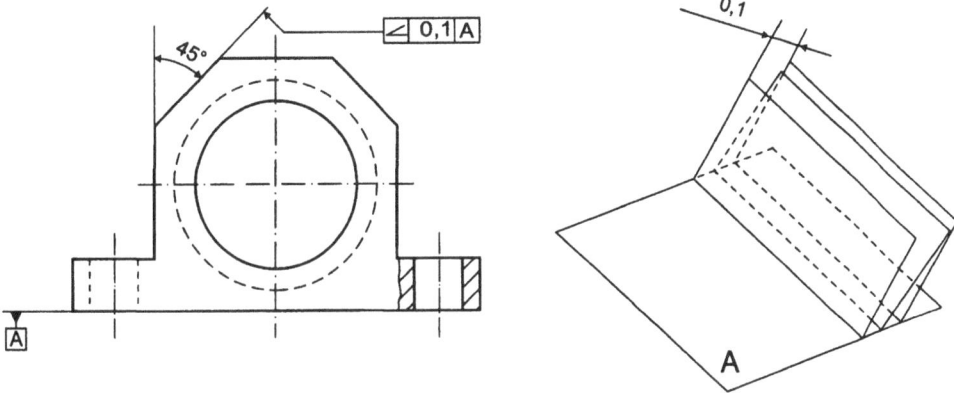

Figure 2.5-17: Angularity

Position tolerance

For the position tolerance, the separate norm DIN EN ISO 5458 "Form- and Position Tolerance" has been created in addition to the description in DIN ISO 1101. It contains comments for the tolerancing of positional tolerances for various applications, explanations of results from tolerance combinations, the interference of dimension and position tolerances for the specification of positions of form elements as well as calculation formulas for determining position tolerances.

According to DIN EN ISO 5458, the position tolerance limits the deviation of a form element as point, line, plane, cube or cylinder from its theoretical exact position. The information about the theoretically exact position is given by theoretically exact dimensions (section 2.5.2.4). The example in **Figure 2.5-18** explains the meaning of the position tolerance. The frame of tolerance, which determines the position tolerance, tolerates the centre axis of the drill hole with the diameter of 10 mm. The tolerance value of 0.05 mm describes the diameter of the cylinder, whose centre axis matches with the axis of the theoretical exact position. The actual axis of the drill hole must lie within this cylinder. The theoretically exact position is defined by theoretically exact dimensions, which give the respective distances to the datum planes "A" and "B".

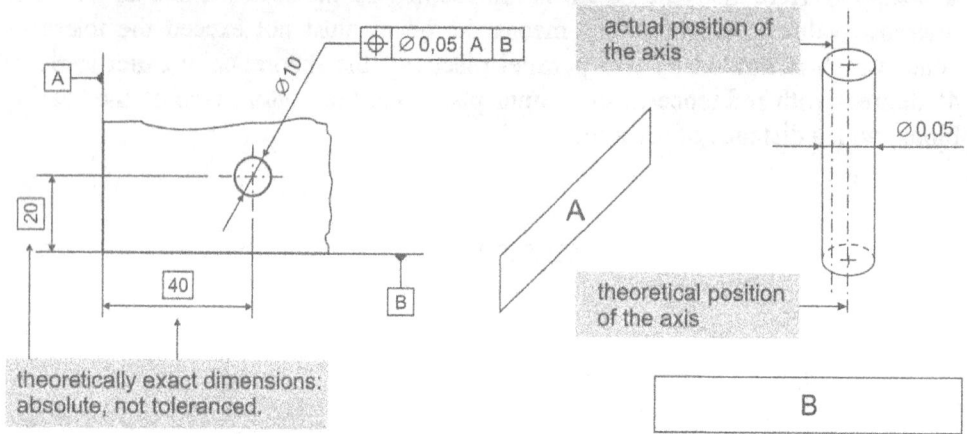

Figure 2.5-18: Position tolerance

Tolerance of concentricity and coaxiality

The term concentricity is used for the relative position of centre points of circular surfaces to one another. The centre point of the toleranced circular surfaces must be within a circular tolerance zone with a centre point that matches the datum centre point.

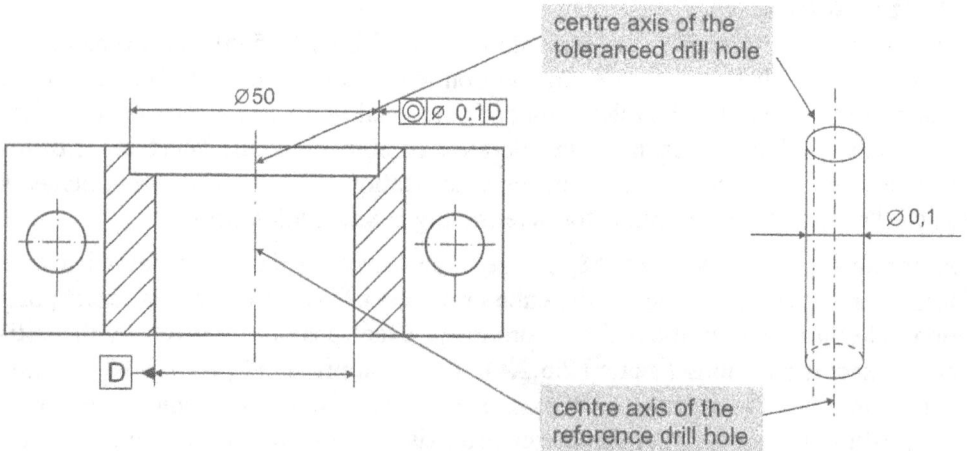

Figure 2.5-19: Concentricity and coaxiality

For cylinder formed geometric elements the term coaxiality is used are called co-axial, if the relative position of two or more cylinders to one another is meant. Similar to concentricity, the position of the cylinder axis is of importance for coaxiality.

At the bearing block, the centre axis of the drill hole with a diameter of 50 mm is toleranced with respect to the of with the drill hole with a diameter of 40 mm for coaxiality (**Figure 2.5-19**). The reference arrow of the tolerance frame, as well as the datum "D", are shown in the extension of the respective dimension lines. The tolerance value is 0.1 mm.

Therefore the position of the centre axis of the 50 mm drill hole is limited to an ideal cylinder around the centre axis of the 40 mm-drill hole, which has a diameter of 0.1 mm.

If the axis of a drill hole is used as datum reference, as in the example shown, the influence of the distance between datum reference and toleranced axis needs to be looked at. The larger the distance, the higher required production accuracy. In some cases, the tolerance of a radial run out can be more useful than the tolerance of coaxiality. However, the radial run-out is limited to cylinder form geometry elements and therefore, for polygon profiles, the coaxiality must be toleranced [Sch 93].

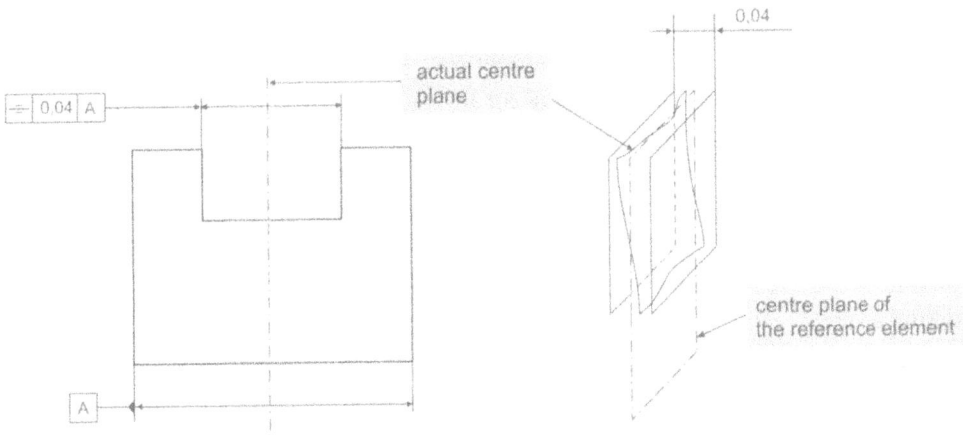

Figure 2.5-20: Symmetry

Symmetry tolerance

A symmetrical element always has a plane or axis of symmetry, also called centre plane or centre axis. For symmetrical tolerances, the centre plane or centre axis of a toleranced element is toleranced with reference to the centre plane of a symmetrical datum reference (**Figure 2.5-20**). The tolerance zone, which should contain the real centre plane of the toleranced element is limited by two planes, which are the tolerance value apart and symmetrical to the datum centre plane.

Tolerance of run-out and total run-out

The tolerance of run out limits the location as well as the form deviations of the toleranced element. Metrologically, the workpiece must be fixed in such a way that the toleranced element can rotate around the reference axis. The distance of the workpiece surface is measured with a measuring device. The maximum difference of the distance during an axial rotation must not be larger than the tolerance value.

Principally, the radial run-out, the axial run-out and the run-out in any desired or specified direction must be distinguished.

The tolerance of the radial run-out is measured vertically to the rotation axis during one rotation. This results in the actual profile of the revolving surface as a section through the workpiece. This has to be within the tolerance zone, which is formed vertically to the axis on each measurement plane by two concentric circles with a distance of the tolerance value, and centre points on the rotation axis. By measuring the radial run-out deviations, the sum of the radial run-out deviations and the double coaxiality deviations are recorded in the respective sections.

The tolerance of radial run out usually refers to a complete rotation, but can be limited to a partial circumference if required.

While for the tolerance of radial run-out the radial deviation of a cylinder formed geometrical element is toleranced, the deviations in axial directions are limited by the tolerance of axial run-out. The geometrical elements, which are toleranced for radial run-out, are for example front faces of cylinders or discs.

The deviation of axial run-out is the largest distance difference between the actual profile of the measured surface and face and a plane, which is perpendicular to the datum axis.

The measurements are taken parallel to the revolving axis, which results in an actual profile for a given radius in the form of a measuring cylinder. This actual profile must be within the tolerance zone, which lies on the measured cylinder and is limited by two circles with a radius of the actual profile, which are apart of by the tolerance value. If the tolerance frame does not have any restricting information, the tolerance is valid for all radiuses, which are within the toleranced plane or surface.

The flatness deviation is not limited by the tolerance of axial run-out, but can in fact assume larger values than the tolerance of axial run-out.

For certain rotationally symmetrical workpieces, e.g. cones, the general definition of the tolerance of run-out is used. There is a distinction between a tolerance of run-out in an arbitrary direction or a tolerance of run-out in a given direction. For the tolerance of run-out in an arbitrary direction, the measured deviation is always vertical to the surface of rotation. By this there results a so-called measurement cone, whose axis is congruent with the datum axis. When measuring the tolerance of run out in a fixed direction, the angle of the measurement of the deviation is explicitly specified and independent of the position or the rotational surface. The tolerance zone, which must be maintained is for both cases on the measurement cone and is limited by two concentric circles, which lie on the barrel line with a distance that equals the tolerance value.

The tolerance of total radial run-out is different from the tolerance of run-out as the deviation is not only limited for an axial or radial section, but for the entire toleranced surface. Therefore, a measuring device is needed, which measures the deviations at multiple points at the same time.

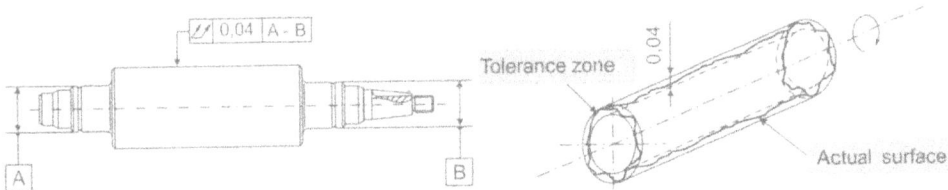

Figure 2.5-21: Tolerances of run-out (total radial run-out)

Total run-out, total radial run-out, total axial run-out and total run-out are also divided into those with arbitrary and those with fixed direction.

For the tolerance of total radial run-out, the vertical deviation to the datum axis must not exceed the tolerance value during multiple rotations and dislocation of the measuring point parallel to the reference axis (**Figure 2.5-21**). The actual profile of the toleranced cylinder must be within the tolerance zone, which is formed by two coaxial cylinders, whose axes match with the reference axis and whose radiuses differ by the tolerance value.

The tolerance of total radial run-out limits the coaxial deviations of the axis of the toleranced element in relation to the reference axis and the cylinder form deviations of the toleranced element [Sch 93].

Contrary to the tolerance of axial run-out for the tolerance of the total radial run-out, the deviation in axial direction is not only taken alongside the graduated cylinder but at multiple radial locations vertical to the reference axis. The largest deviation, which has been arrived at by this method, must not exceed the tolerance

zone, which has been formed by two parallel planes of the deviation of the tolerance value.

Analogous to the tolerance of run-out the tolerance of total run-out can be defined
in random or stipulated direction. For multiple rotations around the reference axis
and for location of the measuring points parallel to the theoretically exact direction, the deviation of the total run-out must not exceed the tolerance value.

2.5.2.4 Tolerance value

The tolerance value defines the size of the tolerance zone, which contains all
points of the element, e.g. points, lines or planes. Marking the tolerance value with
respective symbols determines how to interpret the tolerance value.

Theoretically exact dimensions
Theoretically exact dimensions are only used in connection with a positional, profile or tilting tolerance (**Figure 2.5-22**). They are absolute and therefore not toleranced, which enables the usage of so-called incremental dimensions. These would,
for normal dimensions, lead to an addition of the tolerances. Contrary to the normal dimensions, the exact dimensions are in square brackets [DIN ISO 1101].

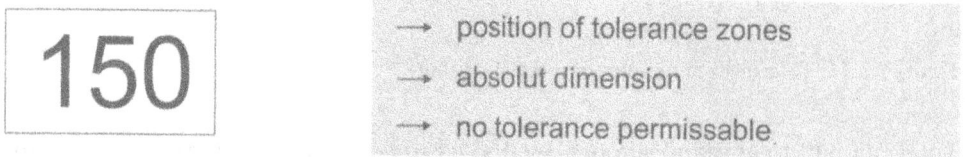

Figure 2.5-22: Theoretically exact dimension

Projected tolerance zone
In cases when the specification of a locational tolerance on the workpiece is not
sufficient to ensure the function or capability of pairing, the tolerance zone can be
shifted, or projected, in the direction of the paired counterpart. In this case the tolerance value is marked by a Ⓟ, the length of the projected tolerance zone must be
given and is also followed by a Ⓟ [DIN ISO 1101]. **Figure 2.5-23** shows an example of a projected tolerance zone. The function requires that the threaded bolt
can be screwed into the workpiece up to the toleranced tilt of the drill hole. The
position of the thread in the bottom section is of secondary importance.

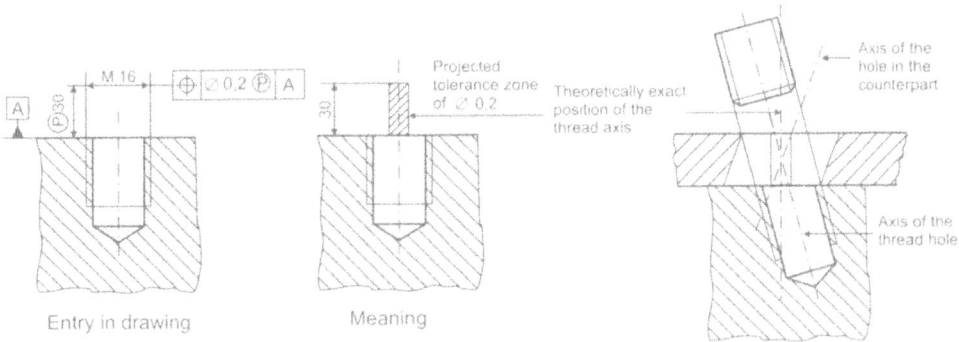

Figure 2.5-23: Projected tolerance

Maximum material principle

If from a functional point of view the possibility of joining a component with a counterpart is important, the maximum material principle according to DIN ISO 2692 can be used to maximise the usable tolerance values for form and location, which relate to the axis and centre planes. The form and location tolerances can be enlarged depending on the degree of utilisation of the measuring tolerances connected to these tolerances, without affecting the assembly. The given tolerances of form and location can, in this case, be increased by the difference between actual dimension and maximum material dimension.

For clarification purposes, a Ⓜ is added to the respective tolerance in the technical drawing when the maximum material principle is used. It can be added to the tolerance, the datum or both, depending on which it relates to.

An example for the usage of the maximum material principle for straightness is shown in **Figure 2.5-24**. The tolerance of straightness of the axis of the shaft in the right section of the figure can be increased from 0.01 mm to 0.03 mm, as the actual diameter being the minimal diameter deviates by 0.02 mm from the material maximum. A function gauge is used to simultaneously check the deviation of straightness and dimension. The diameter of the shaft must be checked separately to ensure that the limits of size are not exceeded.

Minimum material condition

The minimum material condition according to DIN 2692/A1 can be used for tolerances of form and location of axes or centre planes if the desired function allows for. It is usually used when there are certain functional reasons that a minimum measure must not be fallen short of, e.g. for wall thickness. The minimum material condition allows the increase of marked form and location tolerances if the toleranced element deviates from its minimal material dimension towards its maximum material dimension. The marking of the minimum material principle is done with a Ⓛ.

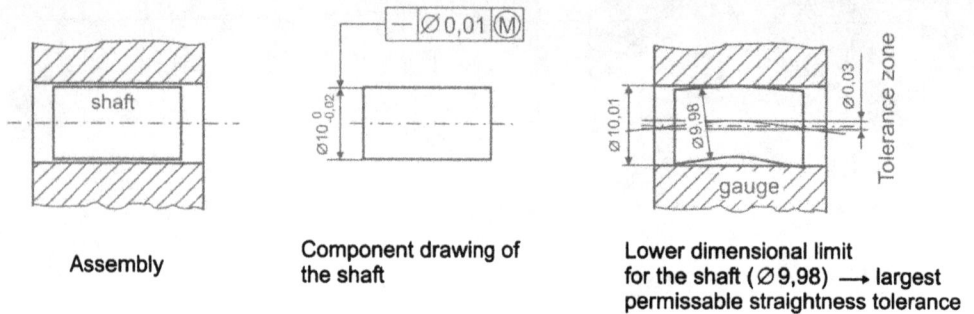

| Assembly | Component drawing of the shaft | Lower dimensional limit for the shaft (∅9,98) ⟶ largest permissable straightness tolerance |

Figure 2.5-24: Maximum material principle

Reciprocity condition
The reciprocity condition is not part of DIN ISO 2692, but of the appendix and therefore may be used. It supplements the maximum and minimum material conditions to the effect that dimension tolerances, connected to a form or location tolerance can be exceeded if the form and position deviation has not been used to the full extent. It is tagged by a Ⓡ after the tolerance value.

Condition of envelope
The condition of envelope demands that the actual form of a component never exceeds the geometrically ideal envelope surface with the maximum-material dimension. By this requirement, the form deviations are also limited by the defined dimensional limit.

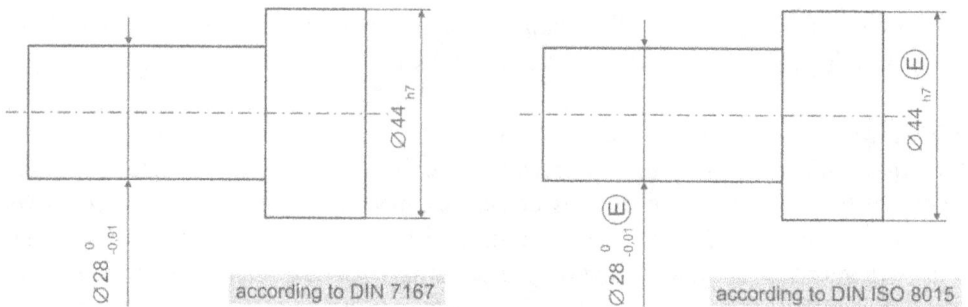

Figure 2.5-25: Condition of envelope

For a shaft, the surface must not penetrate a geometrically ideal cylinder with the maximum dimension, at the same time it must not fall below the minimum dimension, at any position.

A drill hole must not penetrate the smallest cylinder and must not exceed the actual dimension at any position.

According to DIN 7167, the condition of envelope is valid without special tagging on the drawing for all toleranced elements. According to DIN ISO 8015 the validity of the condition of envelope for a feature is marked by adding a Ⓔ to the respective dimension (**Figure 2.5-25**).

Elimination of form deviations for locational tolerancing
According to the American standard ASME Y 14.5 M, the form deviations of a toleranced element can be eliminated for a location tolerance by substituting the toleranced element with an equivalent element (section 2.5.2.3). It is tagged by adding a Ⓣ to the tolerance value.

2.5.3 Tolerancing principles

The necessity of complete dimensioning and tolerancing of work pieces and subassemblies in technical drawings is based on the demand for a functional, productional and qualitative as well as economical production process. To clearly show the relations between dimensions, form and location of elements in technical drawings, tolerancing principles have been defined. A tolerancing principle - *condition of envelope* without explicit inscription in a drawing according to DIN 7167 – has been explained in the last section. According to DIN ISO 8015, the condition of envelope is only one component of the *superposition principles* and therefore tagged by the symbol Ⓔ after the dimension tolerance. In this context, the condition of envelope is also called "old tolerance principle" and the superposition principle according to DIN ISO 8015 as "new tolerance principle".

The following section explains the superposition principle in more detail. After that, the two tolerance principles will be compared.

2.5.3.1 „New" tolerance principle – superposition principle

a) local actual diameter: b) entry on drawing

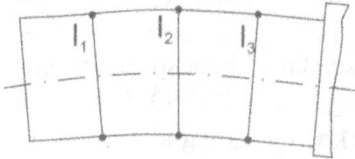

Figure 2.5-26: Inscription of the condition of envelope in drawings for which the   independence
principle generally applies

This tolerance principle assumes that the demands for dimension, form and location have to be met independently of one another. Therefore, tolerance of form and
location can be used to their full extent over it the profile sections of the measured
elements reach maximum material dimensions.

The tolerance of length dimension therefore only limits all actual dimensions,
which have been taken by a two-point measurement (**Figure 2.5-26, a**). This also
limits the cone, saddle and barrel form.

To ensure the function of a component, all necessary tolerances of form and location have to be specified separately. To simplify drawing, general tolerances according to DIN ISO 2768-2 should be used.

The tolerances with regard to the function of a workpiece or the sub assembly are
not necessarily independent of one another. In the drawing, additional symbols,
e.g. Ⓔ for condition of envelope, must be used if a relation exists with regard to
dimension and form for paired elements (**Figure 2.5-26, b**).

2.5.3.2 Comparison of "old and "new" tolerancing principles

To check the condition of envelope theoretically, limit gauging according to the
Taylor principle must be carried out. The transition to the superposition principle
meets the demands of the increasingly applied measuring tests, which gathers the
local actual dimensions. Furthermore, the tolerance details are obvious from the
function requirements of the form elements. **Figure 2.5-27** compares the "old" and
"new" tolerancing principles.

Conditions of envelope	Superposition principle
- function cannot be determined from drawing	- reference to function oriented tolerances by individual entries
- expensive production and inspection	- inexpensive production and inspection
- increased risk of rejection	- reduced risk of rejection
- limitation due to innecessary small tolerances	- no unnecessary small tolerances due to the calibration of individual functional requirements
- low tolerancing effort	- increased tolerancing effort

Figure 2.5-27: "Old" and "new" tolerancing principles

2.5.4   Geometrical product specification and inspection (GPS)

The GPS standards define: dimension, form and location tolerances; characteristics of surfaces and related inspection methods, measuring devices and calibration requirements as well as the uncertainties when measuring geometrical quantities. For all GPS standards, an overview, the so-called masterplan [DIN V 32950], with defined criteria has been developed, into which all new standards are sorted. When reviewing old standards in the next few years, a wide ranging re-structuring is to be expected in the area of form and position tolerancing.

Results from the ISO commission are decisive rules for dimensions. Furthermore, possibilities are being discussed to convert ISO standards into vectorial tolerances, which could influence an ISO standard for vectorial tolerancing.

2.5.4.1 Measurement uncertainty and decision rules

To be able to judge from a measurement if a feature meets the required specifications (tolerances), the decision needs to take into account the extended measurement uncertainty (section 2.3). As decision rule for proving conformance, standard DIN EN ISO 14235-1 can be used for inspection by measuring.

With the example of a double-sided tolerance, **Figure 2.5-28** shows the connections between measurement of result, extended measurement uncertainty U and statements on the compliance of specifications. The area of conformance is given by the reduction of the specification area on both sides by the extended measurement uncertainty. Features, which are measured inside this area meet the specifications in any instance, i.e. they are within the required tolerances. Similarly, the area of the non-conformance lies outside the specification area, which has been enlarged by U. Measured values, which are between these two areas do not allow for a secure statement about meeting the specifications. In this case, it is possible to achieve a statement by a measurement with a lower measurement uncertainty, alternatively special arrangements must be made for this case [Gro 97].

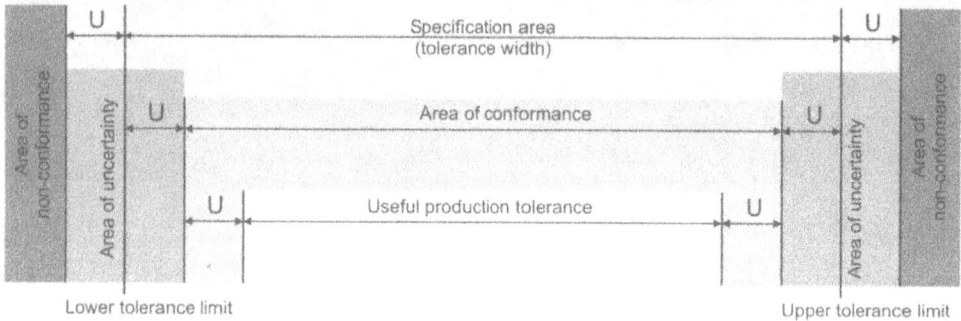

Figure 2.5-28: Relationship between measurement uncertainty and release tolerance

For metrological uses it can be concluded that the useful tolerance zone at the limits is smaller by the extended measurement uncertainty, than the stated tolerance, as it is required that no component outside the tolerance is accepted as a part. If a tolerance is limited in such a way, there is still the probability that parts, which comply with the specification are classified as reject parts as for the worst case the measured value differs from the real value by the amount of the extended measurement uncertainty. To exclude this possibility, the stated tolerance width on both sides must be reduced by double the amount of the extended measurement uncertainty. Therefore, an even smaller effectively usable production tolerance is given [Neu 85].

The description above is related to a double-sided tolerance, in the case of a single sided tolerance, the reduction of the tolerance on the side of the natural limit value is dropped.

2.5.4.2 Vectorial dimensioning and tolerancing

The aim of the vectorial tolerancing is the mathematically unambiguous description of the workpiece form and its geometrical tolerances [Wir 93]. Contrary to conventional tolerancing, a vectorial tolerancing, dimension and location of equivalent elements and the superimposed form deviations are tolerated separately. Equivalent elements are geometrical ideal form elements as e.g. plane, cylinder, sphere, cone and torus. ISO geometrical tolerances, however, are defined as sum tolerances.

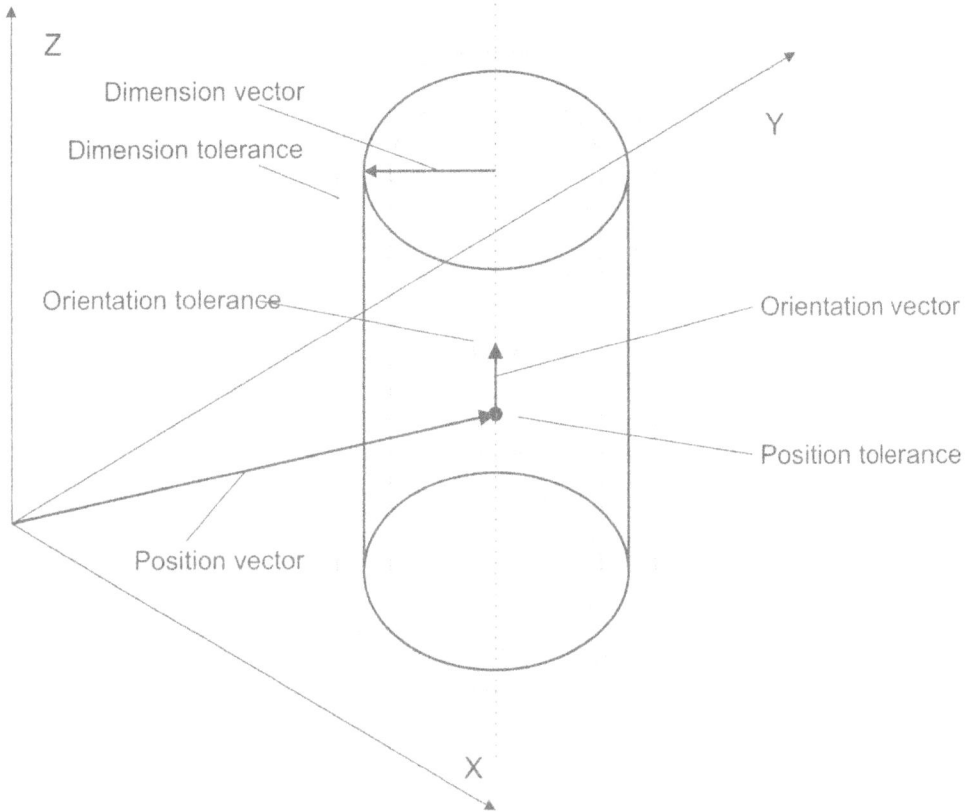

Figure 2.5-29: Vectorial dimensioning and tolerancing

The position of the nominal elements is given by a position vector triple, its orientation by the normal vector. Furthermore, the geometrical ideal form is described by the respective dimensions (**Figure 2.5-29**).

The permitted deviations are separated in a similar data model as for the individual components of the vector triples, form and dimension tolerances (**Figure 2.5-29**).

For vectorial tolerances, the same rules apply as for the vector geometry. That means that e.g. the position of a plane can only be toleranced in normal direction, while the position of a cylinder can only be toleranced vertically to the cylinder axis. Shifting of the position vector within the plane or along the axis does not, by definition, change the position of the elements.

On this basis, a workpiece can be completely described, and deviations or tolerances can be distinguished in dimension, form, orientation and location. Therefore, the vectorial tolerance gives unambiguous information, which can immediately be used for process control [Gro 97]. The position deviation of a drill hole in the -Y-plane, which is known by its components, can for example be used for a correction on a mechanical stage of a pillar drilling machine or a machining centre.

Vectorial dimensioning and tolerancing is especially useful regarding the demands and possibilities of computer aided metrology (e.g. coordinate metrology), which can cost effectively evaluate large amounts of data.

Further reading

[Bag 97]   Baginski, U.: Traditionelle Bemaßungsregeln im Umbruch? DIN-Mitteilungen 76, 1997, NR.: 6, S. 413, Berlin: Beuth-Verlag GmbH

[Ber 68]   Berndt, G.; Hultzsch, E.; Weinhold, H.: Funktionstoleranz und Meßunsi-cherheit. Wissenschaftliche Zeitschrift der Universität Dresden 17 (1968) 2, S. 465-471

[Bet 94]   Betsch, W.; Schatz, M.: Meßräume - Kenngrößen, Klassifizierung, Pla-nung, Ausführung. Qualität und Zuverlässigkeit QZ 39 (1994) 8, S. 884-886

[Bre 93]   Breyer, K.-H.; Pressel, H.-G.: Auf dem Weg zum thermisch stabilen Koordinatenmeßgerät. Neumann, H. J. (Hrsg.): Koordinatenmeßtechnik, Neue Aspekte und Anwendungen. Ehningen: Expert Verlag, 1993

[DIN 94]   Internationales Wörterbuch der Metrologie, International Vocabulary of Basic and General Terms in Metrology, 2. Auflage. Berlin: Beuth-Verlag, 1994

[DIN 95]   Leitfaden zur Angabe der Unsicherheit beim Messen, Deutsche Überset-zung des „Guide to the Expression in Uncertainty in Measurement", 1. Auflage. Berlin: Beuth-Verlag, 1995

[Dut 69]   Dutschke, W.: Zulässige Meßunsicherheit. wt-Z. ind. Fertigung. 59 (1969) 12, S. 630-632

[Dut 96]   Dutschke, W.: Fertigungsmeßtechnik. 3. Aufl., Stuttgart: B.G. Teubner, 1996

[Dut 97]   Dutschke, W.; Braun, M.: Meßunsicherheit - ein Reizwort? Qualität und Zuverlässigkeit QZ 42 (1997) 9, S. 1006-1010

[Gro 97]   Grode, H.-P.: Geometrische Produktspezifikation und -prüfung (GPS), DIN-Mitteilungen 76, 1997, NR.: 7, S. 478-492, Berlin: Beuth-Verlag GmbH

[Her 96]   Hernla, M.: Meßunsicherheit und Fähigkeit, Ein Überblick für die betrieb-liche Praxis. Qualität und Zuverlässigkeit QZ 41 (1996) 10, S. 1156-1162

[Her 97]   Hernla, M.; Neumann, H. J.: Einfluß der Temperatur auf die Längenmessung. Qualität und Zuverlässigkeit QZ 42 (1997) 4, S. 464-468

[Hoi 94]   Hoischen, H.: Technisches Zeichnen - Grundlagen, Normen, Beispiele, Darstellende Geometrie. Berlin: Cornelsen Verlag, 1995

[Kes 95]   Kessel, W.: Meßunsicherheit und Meßwert nach der neuen ISO/BIPM-Leitlinie. Technische Messen tm 62 (1995) 7/8, S. 306-312

[Lem 92]   Lemke, E.: Fertigungsmeßtechnik. Braunschweig/Wiesbaden: Vieweg Verlag, 1988

[Neu 85]   Neumann, H. J.: Der Einfluß der Meßunsicherheit auf die Toleranzausnutzung in der Fertigung, Qualität und Zuverlässigkeit QZ 30 (1985) 5, S. 145-149, München: Carl Hanser Verlag

[Pf 96]   Pfeifer, T.: Qualitätsmanagement. München: Carl Hanser Verlag, 1996

[Pre 97]   Pressel, H.-G.: Genau messen mit Koordinatenmeßgeräten, Grundlagen und Praxistips für Anwender. Renningen-Malmsheim: Expert Verlag, 1997

[Pro 92]   Profos, P.; Pfeifer, T.: Handbuch der industriellen Meßtechnik. München: Oldenbourg Verlag GmbH, 1992

[Pro 97]   Profos, P.; Pfeifer, T. (Hrsg.): Grundlagen der Meßtechnik. München: Oldenbourg Verlag, 1997

[PTB 94]   PTB (Hrsg.): Die SI-Basiseinheiten, Definition, Entwicklung, Realisierung. 1994

[Tra 89]   Trapet, E.; Wäldele, F.: Koordinatenmeßgeräte in der Fertigung - Temperatureinflüsse und erreichbare Meßunsicherheit. VDI-Bericht 751: Koordinatenmeßgeräte als integrierter Bestandteil der industriellen Qualitätssicherung. Düsseldorf: VDI-Verlag, 1989

[Tra 97]   Trapp, W.: Gesetzliche Grundlagen des Meßwesens. Profos, P.; Pfeifer, T. (Hrsg.): Grundlagen der Meßtechnik. München: Oldenbourg Verlag, 1997

[Tru 97]   Trumpold, Beck, Richter: Toleranzsysteme und Toleranzdesign - Qualität im Austauschbau. München, Wien: Carl Hanser Verlag, 1997

[War 84]   Warnecke, H.-J.; Dutschke, W.: Fertigungsmeßtechnik, Handbuch für Industrie und Wissenschaft. Berlin: Springer Verlag, 1984

[Web 93]   Weber, H.: Umgebungseinflüsse auf Koordinatenmessungen. Neumann, H. J. (Hrsg.): Koordinatenmeßtechnik, Neue Aspekte und Anwendungen. Ehningen: Expert Verlag, 1993

[Wec 96]   Weckenmann, A.: Koordinatenmeßtechnik im Qualitätsmanagement. VDI-Bericht 1258: Koordinatenmeßtechnik, sicher-umfassend-zukunftsweisend. Düsseldorf: VDI-Verlag, 1996

[Wir 93]   Wirtz, A.: Vektorielle Tolerierung, das Bindeglied zwischen CAD, CAM
                und CAQ. Neumann, H. J. (Hrsg.): Koordinatenmeßtechnik - Neue As-
                pekte und Anwendungen. Ehningen bei Böblingen: expert-Verlag, 1993

[Wk 95]   Weck, M.: Werkzeugmaschinen, Fertigungssysteme, Band 3.2, Automati-
                sierung und Steuerungstechnik 2. Düsseldorf: VDI-Verlag, 1995

[Zim 94]   Zimmer, M.; Dietzsch, M.: Gestaltung von Meßräumen für Längenmeßge-
                räte. VDI-Bericht 1155: Fertigungsmeßtechnik und Qualitätssicherung.
                Düsseldorf: VDI-Verlag, 1994

Norms and recomendations

DIN 861   DIN 861, Teil 1: Parallelendmaße: Begriffe, Anforderungen,
                     Prüfung. Köln, Berlin: Beuth-Verlag GmbH, 1980

DIN 102   DIN 102: Bezugstemperatur der Meßzeuge und Werkstücke.
                     Berlin: Beuth-Verlag, 1956

DIN 406-10   DIN 406-10: Technische Zeichnungen: Maßeintragung, Begrif-
                     fe, allgemeine Grundlage. Berlin: Beuth-Verlag GmbH, 1992-
                     12

DIN 406-11   DIN 406-11: Technische Zeichnungen: Maßeintragung, Grund-
                     lagen der Anwendung. Berlin: Beuth-Verlag GmbH, 1992-12

DIN 406-11/A1   DIN 406-11/A1 (Norm-Entwurf): Technische Zeichnungen:
                     Maßeintragung, Grundlagen der Anwendung, Änderung 1; Ber-
                     lin: Beuth-Verlag GmbH, 1994-06

DIN 406-11 Beiblatt 1   DIN 406-11, Beiblatt 1 (Norm-Entwurf): Technische Zeich-
                     nungen – Maßeintragung. Teil 11: Grundlagen und Anwendun-
                     gen: Ausgang der Bearbeitung an Rohteilen. Berlin: Beuth-
                     Verlag GmbH, 1996-05

DIN 406-12   DIN 406-12: Technische Zeichnungen: Maßeintragung, Eintra-
                     gung von Toleranzen für Längen- und Winkelmaße. Berlin:
                     Beuth-Verlag GmbH, 1992-12

DIN 406-12   DIN 406-12 (Norm-Entwurf): Technische Zeichnungen: Maß-
                     eintragung, Eintragung von Toleranzen für Längen- und Win-
                     kelmaße. Berlin: Beuth-Verlag GmbH, 1994-06

DIN 1319-1   DIN 1319-1: Grundlagen der Meßtechnik. Teil 1: Grundbegrif-
                     fe. Berlin: Beuth-Verlag GmbH, 1995

DIN 1319-2   DIN 1319-2: Grundlagen der Meßtechnik. Teil 2: Begriffe für
                     die Anwendung von Meßgeräten. Berlin: Beuth-Verlag GmbH,
                     1996

DIN 1319-3   DIN 1319-3: Grundlagen der Meßtechnik. Teil 3: Auswertung
                     von Messungen einer einzelnen Meßgröße, Meßunsicherheit.
                     Berlin: Beuth-Verlag, 1996

DIN 1319-4	DIN 1319-4: Grundlagen der Meßtechnik. Teil 4: Auswertung von Messungen, Meßunsicherheit. Berlin: Beuth-Verlag, 1997
DIN 2257-2	DIN 2257-2: Begriffe der Längenprüftechnik, Fehler und Unsicherheit beim Messen. Berlin: Beuth-Verlag, 1974
DIN 7150-2	DIN 7150-2: ISO-Toleranzen und ISO-Passungen, Prüfung von Werkstück-Elementen mit zylindrischen und parallelen Paßflächen. Berlin: Beuth-Verlag GmbH, 1977-08
DIN 7154-1	DIN 7154-1: ISO-Passungen für Einheitsbohrung, Toleranzfelder, Abmaße in m. Berlin: Beuth-Verlag GmbH, 1966-08
DIN 7154-2	DIN 7154-2: ISO-Passungen für Einheitsbohrung, Paßtoleranzen, Spiel und Übermaße in m. Berlin: Beuth-Verlag GmbH, 1966-08
DIN 7155-1	DIN 7155-1: ISO-Passungen für Einheitswelle, Toleranzfelder, Abmaße in m. Berlin: Beuth-Verlag GmbH, 1966-08
DIN 7155-2	DIN 7155-2: ISO-Passungen für Einheitswelle, Paßtoleranzen, Spiele und Übermaße in m. Berlin: Beuth-Verlag GmbH, 1966-08
DIN 7157	DIN 7157: Passungsauswahl, Toleranzfelder, Abmaße, Paßtoleranzen. Berlin: Beuth-Verlag GmbH, 1966-01
DIN 7157 Beiblatt	DIN 7157, Beiblatt: Passungsauswahl, Toleranzfelderauswahl nach ISO/TR 1829, Berlin: Beuth-Verlag GmbH, 1973-10
DIN 7167	DIN 7167: Zusammenhang zwischen Maß-, Form- und Parallelitätstoleranzen, Hüllbedingung ohne Zeichnungseintragung. Berlin: Beuth-Verlag GmbH, 1987-01
DIN 7168	DIN 7168: Allgemeintoleranzen, Längen- und Winkelmaße, Form und Lage, Nicht für Neukonstruktionen. Berlin: Beuth-Verlag GmbH, 1991-04
DIN 55350	DIN 55350, Teil 13: Begriffe der Qualitätssicherung und Statistik, Begriffe zur Genauigkeit von Ermittlungsverfahren und Ermittlungsergebnissen. Berlin: Beuth-Verlag, 1987
DIN V 32950	DIN V 329950 (Vornorm): Geometrische Produktspezifikation (GPS): Übersicht. Berlin: Beuth-Verlag GmbH, 1997-04
DIN ISO 286-1	DIN ISO 286-1: ISO-Systeme für Grenzmaße und Passungen, Grundlagen für Toleranzen, Abmaße und Passungen. Berlin: Beuth-Verlag GmbH, 1990-11
DIN ISO 286-2	DIN ISO 286-2: ISO-Systeme für Grenzmaße und Passungen, Tabellen der Grundtoleranzgrade und Grenzmaße für Bohrungen und Wellen. Berlin: Beuth-Verlag GmbH, 1990-11
DIN ISO 1101	DIN ISO 1101: Technische Zeichnungen: Form- und Lagetolerierung, Tolerierung, Form-, Richtungs-, Orts und Lauftoleran-

	zen, Allgemeines, Definitionen, Symbole, Zeichnungseintragungen. Berlin: Beuth-Verlag GmbH, 1985-03
DIN ISO 1101	DIN ISO 1101 (Norm-Entwurf): Technische Zeichnungen: Form- und Lagetolerierung, Tolerierung von Form, Richtung, Ort und Lauf, Allgemeines, Definitionen, Symbole, Zeichnungseintragungen. Berlin: Beuth-Verlag GmbH, 1995-08
DIN ISO 1101 Beiblatt 1	DIN ISO 1101, Beiblatt 1: Technische Zeichnungen: Form- und Lagetolerierung, Tolerierte Eigenschaften und Symbole, Zeichnungseintragungen, Kurzfassung. Berlin: Beuth-Verlag GmbH, 1992-11
DIN ISO 2692	DIN ISO 2692: Technische Zeichnungen: Form- und Lagetolerierung; Maximum-Material-Prinzip. Berlin: Beuth-Verlag GmbH, 1990-05
DIN ISO 2692/A1	DIN ISO 2692/A1 (Norm-Entwurf): Technische Zeichnungen: Form- und Lagetolerierung, Maximum-Material-Prinzip, Änderung 1: Minimum-Material-Prinzip. Berlin: Beuth-Verlag GmbH, 1991-05
DIN ISO 2786-1	DIN ISO 2786-1: Allgemeintoleranzen, Toleranzen für Längen- und Winkelmaße ohne einzelne Toleranzeintragung. Berlin: Beuth-Verlag GmbH, 1991-06
DIN ISO 2786-2	DIN ISO 2786-2: Allgemeintoleranzen, Toleranzen für Form und Lage ohne einzelne Toleranzeintragung. Berlin: Beuth-Verlag GmbH, 1991-04
DIN ISO 4291	DIN ISO 4291: Verfahren für die Ermittlung der Rundheitsabweichung, Messen der Radienabweichungen. Berlin: Beuth-Verlag GmbH, 1987-09
DIN EN ISO 5458	DIN EN ISO 5458: Geometrische Produktspezifikation (GPS): Form- und Lagetolerierung, Positionstolerierung. Berlin: Beuth-Verlag GmbH, 1999-02
DIN ISO 5459	DIN ISO 5459: Technische Zeichnungen: Form- und Lagetolerierung, Bezüge und Bezugssysteme für geometrische Toleranzen. Berlin: Beuth-Verlag GmbH, 1982-01
DIN ISO 8015	DIN ISO 8015: Technische Zeichnungen: Tolerierungsgrundsatz. Berlin: Beuth-Verlag GmbH, 1986-06
DIN EN ISO 14253-1	DIN EN ISO 14253-1: Geometrische Produktspezifikation (GPS): Prüfung von Werkstücken und Meßgeräten durch Messung. Teil 1: Entscheidungsregeln für die Feststellung von Übereinstimmung oder Nicht-Übereinstimmung mit Spezifikationen. Berlin: Beuth-Verlag GmbH, 1999-03
ISO 1	ISO 1: Standard reference temperature for industrial length measurements. ISO, 1975

ISO 129	ISO 129: Technical drawings, Dimensioning, General principles, definitions, methods of execution and special indications. Berlin: Beuth-Verlag GmbH, 1985-09
ISO 406	ISO 406: Technical drawings, Tolerancing of linear and angular dimensions. Berlin: Beuth-Verlag GmbH, 1987-10
ISO 14253-2	ISO TR 14253-2 (Draft): Geometrical Product Specifications (GPS): Inspection by measurement of workpieces and measuring instruments. Part 2: Guide to the estimation of uncertainty of measurement in calibration of measuring equipment and product verification. ISO, 1997-04
ASME Y 14.5 M	ASME Y 14.5 M: Dimensioning and Tolerancing. 1994
ASME Y 14.5 1M	ASME Y 14.5 1M: Mathematical Definition of Dimensioning and Tolerancing Principles. 1994
VDI/VDE 3511	VDI/VDE-Richtlinie 3511: Technische Temperaturmessungen. Blatt 1 bis 5, 1994-1996
VDI 2602	VDI/VDE-Richtlinie 2602, Blatt 1 u. 2: Schwingungsisolierung. 1976
VDI 2627	VDI/VDE-Richtlinie 2627, Blatt 1 (Entwurf): Meßräume. 1994

3 Inspection Planning

According to DIN 55350, inspection planning is defined as planning the quality inspection. While quality testing was previously only defined as inspection planning, data acquisition and evaluation, a change has taken place today. Topics such as statistical process control (SPC), inspection equipment monitoring, or even capability tests are now part of the quality inspection. Originally, inspection planning used to be product-orientated, today the manufacturing equipment and production processes are ever more becoming part of inspection planning. It can be expected that even the planning of quality testing for services, product behaviour and the pre-production product development phase (e.g.: marketing, development/design, work preparation) will become part of inspection planning in the future (**Figure 3-1**).

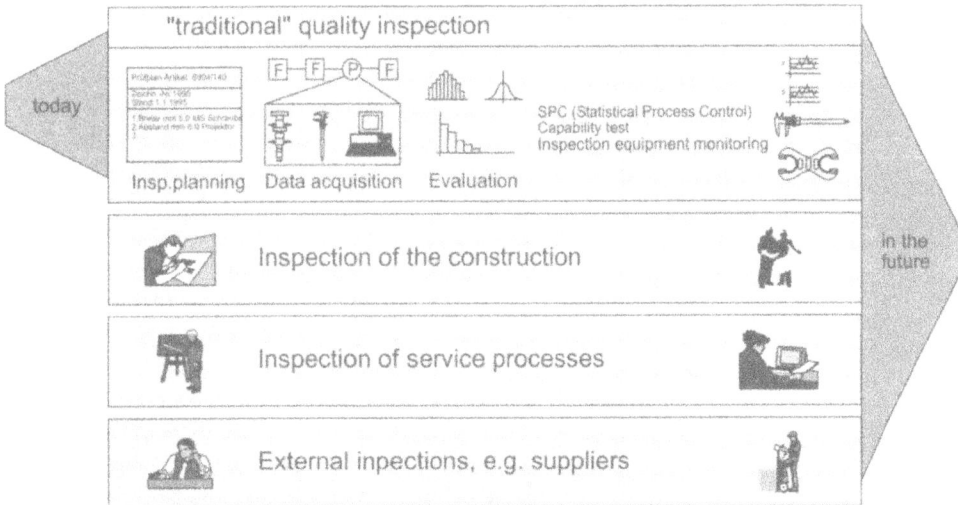

Figure 3-1: Quality inspection

The following will describe inspection planning in its traditional sense, i.e. planning the quality inspection. The main emphasis is on the application-relevant aspects, which are important for production metrology. The steps necessary for test planning are described with the aid of a sample part. An inspection plan is generated step by step. This will show only one possible form of an inspection plan, as each user will develop his/her own layout.

Inspection planning cannot be seen separately from other areas of the production process [Kw 96]. It is imbedded in an overall concept and is drawn up at the same time as the work preparation. Data from the quality planning as well as from development and construction are used. Inspection planning provides organisational and technical methods to carry out quality inspection, which, together with the subsequent data evaluation, is the information source, which completes the quality control loop. The results can be used, for example, to better adapt design specifications to the capabilities of the production process (large quality control loop) or simply to adjust the production equipment to counteract deviations (small quality control loop). A further possibility is to make the inspection dynamic (medium quality control loop, i.e. the control loop between inspection planning and work preparation) (**figure 3-2**).

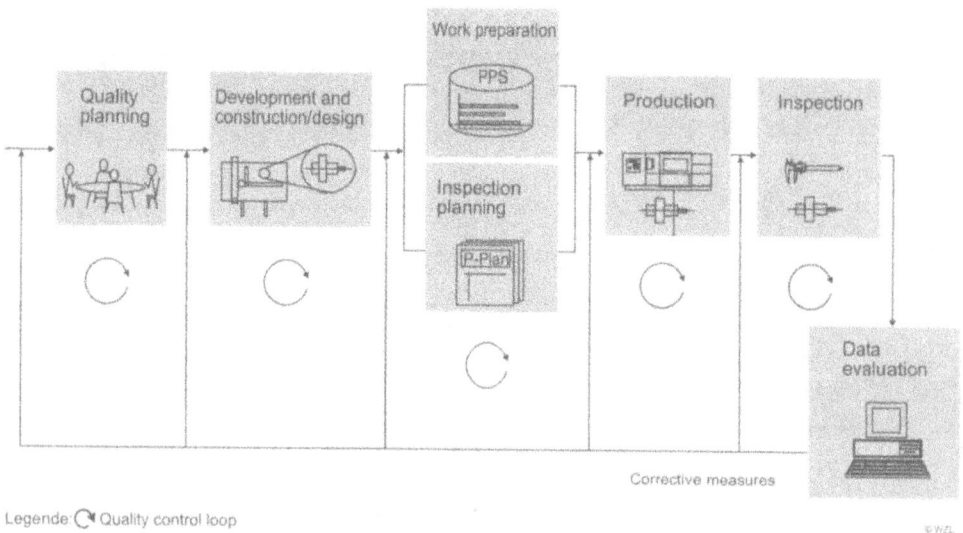

Legende: ◯ Quality control loop

Figure 3-2: Integration within enterprises

3.1   Tasks of Inspection Planning

Inspection planning includes various activities, which are divided into short term and long-term activities. In the short term, an inspection plan, which is a work instruction which regulates the inspection procedures, must be generated. Furthermore, within the inspection plan, consideration must be given to the documentation of results and the processing of data. The programming of the measuring devices is another task of the inspection planning. On the other hand, the inspection method planning, as well as the construction consultation, the staff training, the investment planning and the inspection plan support can only be carried out in the long term.

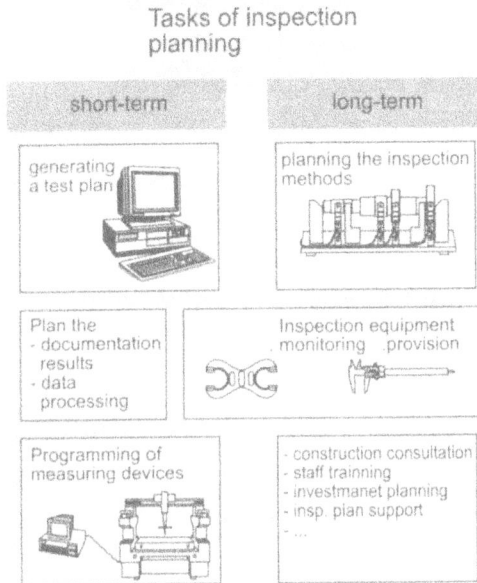

Figure 3.1-1:   Task of inspection planning

More recently, further tasks have been added to inspection planning. Alongside application planning and application of inspection equipment, provision and monitoring are now also part of an inspection planning (**Figure 3.1-1**). The following sections are limited to the actual generating of an inspection plan.

3.2   Procedure for Setting Up an Inspection Plan

The test plan regulates all activities of the quality inspection by type and extent for one product or part. Drawing up a inspection plan requires many different tasks. This should be done early in the product development phase to sensibly integrate the inspection planning into the quality planning, which requires several steps. As these tasks are to some degree dependent on one another, it is important that they are worked through in a logical sequence. The guideline VDI/VDE/DGQ 2619 offers some form of orientation. It describes ten points, which must be adhered to systematically when setting up an inspection plan **(Figure 3.2-1)**.

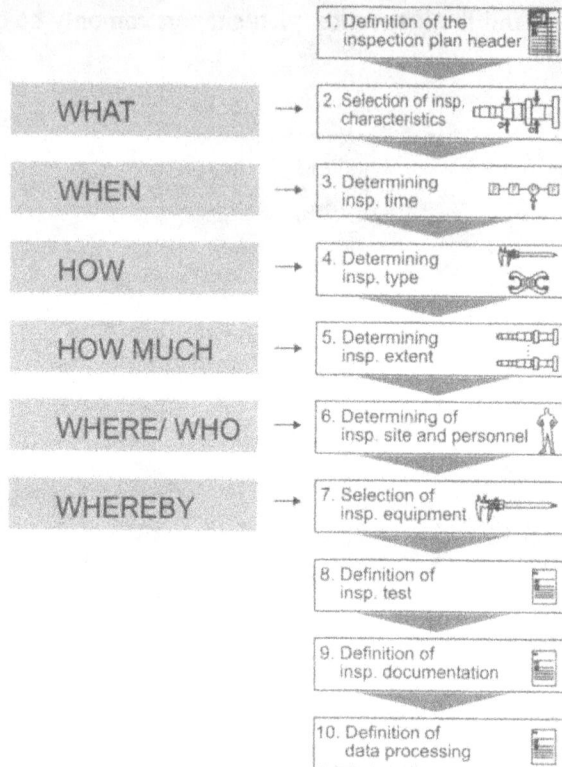

```
                                    ┌────────────────────────────┐
                                    │ 1. Definition of the       │ ▦
                                    │    inspection plan header  │
                                    └────────────────────────────┘
 ┌──────────────────┐               ┌────────────────────────────┐
 │      WHAT        │  →            │ 2. Selection of insp.      │
 └──────────────────┘               │    characteristics         │
                                    └────────────────────────────┘
 ┌──────────────────┐               ┌────────────────────────────┐
 │      WHEN        │  →            │ 3. Determining             │
 └──────────────────┘               │    insp. time              │
                                    └────────────────────────────┘
 ┌──────────────────┐               ┌────────────────────────────┐
 │      HOW         │  →            │ 4. Determining             │
 └──────────────────┘               │    insp. type              │
                                    └────────────────────────────┘
 ┌──────────────────┐               ┌────────────────────────────┐
 │    HOW MUCH      │  →            │ 5. Determining             │
 └──────────────────┘               │    insp. extent            │
                                    └────────────────────────────┘
 ┌──────────────────┐               ┌────────────────────────────┐
 │   WHERE/ WHO     │  →            │ 6. Determining of          │
 └──────────────────┘               │    insp. site and personnel│
                                    └────────────────────────────┘
 ┌──────────────────┐               ┌────────────────────────────┐
 │    WHEREBY       │  →            │ 7. Selection of            │
 └──────────────────┘               │    insp. equipment         │
                                    └────────────────────────────┘
                                    ┌────────────────────────────┐
                                    │ 8. Definition of           │
                                    │    insp. test              │
                                    └────────────────────────────┘
                                    ┌────────────────────────────┐
                                    │ 9. Definition of           │
                                    │    insp. documentation     │
                                    └────────────────────────────┘
                                    ┌────────────────────────────┐
                                    │ 10. Definition of          │
                                    │     data processing        │
                                    └────────────────────────────┘
```

Figure 3.2-1: Sequence of Inspection Plan Generation

The following will describe these points. In parallel with describing these points, an inspection plan for a sample workpiece will be drawn up in order to provide an easier understanding of the process.

An inspection plan is based on standards and guidelines, legal obligations and company procedures, technical documents as drawings, parts lists, work schedules, knowledge of the processes and of the customer's demands.

3.2.1   Definition of the Inspection Plan Header

The inspection plan header is important for organisational reasons and is specific to a company. It includes identity number, workpiece number, drawing number, work schedule number, responsible member of staff and other related information.

Figure 3.2-2 shows the inspection plan header for the sample company WZL and the sample workpiece shaft.

WZL TH AACHEN	Inspection plan	*Insp. plan no.: 4711*	Page 1 of 1
Drawing no.: 8904-140	Component test: shaft	Contact person F. Lesmeister	Date 01.12.1999
PFO-Nr	Insp. charact.	Limit values	Insp.equip.	Insp. extent	Comments

Figure 3.2-2: Inspection plan header

3.2.2 Selection of Inspection Characteristics

The selection of the inspection characteristics has an important role within the set-up of the test plan as it influences all subsequent actions. Inspection characteristics must be identified and evaluated according to their inspection necessity. To identify potential inspection characteristics, data from construction drawings, construction or process FMEA, machine performance, work plans, etc. is analysed.

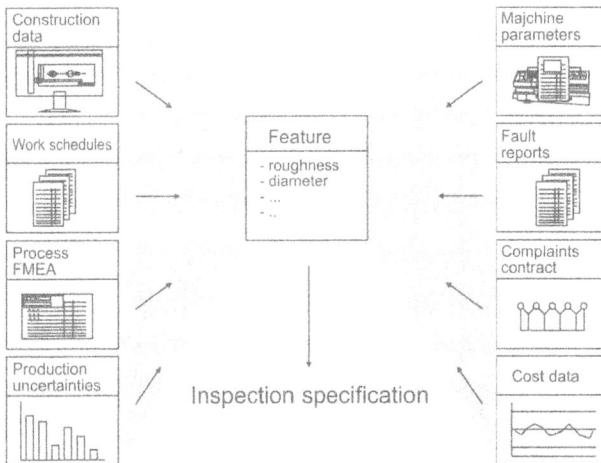

Figure 3.2-3: Selection of the Inspection Characteristics

Additional knowledge can be derived from discussions with designers, manufacturers and users of the parts **(Figure 3.2-3)**. Usually, the need for information about potential inspection characteristics grows with the complexity of the component.

Once the characteristics have been identified, the necessity of an inspection is evaluated. The aim is to minimise the inspection costs but also to ensure product quality. Safety-relevant features are always necessary inspection characteristics. In addition to ensuring functionality of a feature, the decision on whether a feature is inspected is based on customer requirements, stability of the production process, the intended utilisation of the workpiece, the production site (within own company or third party) and costs (e.g. follow on cost for tolerance excesses and testing costs).

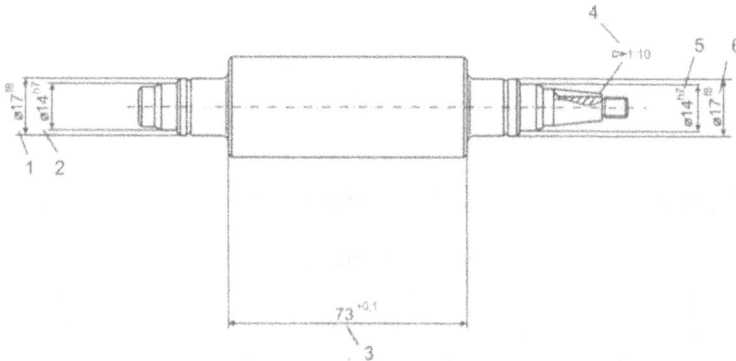

Figure 3.2-4: Sample workpiece shaft – selection of inspection characteristics.

The inspection characteristics can be physical, chemical, functional or optical. The result of these steps is the characteristics to be inspected with their desired values and allowed tolerances.

For choosing the inspection characteristics for the sample shaft, the construction drawing and the design engineer were consulted. There are six relevant inspection characteristics **(Figure 3.2-4)**.

All but these six dimensions have been left out of the construction drawing to simplify the drawing. The functioning of the shaft (bearing seat) is the decisive criterion for the choice of the inspection characteristic. Therefore, the fits of the shaft (1, 2, 5, and 6), length (3) as well as the angle of the cone (4) must comply with the according tolerances. These characterisitcs are entered into inspection test plan with the respective numbers and the limit values **(Figure 3.2-5)**.

WZL TH AACHEN	Inspection plan	Insp. plan no.:4711	Page1 of 1
Drawing no.: 8904-140	Test component: shaft	Contact person	Date
		F. Lesmeister	01.12.1999
PFO-Nr	Insp.charact.	limit values	Insp.equip.	Insp.extent	Comments
1	diameter	16,984 16,957			
2	diameter	14,000 13,982			
3	interval	73,1 73,0			
4	cone				
5	diameter	14,000 13,982			
6	diameter	16,984 16,957			

Figure 3.2-5: Sample workpiece shaft – insoection plan

3.2.3   Determining Inspection Time

This step determines the inspection time, i.e. when to inspection the characteristics. A timely inspection can contribute to not working on "scrap material" in the ensuing production process. If this intermediate inspection is taken too far or carried out too late, unnecessary costs are incurred, e.g. by long storage or buffer times until inspection and long delivery periods. It also must be taken into account that some parts cannot be reached once they have been assembled and therefore cannot be inspected.

The choice of the inspection time is hence reliant on a number of criteria such as inspection costs, damage risk, accessibility of the characteristic, increase in the product value, etc **(Figure 3.2-6)**.

The inspection time for the shaft can only be determined if the production process is known, e.g. on the basis of allocated capability values, the number of units and the application field. The description of these is too complex for this chapter. As a simplification, the shaft will only be tested during a final inspection.

3.2.4   Determining Inspection Type

There is a principal distinction between an inspection by attributes and by variables. The inspection by attributes is a good/bad test, whereas the inspection

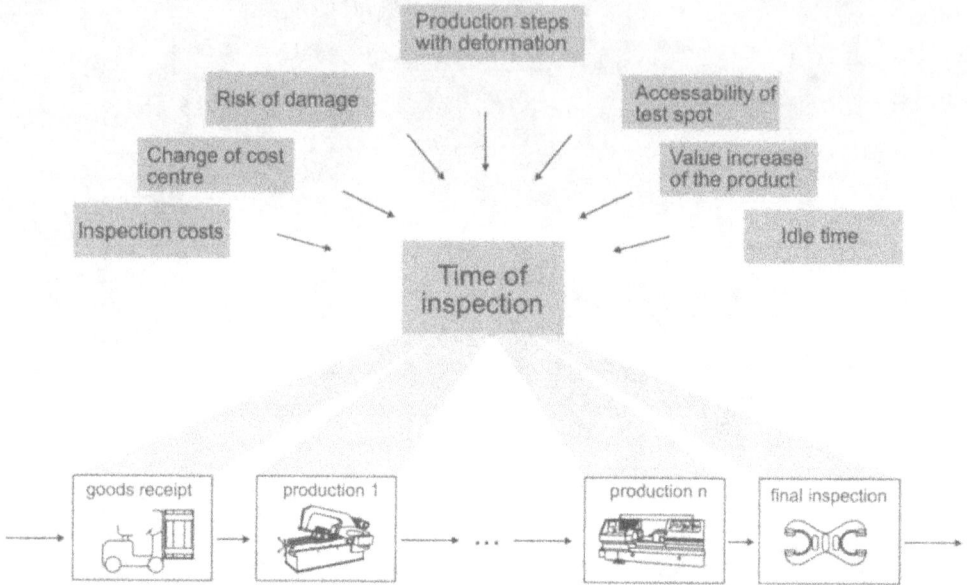

Figure 3.2-6: Determining Inspection Time

by variables is used to determine quantitative characteristics. The decision for one or the other inspection type is made by considering cost aspects, the applicability of inspection equipment and, of course, the characteristic to be inspected. Principally, the inpection by variables is superior to the inspection by attributes as the result is of higher statistical value **(Figure 3.2-7)**.

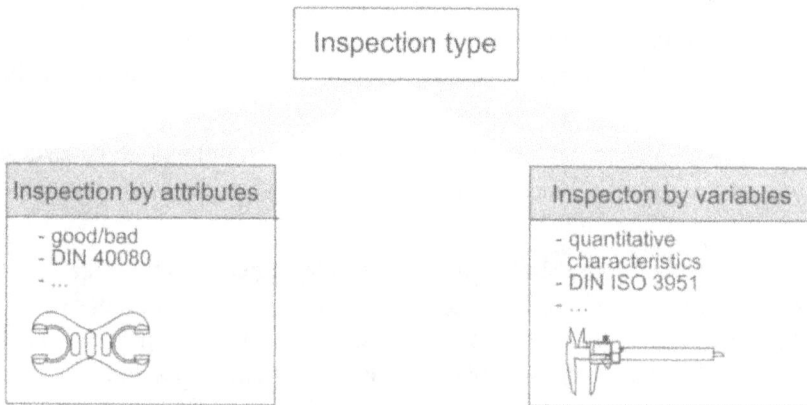

Figure 3.2-7: Determining Inspection Type

With regard to the shaft, the choice of test type depends on the batch size (**Figure 3.2-8**).

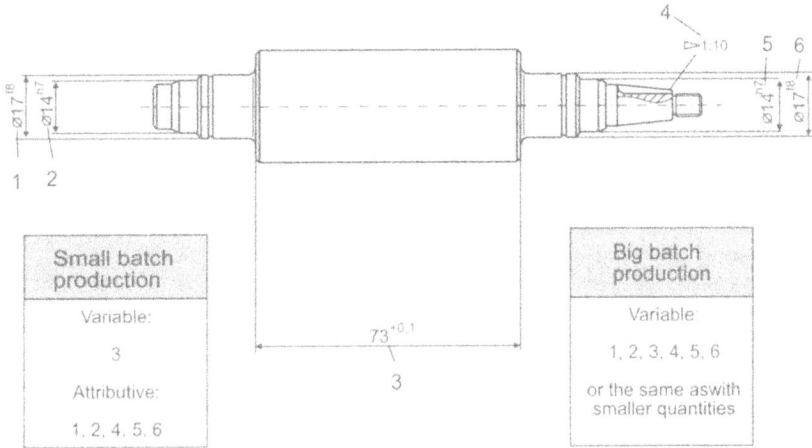

Figure 3.2-8:   Sample workpiece shaft – choice of inspection type

In principle, it is sufficient to perform an inspection by attributes on the features 1, 2, 4, 5, and 6. An inspection by variables is carried out for feature 3 as, although an inspection by attributes would be sufficient, a new gauge would have to be built, because it is a length for which no standardised gauge is available for the characteristics 1, 2, 4, 5 and 6, which are diameters or cones. Appropiate gauges are available for a larger batch size, or if a statistical evaluation of the results is needed, the inspection by variables should be chosen over an inspection by attributes. For this, special inspection equipment is needed. The working example is based on a small bath production.

3.2.5   Determining the Inspection Extent

The inspection plan defines the test extent (inspection frequency, inspection level). This directly influences inspection and error cost. It ranges from random and intermittent tests all the way to a 100% test. Within this range are a number of test focuses.

The 100% inspection considers all parts. This is especially important for safety-relevant parts. In mass production, it is only economical when using automated inspection equipment. Additionally, the error quota is relatively high in manual testing of mass produced parts, as the activity is highly monotonous. The manual inspection of random samples does not carry such a high error rate with it (**Figure 3.2-9**).

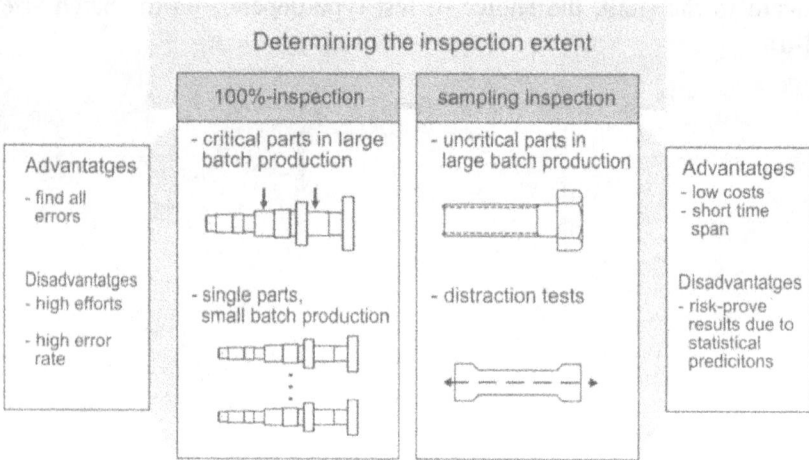

Figure 3.2-9: Selection of the test scope

Sample inspection is carried out according to externally valid standards or company internal regulations [DIN ISO 3951], [DIN ISO 2859]. An extension to this is the dynamic inspection extent definition [DIN ISO 2859], for which the sampling inspection extent varies according to the test results. Additional methods to determine the inspection extent, such as SPC, are explained later on (section 5.2). When choosing the inspection extent, the importance of the feature for the quality of the product (critical feature, main or ancillary feature) and the process capability are important.

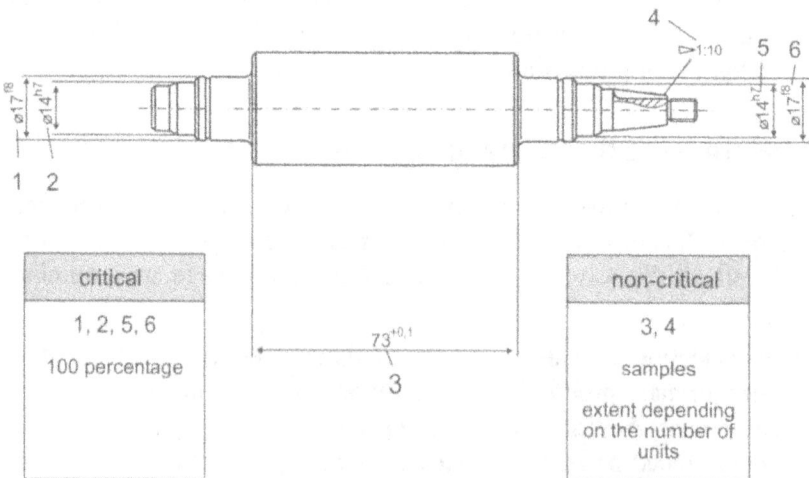

Figure 3.2-10: Sample workpiece shaft – determining the inspection extent

For our sample shaft, which is only produced in small quantities, the critical features are the fits (1,2,5,6). These have to be inspected by a 100% test, to ensure correct production behaviour. The other two features (3, 4) are not critical. For these sampling inspection is sufficient, whose extent has to be defined depending on the batch size **(Figure 3.2-10)**.

3.2.6   Selection of Inspection Site and Personnel

Defining the inspection site is reliant on the inspection characteristics, the inspection equipment, the production flow and the dimensions of the parts **(Figure 3.2-11)**. The production flow should not be disrupted by the inspection steps. This can be achieved by co-ordinating the inspection steps with the production flow or by intermediate storages.

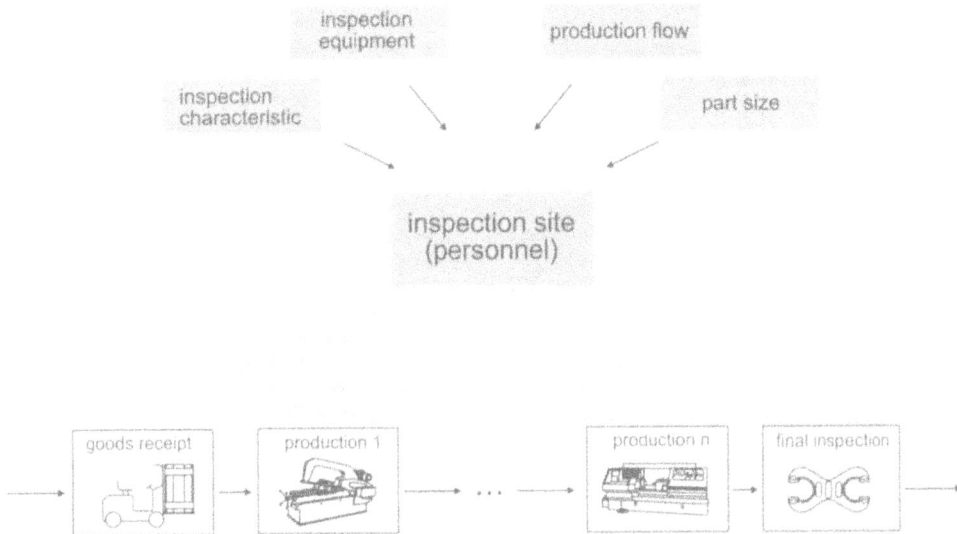

Figure 3.2-11: Determining of inspection site and personnel

The question of the inspection personnel is usually answered by the selection of the inspection site. Complex measurement devices, e.g. coordinate measuring machines can usually only be handled by specialised staff. Otherwise the production staff should carry out the inspection as they can correct errors immediately. In addition to this, the staff are given more responsibility and identify better with the product.The machine operator should therefore inspect the shaft immediately after production.

3.2.7 Selection of Inspection Equipment

After the inspection characteristics have been defined and extent, time, type and site of inspecion have been determined, suitable inspection equipment needs to be chosen. For this, the question of availability and suitability of inspection devices must be clarified. An inspection equipment database can help in this decision, as it

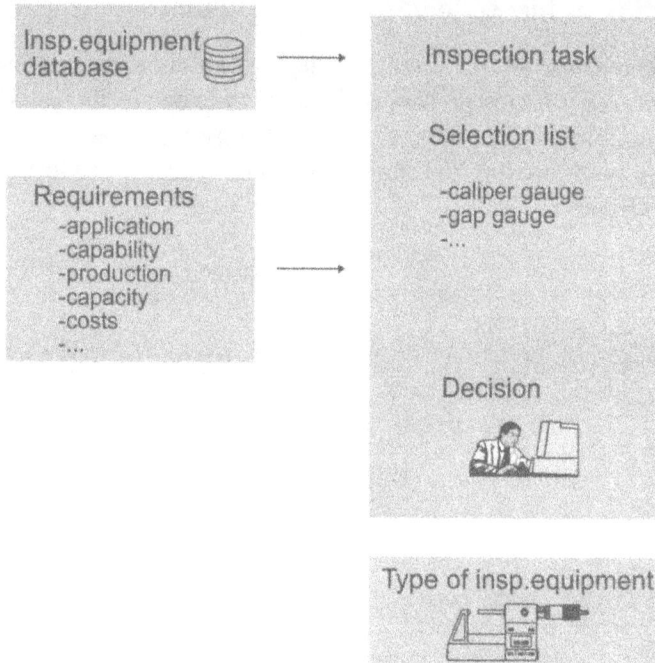

Figure 3.2-12: Selection of inspection equipment

lists all available measurement devices. The list can be produced by a computer, which takes into account the inspection task. The decision of which inspection equipment should be used can be made manually or with the aid of a computer. The capability of the inspection device (measurement uncertainty) and the tolerance of the inspected characteristic are the main selection criteria. Furthermore, acquisition cost, capacity and other elements are taken into account. The result of the inspection device selection, the type of inspection equipment, is taken into the inspection plan **(Figure 3.2-12)**.

For the sample shaft, a selection list is compiled for the measurement tasks **(Figure 3.2-13)**. The selection process for our example seems easy and is done manually. For other parts, it can become very complex, in which case the use of a computer becomes necessary. For the characteristcs 1, 2, 5 and 6, a gap gauge can be used, which can quickly check diameters. A cone gauge and a calliper gauge can measure characteristics 3 and 4. Principally, the use of a coordinate measuring

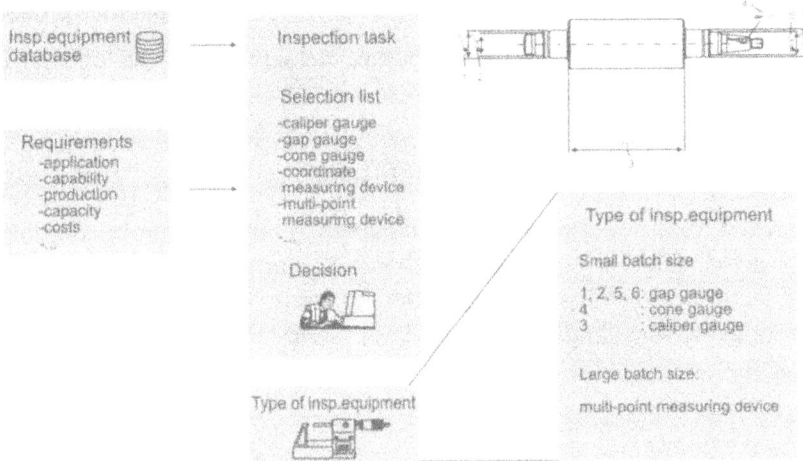

Figure 3.2-13: Sample workpiece shaft – selection of inspection equipment

machine is feasible, but would incur higher costs and is therefore not chosen. If the shaft is produced in a larger batch, it can be effective to use a multi-point measuring device.

WZL TH AACHEN	Inspection plan	Insp. plan no.: 4711	Page 1 of 1
Drawing no.: 8904-140	Test component: shaft	Contact person	Date
		F. Lesmeister	01.12.1999
PFO-Nr	Insp.charact.	limit values	insp.equip.	Insp.extent	Comments
1	diameter	16,984 16,957	gap gauge	100 %	
2	diameter	14,000 13,982	gap gauge	100 %	
3	interval	73,1 73,0	caliper gauge	3 each batch	
4	cone		cone gauge	3 each batch	
5	diameter	14,000 13,982	gap gauge	100 %	
6	diameter	16,984 16,957	gap gauge	100 %	

Figure 3.2-14: Sample workpiece shaft – inspection plan

This has the advantage that it automatically inspects all characteristics in one run and also records them. It is therefore a good time saver and provides additional information. The sample of the shaft is however based on small quantity production.

A suitable inspection plan with manual inspection equipment can be seen in **Figure 3.2-14**.

3.2.8   Inspection Text and Documentation

The inspection text should incorporate all necessary information in addition to the inspection tasks. This is especially advisable for complex inspection characterisitcs and facilitates the inspection procedure **(Figure 3.2-15)**. For example, a simple inspection drawing simplifies the work for the tester.

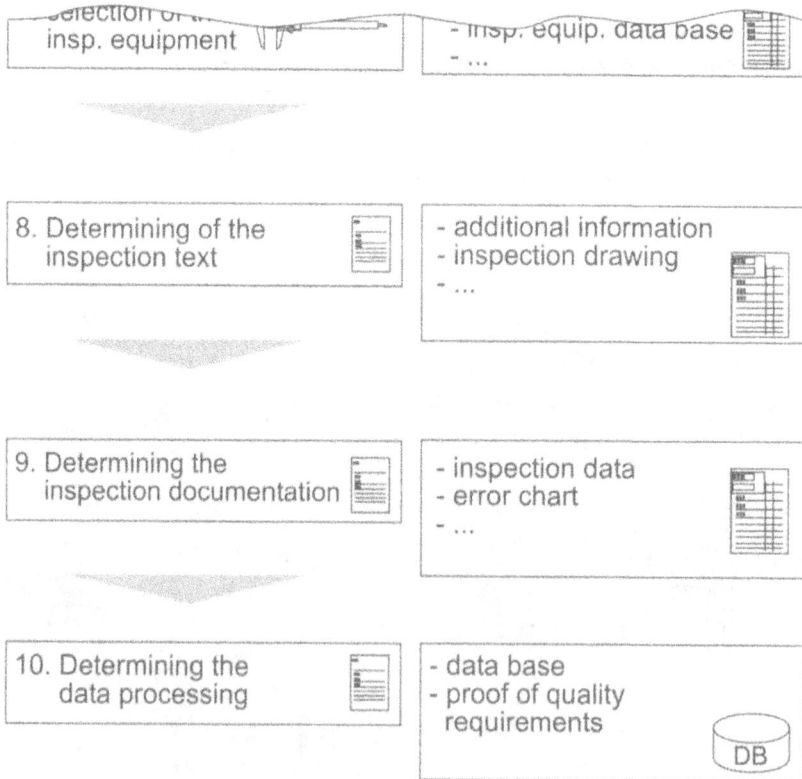

Figure 3.2-15: Inspection text and documentation

An inspection without documentation and evaluation does not make sense, as the data would be lost without being used and is not available for future use.

3.3   Utilisation of Results

The inspection data evaluation is an important basis for realising internal quality control loops. For this, the data must be processed in a useful manner and, in order to avoid having too much data. A visualisation can simplify the utilisation of the data (**Figure 3.3-1**).

The data must be documented to comply with legal regulations or those imposed by the customer. Especially stringent regulations are predominant in the area of aviation, automotive industry or generally for safety relevant parts. For these areas, inspection results are archived for many years so that the manufacturer or end user is able to call upon the results in the case of a damages claim. Documentation is also required for internal use in order to be able to follow production/inspection steps. Inspection results can be the basis for future improvements.

Further use for the results can be found during inspection evaluation, as the data can be used for process control or to adapt the inspection extent.

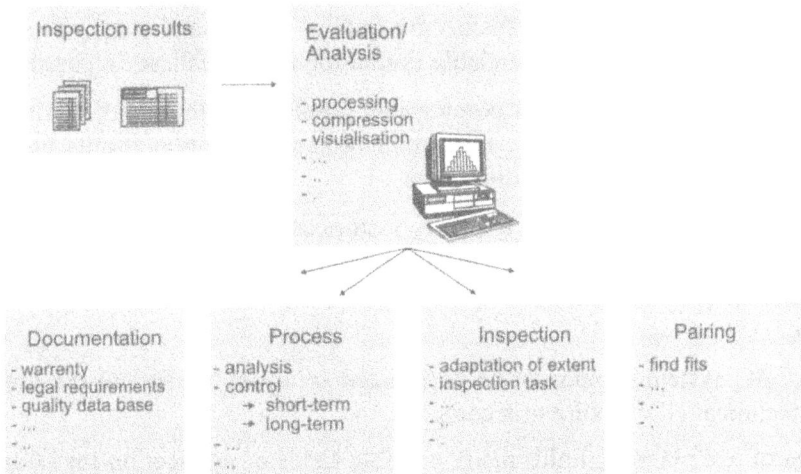

Figure 3.3-1: Inspection data evaluation

3.4   Possible fields of application

Setting up an inspection plan, as described above, includes a number of activities. Especially when the tasks are complex, it is possible that partial aspects are overlooked or the activities cannot be carried out due to missing information. For this reason, IT-systems are increasingly used for inspection planning. These cannot be looked at on their own, but need to be viewed in the context of the overall quality control. Only in combination with other quality assuring functions and other internal IT systems, can they be of use for inspection planning.

The use of computer-aided solutions in the area of quality assurance has the following aims [Mas 94]:

- To rationalise and simplify planning by suitable functions.

- To optimise the inspection process, which aims at the inspection extent and the necessary activities.

- To channel, bring together and make available the data, which is collected from various areas of a company.

- To speed up the usage of data, free it of manual tasks and extend this through additional possibilities.

- To be able to use the data sensibly for quality control and quality reporting, it must have adaptable and extendable evaluation and visualisation possibilities.

- The CAQ system must offer comfortable possibilities to administer inspection equipment, inspection plans, inspection orders, inspection results, as well as part, customer, supplier and other data.

- The CAQ system should be a quality assurance system as well as a quality information system, to take account of the increased information requirement and the necessity to make available the required information at the right time and place.

- The CAQ-system should not be an isolated solution, but fit into the functional and technical IT structure of a company.

The area of test planning holds many possible fields of application for IT systems. An overview can be seen in **Figure 3.4-1**. There are many systems, but some of them only work on partial aspects. Some software is in need of improvements. Data maintenance, especially for technical changes, is one of the areas requiring improvement. Much care must be taken when choosing the software that is suitable for a company or special task.

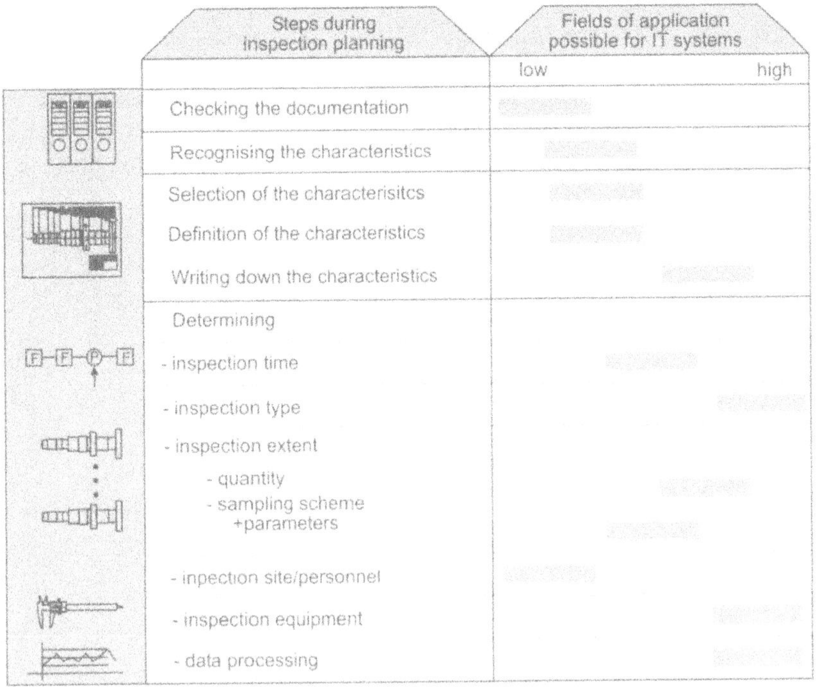

Steps during inspection planning	Fields of application possible for IT systems	
	low	high
Checking the documentation		
Recognising the characteristics		
Selection of the characterisitcs		
Definition of the characteristics		
Writing down the characteristics		
Determining		
- inspection time		
- inspection type		
- inspection extent		
- quantity		
- sampling scheme +parameters		
- inpection site/personnel		
- inspection equipment		
- data processing		

Figure 3.4-1:   Application possibilities of IT systems for test plan creation

Further reading

[Pf 96]   Pfeifer, T.: Qualitätsmanagement: Strategien, Methoden, Techniken. 2. Auflage, Carl Hanser Verlag, München, 1996.

[Kw 96]   Kwam, A.: Methodik zur Integration der Prüfplaung in die Qualitätspalung. Dissertation RWTH Aachen, 1996.

[Mas 94]   Masing, W.: Handbuch der Qualitätssicherung. 3. Auflage, Carl Hanser Verlag, München, 1994.

[Dut 96]   Dutschke, W.: Fertigungsmeßtechnik. 3. Auflage, B. G. Teubner, Stuttgart, 1996.

Norms and Guidelines

VDI/VDE/DGQ 2619   Richtlinie zur Prüfplanung. VDI/VDE/DGQ 2619, Beuth Verlag, Berlin, 1985.

DIN 55350   DIN 55350 Teil 11ff.: Begriffe der Qualitätssicherung und Statistik. Beuth Verlag, Berlin 1995.

DIN ISO 2859   DIN ISO 2859: Annahmestichprobenprüfung anhand der Anzahl fehlerhafter Einheiten oder Fehler (Attributprüfung) / AQL. Beuth Verlag, Berlin, 1993.

DIN ISO 3951   DIN ISO 3951: Verfahren und Tabellen für Stichprobenprüfung auf
                          den Anteil fehlerhafter Einehiten in Prozent anhand quantitativer
                          Merkmale (Variablenprüfung). Beuth Verlag, Berlin, 1992.

4 Test Data Acquisition

After describing the fundamentals of production metrology and test planning in both of the preceding chapters, the subject of the test data acquisition will now follow. The goal of the chapter on test data acquisition is to provide the reader with an overview of the most diverse inspection devices and testing methods, in order to be able to select and correctly implement a suitable testing device, in accordance with specific demands on measurement uncertainty and the additional boundary conditions, which arise from the direct environment of test data acquisition, whether testing must take place directly within the production process or be transferred to an acclimatised measuring room. Beginning with simple inspection devices, which are found on every shop floor, right up to complex 3D inspection arrangements and optical measuring systems which can determine positions and geometry with a high level of accuracy, this chapter provides a solution for nearly every metrological problem which arises in the area of production technology.

4.1   Shop floor Inspection Equipment

Under the generic term, shop floor inspection equipment, measuring instruments for the recording of one-dimensional features are summarised below, i.e. for the measurement of

- outer, inner and recess dimensions,
- diameters, widths and thicknesses,
- heights and depths,
- angles

and similar dimensions.

The following sections deal with the most commonly used handheld inspection devicesfor the manual recording of individual dimensions. These handheld inspection devices are sold by numerous manufacturers as complete units and in large numbers and usually offer a complete solution for simple measuring tasks.

The hand-guided measuring equipment, also referred to as measuring instruments **(Figure 4.1-1)** are of particularly great importance for the shop floor area as they are, on the one hand, universally applicable and, on the other hand, relatively easy to use. They probably rank among the oldest measuring devices and are almost exclusively mechanically constructed. Today, apart from these mechanical versions, which are still used due to their robustness, low cost and partly also simpler handling, electronic versions are increasingly also being used. These often have technical data interfaces for the transmission of measurement results, in addition to a more easily readable display and a higher degree of accuracy and functionality.

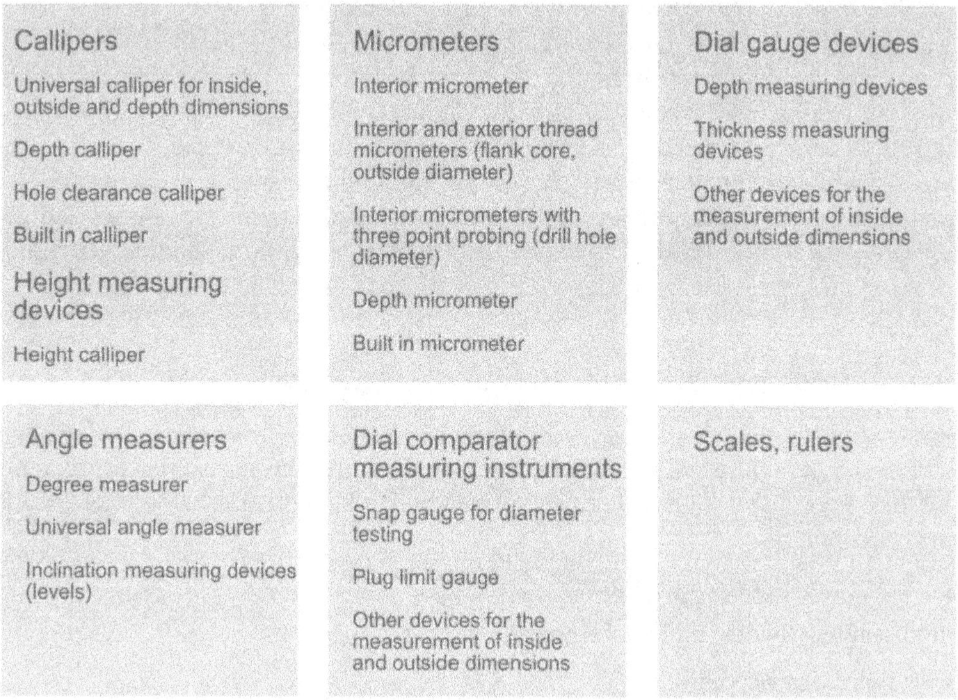

Callipers	Micrometers	Dial gauge devices
Universal calliper for inside, outside and depth dimensions	Interior micrometer	Depth measuring devices
Depth calliper	Interior and exterior thread micrometers (flank core, outside diameter)	Thickness measuring devices
Hole clearance calliper	Interior micrometers with three point probing (drill hole diameter)	Other devices for the measurement of inside and outside dimensions
Built in calliper		
Height measuring devices	Depth micrometer	
Height calliper	Built in micrometer	

Angle measurers	Dial comparator measuring instruments	Scales, rulers
Degree measurer	Snap gauge for diameter testing	
Universal angle measurer	Plug limit gauge	
Inclination measuring devices (levels)	Other devices for the measurement of inside and outside dimensions	

Figure 4.1-1: Examples of hand-guided shop floor inspection equipment for length and angle measurement

4.1.1 Callipers and Height Measuring Devices

Callipers **(Figure 4.1-2)** rank among the most common handheld measuring instruments in length metrology. The standard versions can be used for outside, inside and depth measurements. DIN 862 describes the specifications for callipers, inspection of them and even their design. The calliper consists of a beam and a relative moveable slider. For measuring outside and inside diameter dimensions, the beam has a fixed measuring jaw with a corresponding mobile measuring jaw on the slider. The measurement value is read off of a vernier calliper from the main scale. Frequently there are two scales, one in inches and one in millimeters. For the metric scale, vernier values of 1/10 mm, 20 mm and 1/50 mm are usual. The readability of the vernier calliper depends on the vernier length, whereby a greater vernier length provides better readability. While in former times, a 9 mm scale was selected for the 1/10 mm vernier, a 19 mm scale is common today and a length of 39 mm for the 1/20 mm vernier. The measuring range of callipers is standardised up to 2000 mm, however, commercial versions are usual of up to 3000 mm. Smaller versions of up to 160 mm are known as pocket callipers.

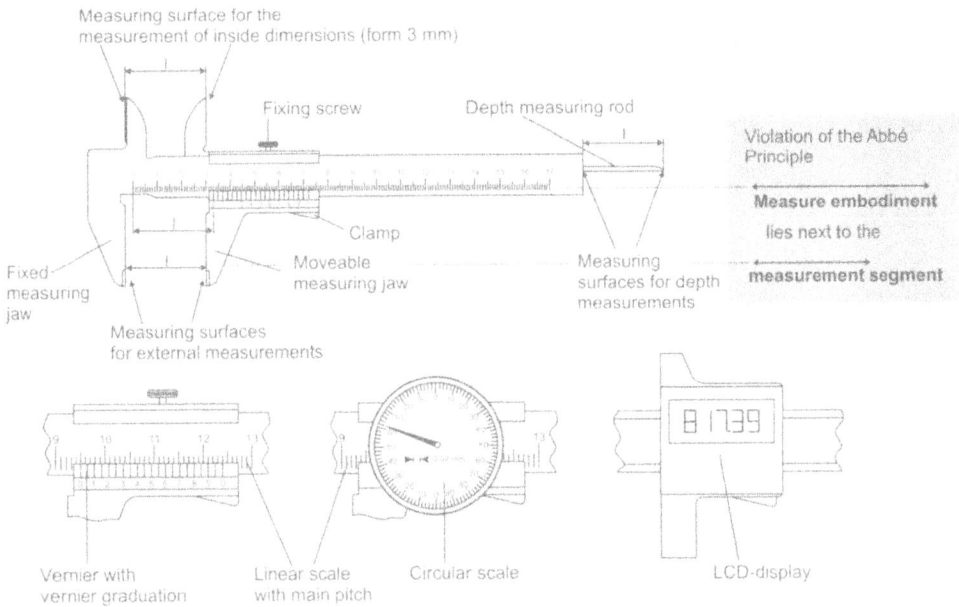

Figure 4.1-2: Callipers

In addition to callipers with a linear scale and vernier reading, devices with a circular scale are increasingly being used, which enable a more rapid and dependable reading. Many pocket calliper gauges are therefore equipped in this way with circular scales (scale graduation 0.02...0.1 mm). Digital callipers are also used with incremental measurement systems, e.g. optical grid scales or capacitive measurement systems and digital display. Apart from better readability, these usually also have additional functions, e.g. resetting the display: it can be set to zero at any position and thereby enables difference measurements, where the deviation is displayed in relation to a reference value.

When looking at measurement uncertainty, it must be considered that callipers, with the exception of depth callipers, do not conform to the Abbé principle, due to their design: the measure embodiment and measuring path do not lie behind one another, but next to one another. Slackness in the track of the slider and strong pressure from the moveable measuring jaw on the inspected object cause tilting of the slider and flexible bending of the beam. This results in measurement errors which feed directly into the measurement result (see Section 2.3).

In addition to the most common version shown in **Figure 4.1-2**, there are numerous special versions of callipers in use, for example, for measuring drill hole intervals or depth and height dimensions. Height measuring devices operate on the same principle as the calliper described above, with the difference that the measurement is almost always taken with reference to a surface through 1 point probing. While the relationship to a calliper is most obvious with a simple, mechanical height calliper, with the more complex height measuring devices, often only the same principle is still recognisable **(Figure 4.1-3)**.

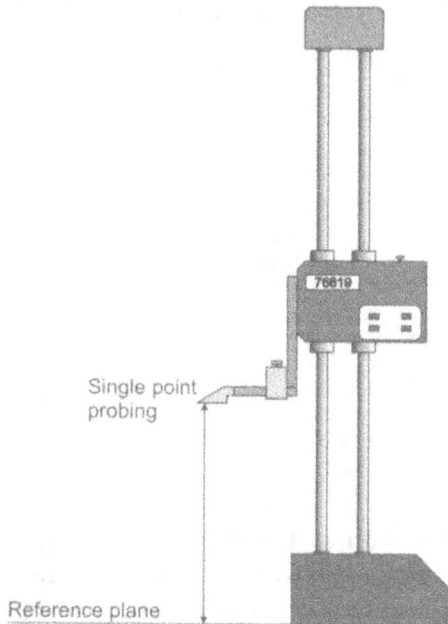

Figure 4.1-3:   Schematic diagram of a height measuring device with digital display

4.1.2 Micrometers

In contrast to the calliper, with the micrometer **(Figure 4.1-4)** the measure embodiment and measuring path are aligned, e.g. the Abbé principle is fulfilled here. The thread gradient of a screw serves as a measure embodiment. The screw is guided into an internal thread, the measuring spindle, to experience a defined forward shift. In order that the measuring force does not become too large through the manual operation of the spindle drive, micrometers are generally provided with a slip clutch, which limits the torque and thereby the measuring force. Depending on the model, measurement values can be recorded into the micrometer range.

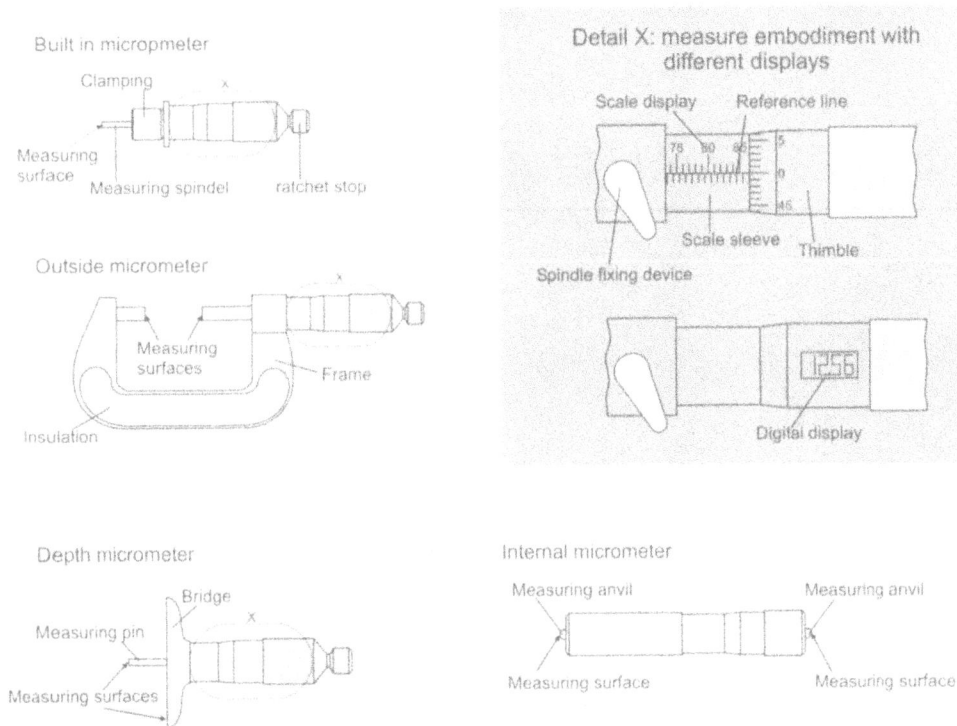

Figure 4.1-4: Micrometers

The actual core of all micrometers, which is also individually traded as a built-in micrometer, is only suitable for 1 point probing, e.g. it must be expanded to a measuring device by appropriate structures. There are numerous possibilities for a simple adaptation to special measuring tasks and thus a broad spectrum of commercial micrometers. The most common include the micrometer gauge, for which structure, specifications and testing are described in DIN 863, Part 1. They are primarily used for external measurements such as the measurement of numerous thickness and diameter dimensions, external thread diameters or tooth widths. The various versions of micrometer gauges are differentiated, not only through the structure of the bow, but more particularly through the measuring surfaces which are specially adapted to the measuring tasks **(Figure 4.1-5)**. For example, with a thread inspection for the measurement of pitch diameter, core diameter and outside diameter a special exchangeable pair of measuring inserts are used in each case.

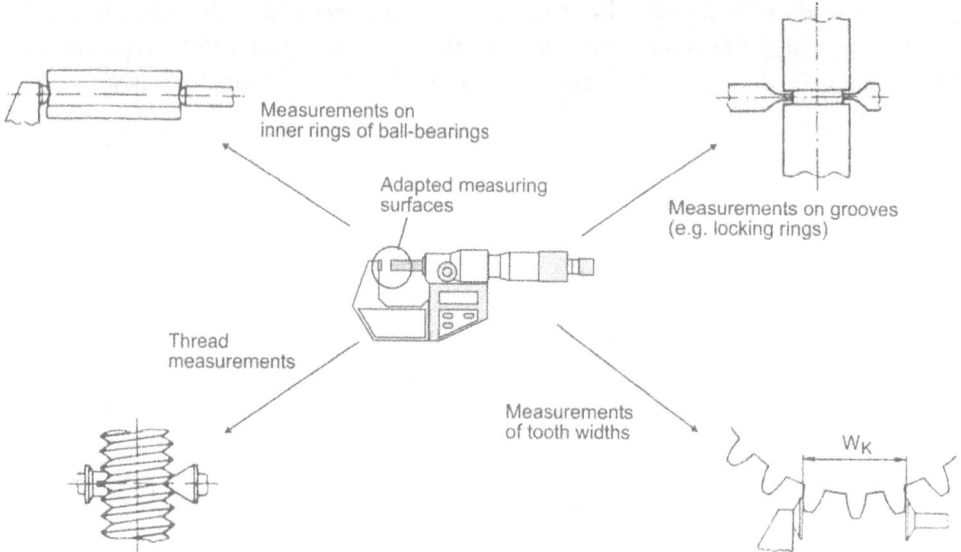

Measurements on inner rings of ball-bearings

Adapted measuring surfaces

Measurements on grooves (e.g. locking rings)

Thread measurements

Measurements of tooth widths

W_K

Figure 4.1-5: Typical measuring tasks for micrometer gauges

As with the calliper, apart from external measurements, the most diverse range of internal measurements can be carried out with special micrometers. Examples are micrometers for internal thread, nut width, depth or drill hole measurement.

A widely used instrument for drill hole measurement is the three-point inside micrometer gauge **(Figure 4.1-6)**, where three anvils probe the wall of the drill hole in a radial direction with their symmetrical rotary distribution having a self-centring effect. The measuring heads, which contain the respective deflection mechanism are often exchangeable, so that drill hole diameters in a typical region of 20 mm to over 300 mm can be measured with several heads. However, as the exchanging of heads itself causes errors (dirt, different assembly forces, etc.) the micrometer must be calibrated at least once after each change of head, whereby every head or measuring range has a particular adjusting ring or master adjuster.

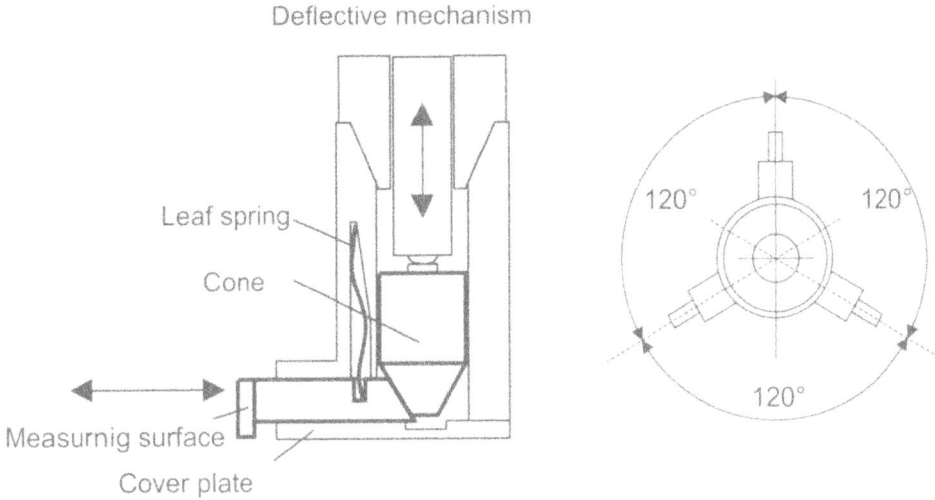

Deflective mechanism

Leaf spring

Cone

Measurnig surface

Cover plate

120°   120°

120°

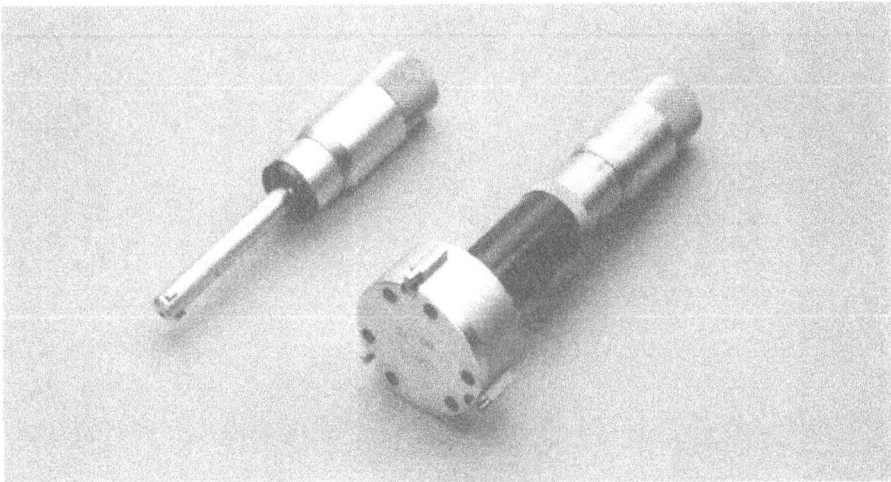

Figure 4.1-6: Three-point internal micrometer gauge

4.1.3 Display Sensors with Mechanical Transmission

Measurement reading recorders with indicators include dial gauges, dial comparators and indicators **(Figure 4.1-7)**. Similar to integrated micrometers, they can not
be used on their own due to one-point probing, but must be supplemented by
mechanisms which are adapted to the particular measuring task. Here, the display

is integrated into the sensor and connected to the moveable spindle or lever through mechanical coupling.

A dial gauge is essentially made up of four modules: the housing with a clamping shaft, the spindle with a measuring probe and the gearing as well as the scale face with the accompanying sensor, pointer, glass casing and the tolerance labels. For the housing with the clamping shank, the most important installation dimensions are given in DIN 878. The measuring probe (probe tip) with likewise standardised connection dimensions is exchangeable, in order to adapt the shape of its measuring surface to the surface which is being probed. The measuring path taken up by the spindle contact is transferred to the gear wheel mechanism via a rack. Dial gauges normally have display ranges of 3 mm or 10 mm with a scale interval of 10 µm.

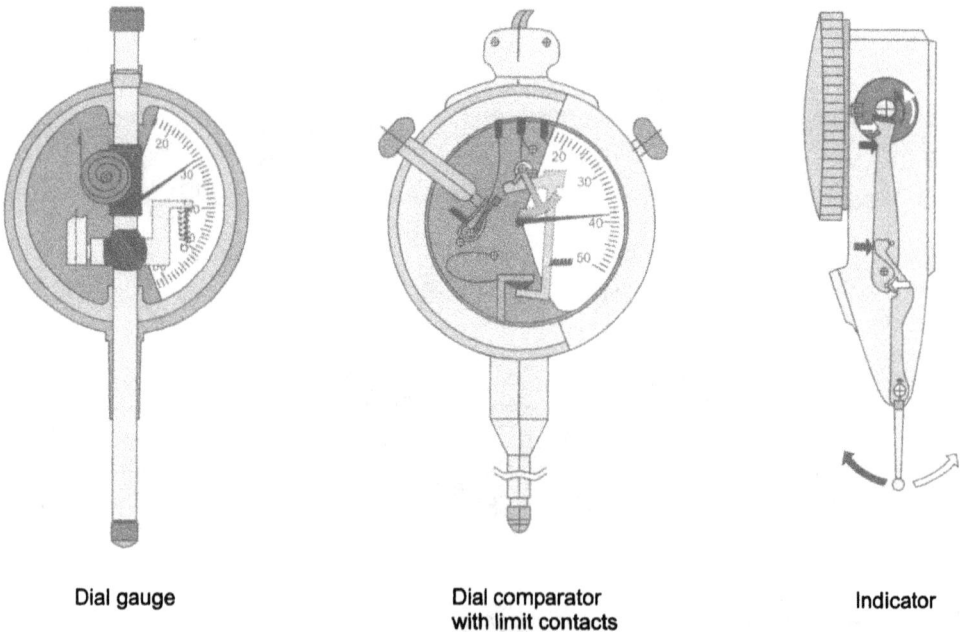

| Dial gauge | Dial comparator with limit contacts | Indicator |

Figure 4.1-7: Display sensors with mechanical transmission

The dial comparator is a length measuring device, standardised according to DIN 879 where the pointer has an angular deflection of less than 360 degrees. It mainly differs from the dial gauge through higher demands on precision and other mechanical transmission elements, which are generally realised through combinations of levers and gear segment pinion systems.

With their various types of applications, the dial comparator and dial gauge can, in a certain way, be regarded as a supplemental to gauge inspection. With the aid of a dial comparator or a dial gauge, the body material of gauge construction can be-come measuring devices which can be used to carry out relative measurements, e.g. with the comparator snap gauge

In contrast to the dial gauge, dial comparator and indicator, the functioning of a test indicator, standardised according to DIN 2270, is based on a swivelling sensor. The movement of the sensor is transmitted via a combination of levers and a gear to the pointer. From its initial position, the sensor can be moved around an axis, turning in two directions and thereby measuring in both directions. Similar to dial comparators, indicator probes are mainly used for comparative measurements so that their use is particularly common in a height measuring device. Their meas-urement uncertainty corresponds to that of dial gauges, while indicators have only a very small measuring range.

In addition to these mechanical dial gauges, comparators and indicators, electronic versions are also available with analogue and incremental measurement systems. Their structure will not be described further here. **(Figure 4.1-8)**.

Source: Mitutoyo

Figure 4.1-8: Dial gauge with incremental position indicator and digital display

4.1.4 Angle Measuring Devices

The simple angle measurer **(Figure 4.1-9)** permits the measuring of angles in degrees. A half or quarter degree can still be estimated with good devices.

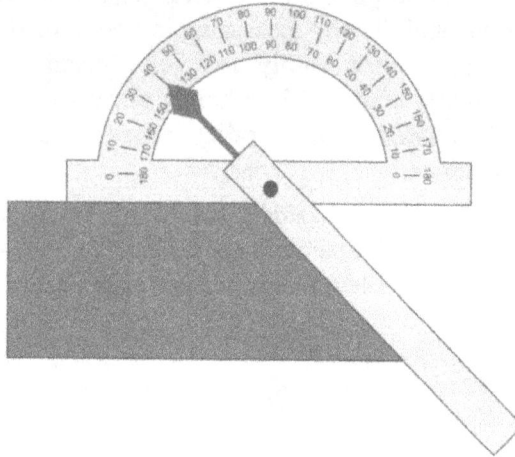

Figure 4.1-9: Simple angle measuring device

An improved form of the simple angle measurer is the universal angle measurer. It consists of two fixed and one moveable measuring arm, a complete circular scale, a twelve part vernier to the left and right of the zero line and a locking nut. The reading precision results from the difference between the graduation of the main scale and the graduation of the vernier and is usually $1/12°$. The moveable measuring arm is also adjustable in its longitudinal direction and has measuring edges at its ends of 45° or 30°. With optical universal angle measurers, the reading takes place via a lens system, which makes the magnified adjusted angle value visible on a ground glass plate **(Figure 4.1-10)**. With 30 times magnification, the angle size can be read off to $1/12°$ without the aid of a vernier [App 77], [VDI/VDE/DGQ 2618].

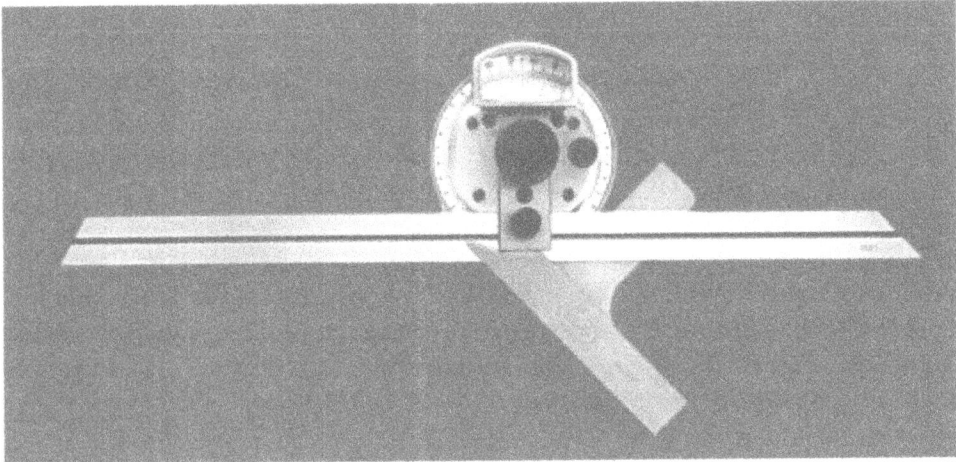

Source: Mahr

Figure 4.1-10: Optical universal angle measurer

4.2   Measuring Sensors

In this section, base sensors will be introduced which are used for length or angle measurements in the area of production metrology. Base sensors are understood to be measuring sensors integrated into the machinery or near to production which can be connected to manufacturing devices or be part of a complex measuring device to record data. Such sensors convert the quantity to be measured into a analogue electrical quantity, using the ohm, inductive or capacitive principle. On the other hand, sensors are often used for length measurement, which, via an integrated scale, either determine the measurand by counting known path increments, or take readings directly from the scale of so-called encoded material measures.

4.2.1 Potentiometer Sensors

Potentiometer sensors represent a simple and economical possibility for recording linear and angular displacements and converting them into an electrical quantity. In its most basic form, a potentiometer consists of a tensioned wire to which a moveable sliding contact is attached as a pick-up **(Figure 4.2-1)**. The resistance R between the pick-up and the end of a wire is proportional to its distance s. With a total resistance R_0 and a total wire length s_{max}, the picked up resistance is a function of the slider position to:

$$R = \frac{s}{s_{max}} R_0 \tag{4.2-1}$$

For a circular wire, the angle position α of the slider similarly results in:

$$R = \frac{\alpha}{\alpha_{max}} R_0 \tag{4.2-2}$$

In order to transform the resistance into a voltage which is easily measurable and thus displayable, the potentiometer sensor is operated as a voltage divider (**Figure 4.2-1**). The output voltage U_A is then dependent on the slider position s. Accordingly, the potential divider principle (equation 4.2-3) applies:

$$\frac{U_A}{U_0} = \frac{\dfrac{s}{s_{max}}}{1 + \dfrac{R_0}{R_B} \dfrac{s}{s_{max}} \left(1 - \dfrac{s}{s_{max}}\right)} \tag{4.2-3}$$

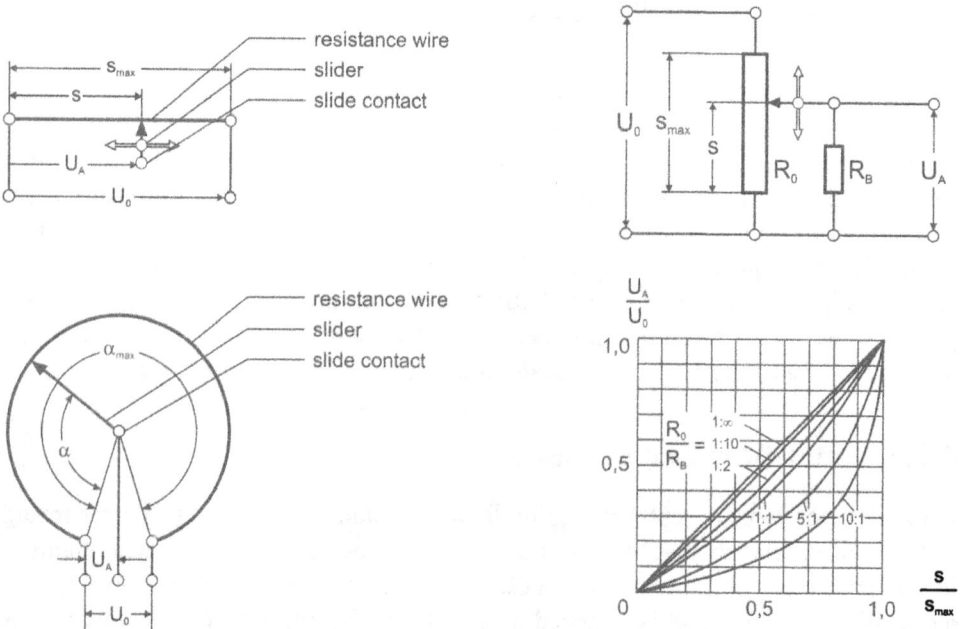

Figure 4.2-1: Potentiometer principle and voltage divider circuit

If the input resistance R_B of the downstream device, e.g. an amplifier, is sufficiently high, the linearity error can be kept below 0.1%, however this rises substantially with insufficient resistance (**Figure 4.2-1**).

4.2.1.1 Embodiments of Resistance Sensors

Basically there is a differentiation between the following types of potentiometer systems, depending on the manufacturing process:

- wire-wound potentiometers
- coated potentiometers

Wire-wound Potentiometers
In the simplest case, wire-wound potentiometers consist of a tensioned resistance wire with a pick-up slider. These sensors have a high resolution potential. It is a disadvantage that there is a limit as to how thin the wire can be, so that for R_0,values under 10 Ω usually result. Apart from that, the total resistance R_0 is highly wear-dependent.

Larger resistance values are achieved through a spiral arrangement of the resistance wire on an insulator (**Figure 4.2-2**), thereby also reducing the wear-dependent change in resistance. However, the disadvantage that the resolution is limited through the individual windings must be considered here.

Helical potentiometer   Coated potentiometer

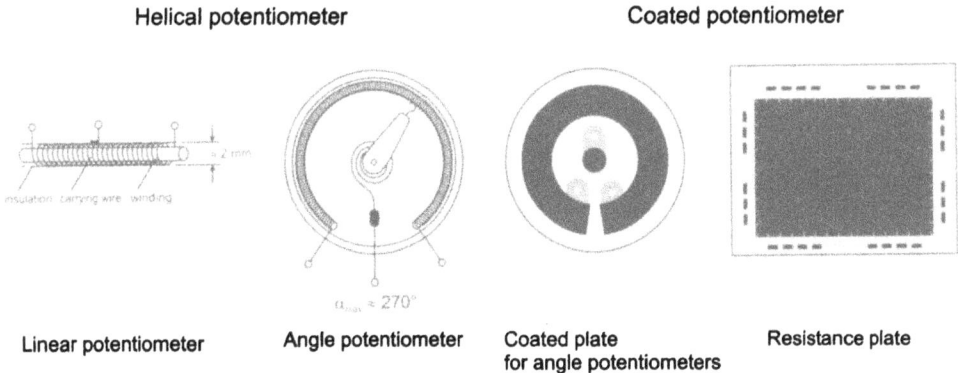

Linear potentiometer   Angle potentiometer   Coated plate   Resistance plate
                                                      for angle potentiometers

Figure 4.2-2:   Helical and coated potentiometers

Wire potentiometers have only become generally accepted for certain special applications in metrology.

Coated Potentiometers

Coated potentiometers consist of a resistance coating which is applied on a suit-able substrate. The type of manufacture depends on the application requirements [Ber 93]. For metrological use, potentiometers with a polymer coating have be-come generally accepted. In special cases, potentiometers with a metal oxide glass linkage coating or thin film metal systems can also be used.

The resistance layer of the potentiometer with a polymer layer consists of a lacquer resin system pigmented with soot or graphite. This achieves very smooth and non-abrasion-proof surfaces which permit a high number of operation cycles. Precious metal alloys are increasingly being used as slider materials. With these potenti-ometers, the resistance R_O can be set and compared in a wide area. The resolution of these sensors is very high, but the temperature coefficient is higher than with wire potentiometers, resulting in a higher error influence through temperature variations.

Figure 4.2-3 shows the application of a potentiometer sensor for position meas-urement of a machine table.

Conventional resistance sensors can only be used to evaluate a coordinate such as a distance or angle. A resistance plate, however, also enables the recording of co-ordinates on one level **(Figure 4.2-2)**. It consists of a resistance layer with embed-ded conductive strips. The achievable resolution is primarily only limited by the pigment size. Very hard and abrasion-proof surfaces can be attained with a special type of pigment in the resistance layer, which enables mechanical probing, e.g. with a "writing stylus" [Ber 93].

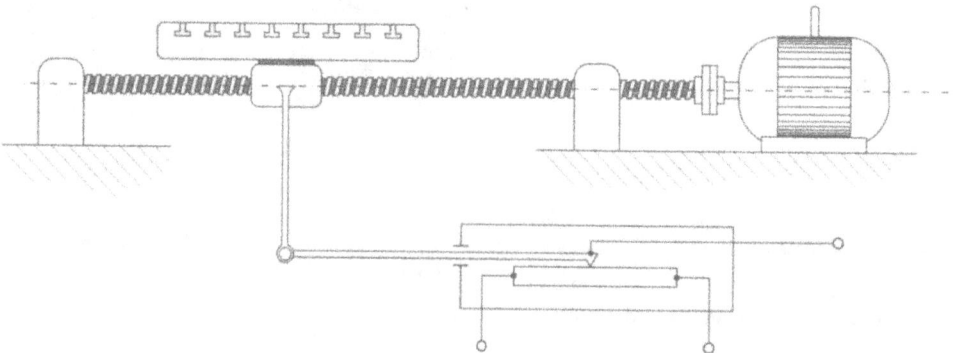

Figure 4.2-3: Application of a potentiometer sensor for the position measurement of a machine ta-ble

4.2.2 Inductive Sensors

The basis of all inductive measuring procedures is the induction law of electrodynamics:

$$U = -L \cdot \dot{I} \qquad (4.2\text{-}4)$$

Derived from the Maxwell equations, this law states that the magnetic field of a coil subjected to an alternating current I, induces a voltage U in the coil. According to Lenz's Law, this induction voltage always exactly opposes the change producing it – the excitation voltage. As the induced voltage is proportional to the temporal   change   in   the   current

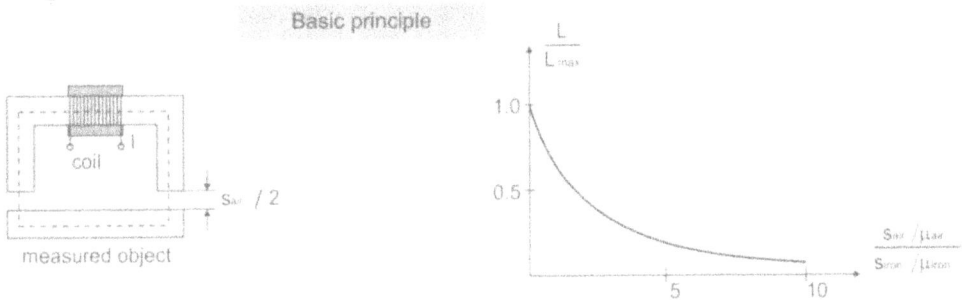

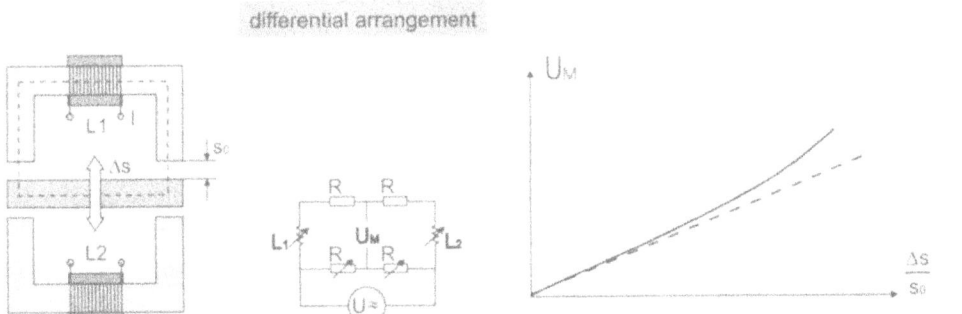

Basic principle

coil

measured object

$S_{av} / 2$

differential arrangement

L_1

Δs

L_2

(Equation 4.2-4), inductive sensors can only be operated with alternating current. The proportionality L is called *inductivity* and embodies the measurand for all inductive measuring procedures.

Inductivity is dependent on geometrical quantities (coil length, coil diameter), as well as the number of windings and the material inside the coil.

For a simple longitudinal coil:

$$L = \mu_0 \cdot \mu_r \cdot w^2 \cdot \frac{A}{l} \qquad (4.2\text{-}5)$$

applies

with   A = cross section area of coil, l = coil length, w = number of windings,

μ_0 = absolute permeability of the vacuum, μ_r = relative permeability

It is common to all forms of inductive sensors that the inductivity is in some form changed through interaction with the measured object and from this change, an inference can be made about the actual measurand – e.g. the path or angle.

4.2.2.1 Operating Principles of Inductive Sensors

Inductive measuring sensors are suitable for recording paths or angles, with a measurement range from micrometers to several meters. Therefore, inductive path sensors can be used to measure small as well as medium paths, depending on the design. Additionally, dynamic path changes can also be recorded with inductive sensors. The following sensor principles can be differentiated with regard to object probing and the operating principle:

- transverse anchor sensor

- plunger coil sensor

- eddy current sensor

As additional auxiliary voltage is required to operate these sensors, they are also referred to as *passive* measuring sensors.

Transverse Anchor Sensors
The main field of application for transverse anchor sensors is non-contact interval specification . However, the application of these sensors is limited to ferromagnetic materials, as the magnetic lines of force are completed by the measured object itself. The measurement value is thus influenced directly by the workpieces. Therefore it is of great importance for the instrumentation application of the transverse anchor sensor as a non-contact measuring sensor, that the magnetic properties of the workpiece are spatially constant. Through the change in size of the air gap between the sensor and the workpiece surface, the inductivity of the magnetic circle, which is formed by the coil, the ferromagnetic core and measured object, changes. As the core and the anchor (workpiece) possess a very high relative permeability, the lines of force pass in a straight line through the air gap **(Figure 4.2-4 top)**. With regard to the maximum inductivity, with $S_{Air} = 0$ the following non-linear analytical relationship results between inductivity and the size of the air gap:

$$\frac{L}{L_{max}} = \frac{1}{1 + \dfrac{S_{Air}/\mu_{Air}}{S_{Iron}/\mu_{Iron}}} \tag{4.2-6}$$

A linearisation of the characteristic curve can result with the aid of so-called differential circuits, which are very often used in practice. Through the detuning in opposite directions of two identically constructed sensors, with their output signals switched differentially, an extensive linearisation of the characteristic curve takes place **(Figure 4.2-4 bottom)**. Such differential arrangements are usually operated in a bridge circuit using the deflection procedure. For the bridge circuit, the size of the air gap is considered a good approximation as a function:

$$U_M \approx const \cdot \frac{\Delta s}{s_0} \cdot U \qquad\qquad (4.2\text{-}7)$$

An important prerequisite for accurate measurement results is a straight-line process of the lines of force in the air gap. For this reason, the transverse anchor sensor can basically only be used for the measurement of small to very small paths. The maximum measuring path with a linearity error <1% is 0.7 mm. With an increasing distance between the anchor and the core, the lines of force are distorted within the air gap until they finally tear off, at which point the prerequisites for a distance measurement are no longer given.

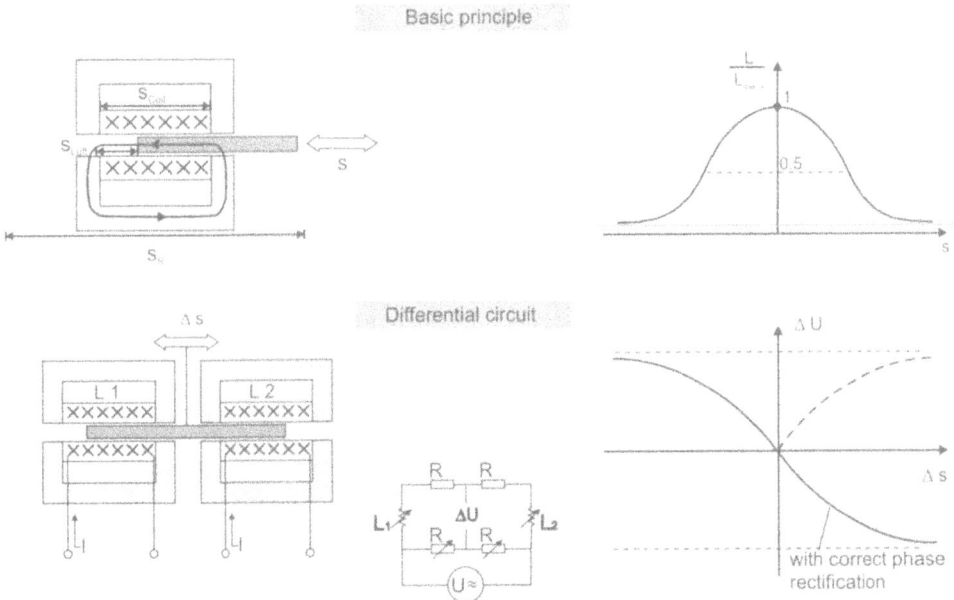

Figure 4.2-4: Principle diagrams and in principle characteristic curve path of inductive transverse anchor sensors

Plunger Coil Sensors
In contrast to the transverse anchor sensor, the so-called down hole anchor sensor is suitable for the measurement of smaller and medium sized length. The coupling of the sensor to the workpiece takes place by tactile means through a probe which is rigidly connected with the plunger coil anchor. During tracer deflection, the plunger coil anchor is moved within the coil, thereby changing the inductivity of the coil. The principle structure of a plunger coil anchor sensor is represented in **Figure 4.2-5**.

The following analytical relationship applies for the characteristic curve of this basic arrangement:

$$\frac{L}{L_{max}} \approx \frac{1}{1 + \dfrac{S_{Air}/\mu_{Air}}{S_{Fe}/\mu_{Fe}}} \tag{4.2-8}$$

with $S_{Fe} = S_R + S_{Coil} - S_{Air}$ and μ_{Fe} = relative permeability of iron.

As the relationship between inductivity and measuring path runs strongly non-linearly like the transverse anchor sensors, the plunge coil anchor sensor is also usually operated in a differential arrangement.

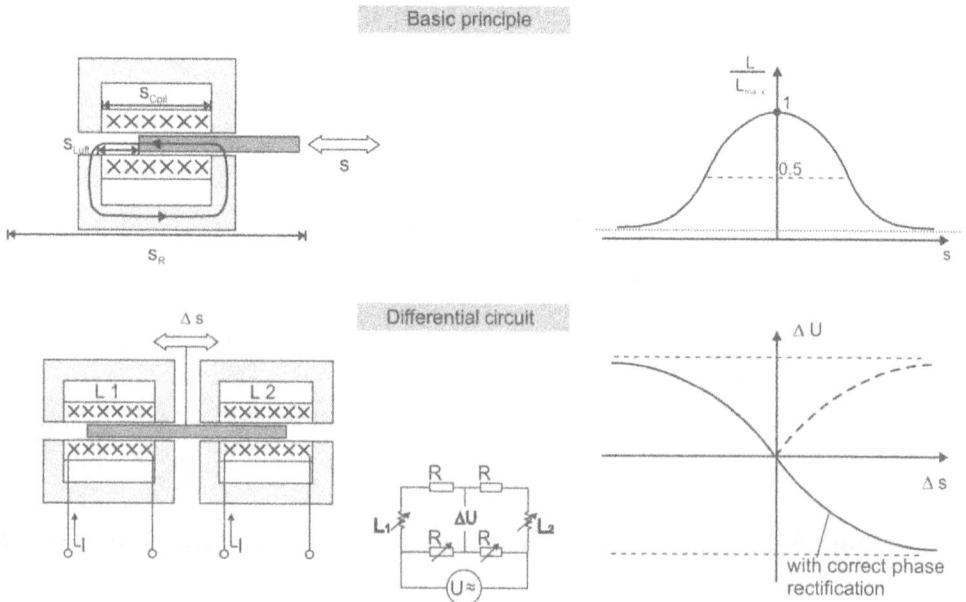

Figure 4.2-5: Principle diagram and characteristic curve path of an inductive down hole anchor sensor

Eddy Current Sensors

With the inductive sensors described so far, the change of inductivity was caused by the change of the magnetic field in a coil by shifting a ferromagnetic anchor. With eddy current sensor, however, the change in inductivity required for measurement through the physical effect results from the formation of eddy currents in electrically conductive, non-ferromagnetic materials. These eddy currents develop if the material is exposed to a high frequency magnetic field e.g. with the aid of a coil having an equally high frequency magnetic field flowing through it. The developing eddy currents cause a magnetic flow in the coil, which exactly opposes the magnetic flow in the coil and therefore partially compensates it. As the effect depends on the distance between the coil and the workpiece surface, such sensors can be used for non-contact distance measurement. The characteristic curve of the sensor is strongly non-linear, as with the transverse anchor or plunger coil sensor, so that eddy current sensors are usually operated in a differential circuit (**Figure 4.2-6**).

The difference to the inductive transverse sensor consists of the fact that eddy current sensors are particularly suitable for non-ferromagnetic materials such as copper or aluminium, while the transverse sensors can basically only be used with ferromagnetic materials. As, with ferromagnetic sensors, the inductivity of the total arrangement rises as the sensor approaches and, as described, falls with non-ferromagnetic materials, eddy current sensors provide a simple way of e.g. separating iron parts from non-magnetic metals [Tr 89].

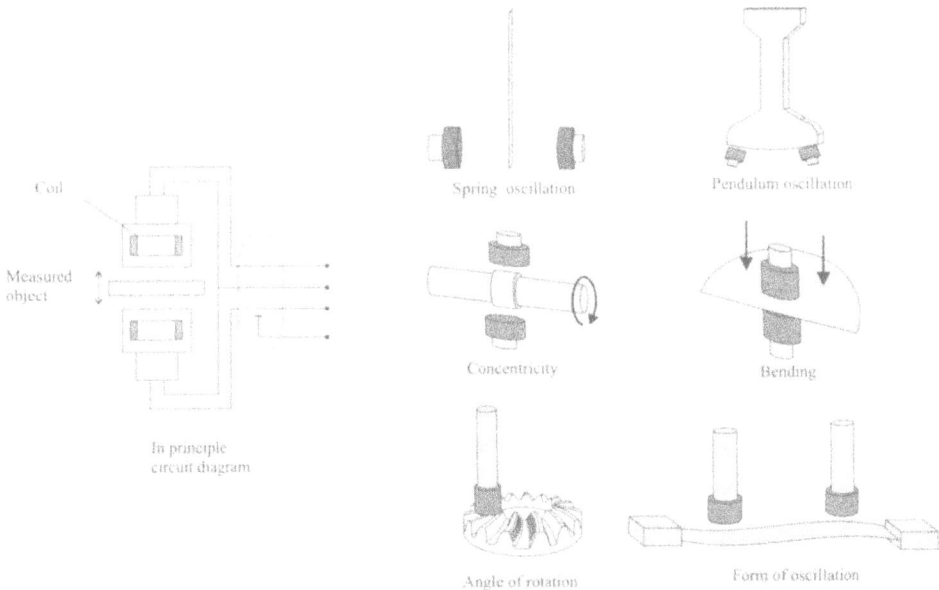

Figure 4.2-6: Circuit diagram and application examples for eddy current sensors

A further field of application of eddy current sensors is non-destructive material testing [St 88]. These sensors are particularly suitable for detecting shrink holes and tears, as the local eddy current formation is obstructed in the material by these material defects and no compensating magnetic fields are formed in these places.

4.2.2.2 Technical Embodiments of Inductive Displacement Sensors

The following sections describe several technical embodiments of inductive displacement sensors, together with their typical fields of application.

Inductive Measuring Touch Probe

Contact inductive plunger coil systems are very common within the field of production metrology. There are two technical embodiments of these sensors, with a differential throttle or with hybrid coil **(Figure 4.2-7)**.

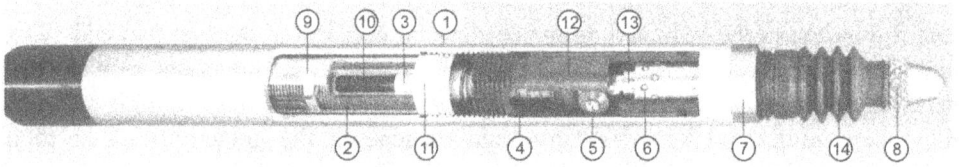

1	cylindrical clamping shaft	6	ball retainer
2	subdivided field coil	7	adjustable lift limitation
3	insulation element for equalising thermal influences	8	exchangeable measuring probe
4	measuring force spring	9	shield
5	torsion protection guide	10	ferrite core

11	mpact of the measuring force spring
12	guiding bushing
13	spindle
14	sealing cover

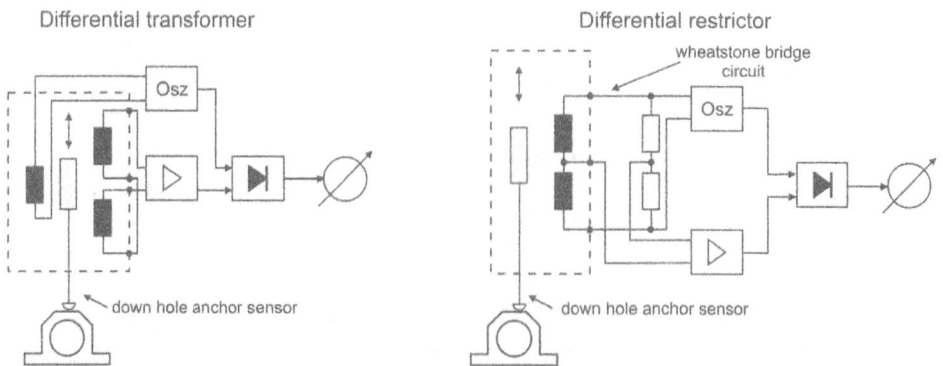

Figure 4.2-7: Structure and operating principles of an inductive measuring sensor

A *hybrid coil* consists of a coil system of one primary coil and two secondary coils. A high frequency alternating current is set on the primary coil which induces voltage in both of the secondary coils, depending on the position of the plunger coil. When the plunger coil is in a central position, the induced voltage is the same in both of the secondary coils. However, if the plunger coil is shifted in one direction, the induced voltage in the secondary coil, which is positioned in the appropriate direction, increases and the induction voltage in the other secondary coil decreases. A phased rectification of the difference of both output voltages produces a voltage level which is proportional to the measurand.

With the *differential throttle* both coils are interconnected in a Wheatstone bridge circuit in the form of a half bridge, which is completed by the ohm half bridge in the amplifier. This technical circuiting arrangement produces a bridge voltage proportional to the shift of the plunger.

Resolvers
One of the oldest applications of the inductive measurement principle is the resolver. It consists of a revolving mounted so-called *rotor coil* and several fixed *stator coils*. With three stator coils, they are usually arranged against one another at an angle of 120° around the rotor coil (**Figure 4.2-8**).

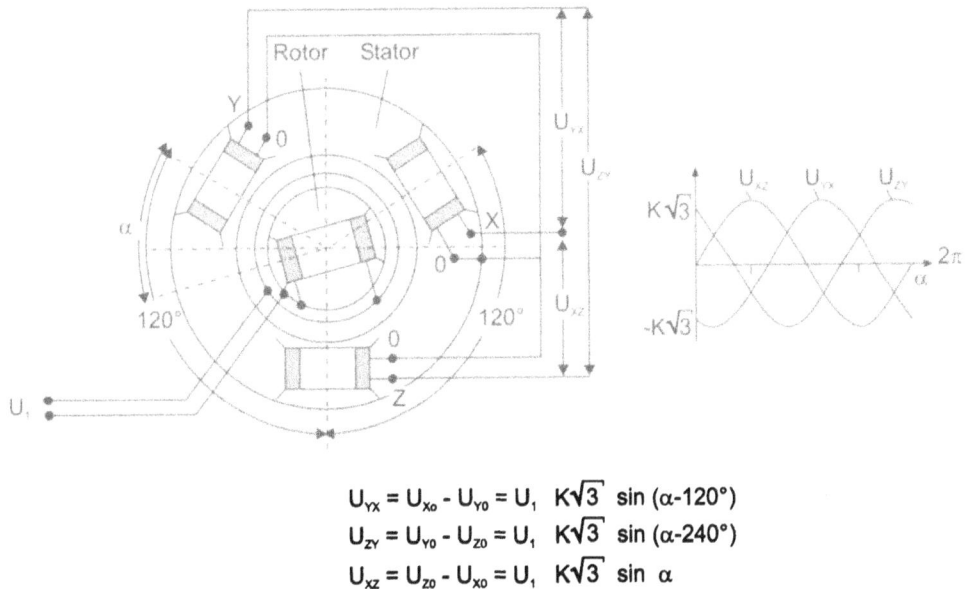

$$U_{YX} = U_{X0} - U_{Y0} = U_1 \ K\sqrt{3} \ \sin(\alpha - 120°)$$
$$U_{ZY} = U_{Y0} - U_{Z0} = U_1 \ K\sqrt{3} \ \sin(\alpha - 240°)$$
$$U_{XZ} = U_{Z0} - U_{X0} = U_1 \ K\sqrt{3} \ \sin \alpha$$

Figure 4.2-8: Principle diagram of an inductive resolver

The rotor coil is on a ferromagnetic core and is fed with an alternating current, e.g. through slip rings. An alternating voltage is induced in the stator coil caused by the magnetic field in the rotor coil, with the amplitude depending on the angle of rotation, as the coupling of the rotor coil – stator coil transducer periodically changes with the turning of the rotor. If the stator coils are arranged as in **Figure 4.2-8**, a 120° phase shifted induction voltage results in each of the three coils. The calculation of the angle of rotation and direction of rotation can now be carried out in a simple manner from the measured voltage amplitudes or the rms-values of the voltages.

4.2.3 Capacitive Sensors

The capacity C of an arrangement of two plates with area A, which face each other with distance d, results in:

$$C = \varepsilon_0 \cdot \varepsilon_r \frac{A}{d}$$
(4.2-9)

ε_0 : dielectric constant of the vacuum

ε_r : relative dielectric constant

    (air : $\varepsilon_r \approx 1$; oil : $\varepsilon_r \approx 2,5$; water : $\varepsilon_r \approx 80$)

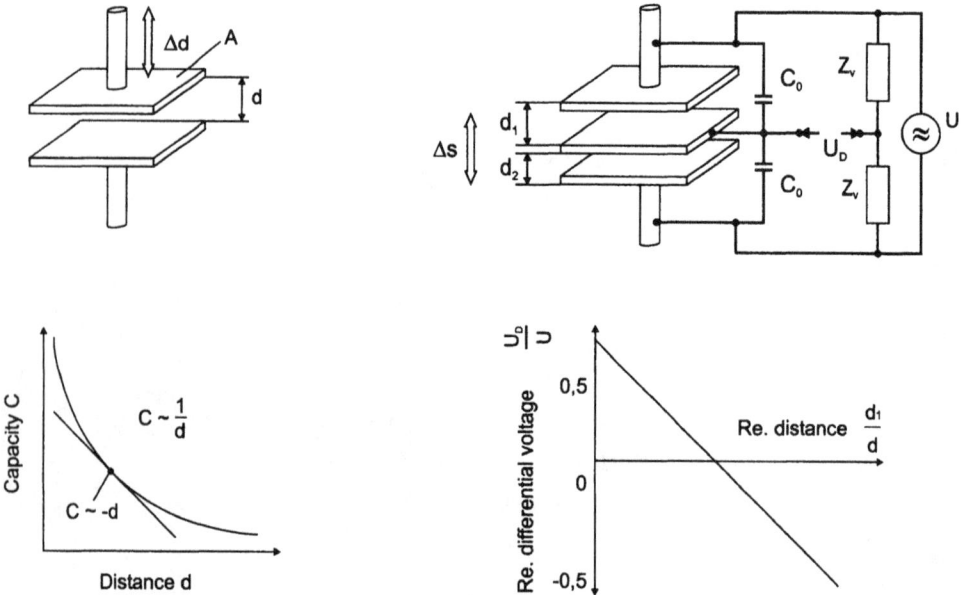

Figure 4.2-9: Capacitive distance sensor in basic and differential arrangements

Capacitive sensors are insensitive to magnetic interference fields, but the measurement result is directly influenced by the dielectric constant ε_r of the medium between the plates. Interferences through oil, water, etc. must also be considered.

4.2.3.1 Embodiments of Capacitive Sensors

Distance Sensors

When using distance sensors, it must be considered that the relationship between capacity and distance is inverse proportional and thus non-linear. Linearity can be achieved by arranging the object to be probed as a capacitor plate between the plates of the condenser. This produces a capacitive differential sensor which supplies an output voltage U_D proportional to the plate distance d_1 when the bridge circuit is appropriately tuned **(Figure 4.2-9)**:

$$U_D = U\left(\frac{1}{2} - \frac{d_1}{d}\right); d = d_1 + d_2 \qquad (4.2\text{-}10)$$

Capacitive distance sensors measure without contact and are independent of the magnetic characteristics of the probed workpiece. With a measuring range of 1 mm and a relationship of the change in distance to the plate distance of 0.1, the linearity error is 1% [Pro 92].

Surface Sensors

The surface sensor uses the linear relationship between the capacity and the capacitor surface. With a fixed plate distance and fixed plate width b, and coverage s of both plates, the capacity results in:

$$C = \varepsilon_0 \cdot \varepsilon_r \frac{b}{d} s \qquad (4.2\text{-}11)$$

With a simple capacitor configuration, an error must be considered, which results due to a transverse shift of the plates which is not completely unavoidable, with the factor

$$\frac{1}{1 - \dfrac{\Delta d}{d}} \qquad (4.2\text{-}12)$$

A decrease in the transverse sensitivity can be achieved by executing a condenser plate as a double plate. The arrangement according to **Figure 4.2-10** is particularly suitable for small paths Δs.

The sensitivity of the capacitive surface sensor can be increased through the arrangement shown on the top right of **Figure 4.2-10**. As a result of a parallel connection of several condensers a larger capacity change ΔC results from an equal shift Δs, as the surface change is larger by the number of surface pairs, than with the single plate arrangement.

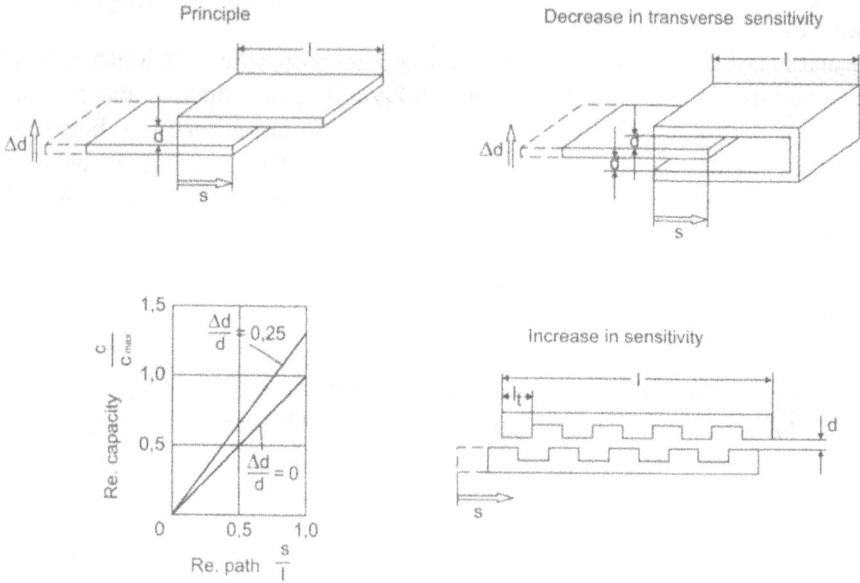

Figure 4.2-10: Capacitive surface sensor

4.2.4 Pneumatic Sensors

Pneumatic displacement sensors are based on the principle that the volume flow through a flow channel is limited by the narrowest diameter. In a nozzle-baffle system, the narrowest diameter through the annular gap surface is given between the nozzle and baffle plate on the nozzle opening. As long as the annular surface remains smaller than the nozzle diameter, the volume flow v through the gap is proportional to the annular gap surface and thus to the distance s between the nozzle and baffle plate.

In order to display the measurand, the flow, pressure and velocity measuring methods can be implemented **(Figure 4.2-12)**.

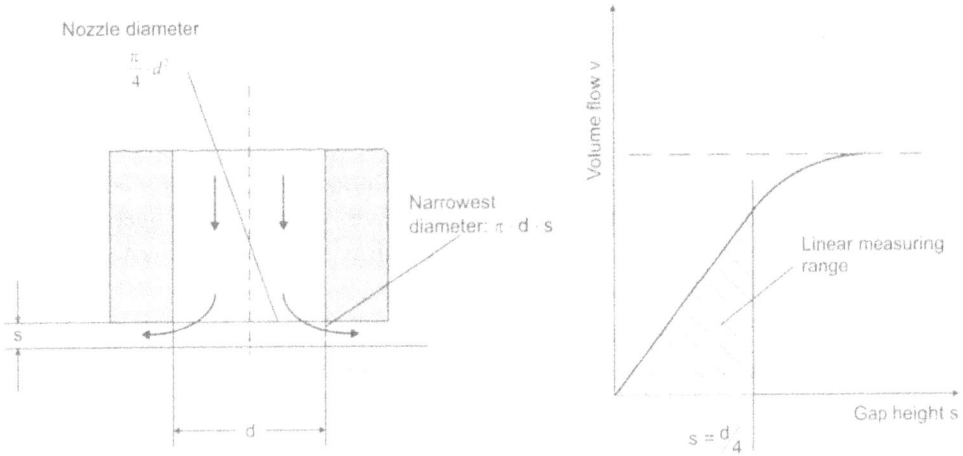

Figure 4.2-11: Nozzle-baffle plate system

With the *flow measuring method* the flow volume v is measured directly through the nozzles, where the air flows through a vertically arranged spherical glass tube, in which an also spherical, very light float is suspended on the airflow. The height of the float is directly dependent on the measurand, so that this can be read off of a scale next to the measuring tube.

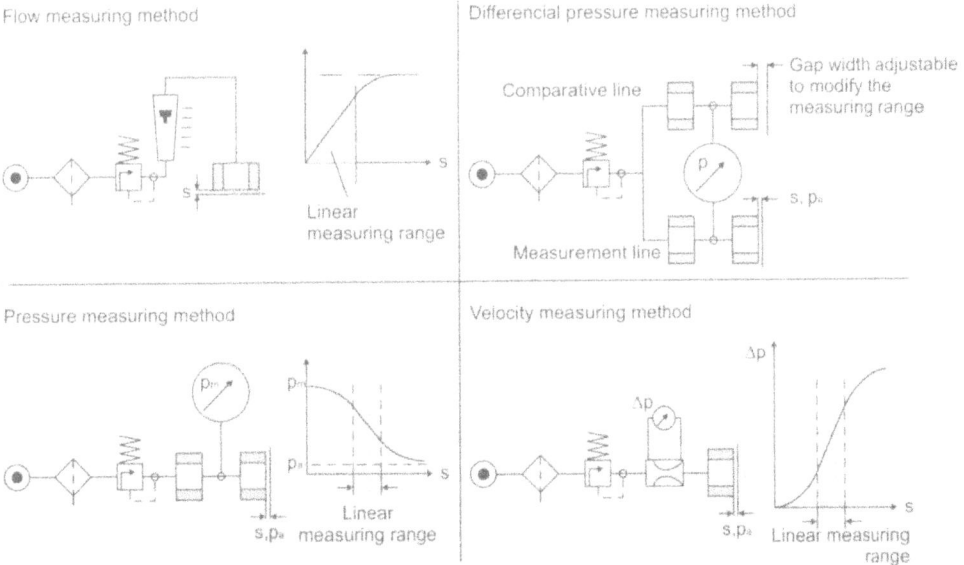

Figure 4.2-12: Measuring methods for displaying the measurand

The *pressure measuring method* makes use of the effect that the pressure p between a base nozzle and the measuring jet is directly dependent on the volume flow and thus on the height of the gap. This is displayed using a manometer, which is usually easier to take readings from than liquid manometers.

With *differential pressure measurement* the difference in pressure between a measuring sensor and a reference sensor is measured, whereby both operate using a pressure measuring method. By changing the gap width on the reference sensor, the measuring range of the sensor can be adjusted.

With the *velocity measuring method*, the flow volume is converted into a velocity change using a throttle. The pressure drop at the throttle again depends on the flow velocity and can be used to record the distance s. The display takes place on a manometer.

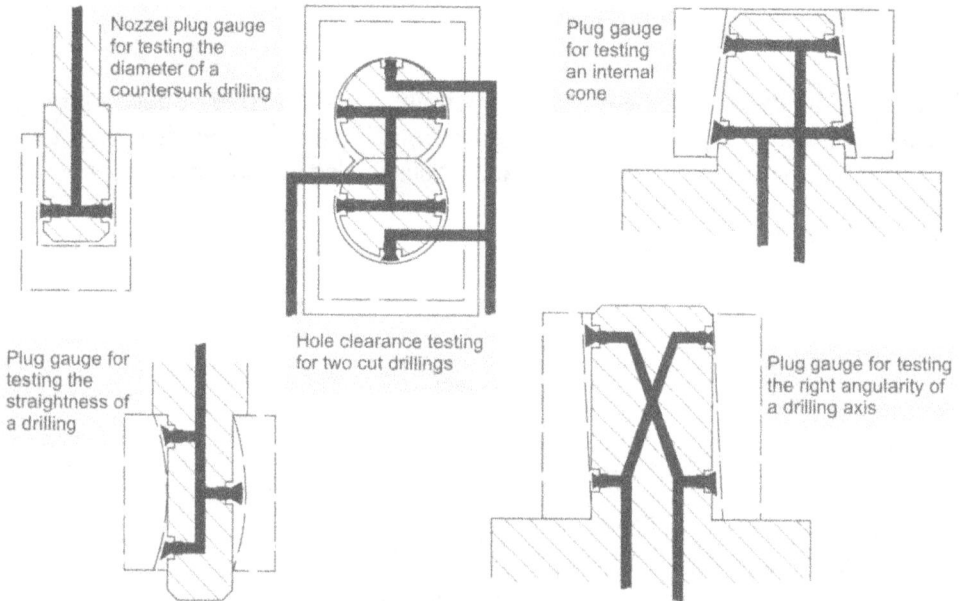

Nozzel plug gauge for testing the diameter of a countersunk drilling

Plug gauge for testing an internal cone

Plug gauge for testing the straightness of a drilling

Hole clearance testing for two cut drillings

Plug gauge for testing the right angularity of a drilling axis

Figure 4.2-13: Diameter, form and positional inspection with pneumatic plug gauges

Today only the flow and pressure measuring methods are usually used. According to the supply pressure behind the pressure control valve, there is a differentiation between low pressure and high pressure methods. For the low pressure method, the supply pressure must be below 0.1 bar. Today, however, the high pressure method has become generally accepted, where the supply pressure is above 0.5 bar, so that the sound velocity occurs in the narrowest diameter [Dut 96].

This measurement principle can be used for both contact and non-contact work-piece measurement. With contact measurement, the workpiece is touched by a probe element which thereby changes its position in the sensor and reduces the annular gap of the internal nozzle-baffle plate system. With non-contact inspection, the workpiece itself represents the baffle plate. The advantage of this method is the self-cleaning effect created by the outflow of compressed air.

Depending on the arrangement of the nozzles in the measuring sensor, measurements of lengths or diameters, or form inspections can be carried out. Through a pneumatic switching of the nozzle arrangement, a measuring sensor can also be successively used for dimensional and form testing. Through the mechanical and pneumatic coupling of several sensors, complex testing functions, e.g. determining the position of two drill holes rapidly and highly precisely, can be executed in one procedure.

Pneumatic length measurement is particularly suitable for recording small paths or length differences in mass production, as the initial costs for pneumatic measuring equipment are quite high, while the flexibility is rather small. Pneumatic length measurement also puts high demands on the compressed air network. Due to the relatively long response times of 0.5 ... 3 s, these sensors are unsuitable for the measurement of dynamic effects [Dut 96].

Figure 4.2-14 shows different embodiments of contact and non-contact pneumatic measuring sensors for production metrology.

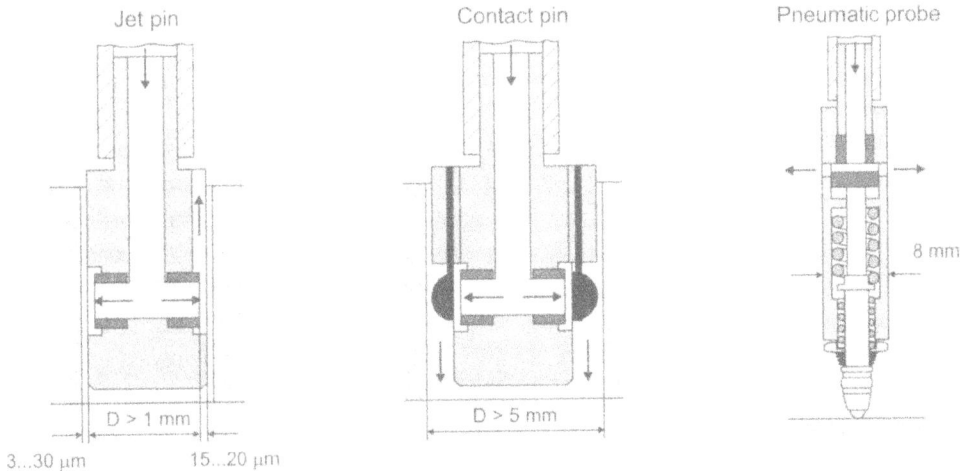

Figure 4.2-14: Pneumatic measuring sensors

4.2.5 Ultrasonic Measuring Methods

The term "ultrasonic" refers to mechanical waves in a frequency range in the upper threshold of human hearing at approximately 20 kHz up to the gigahertz range.

In industry, ultrasonic sensors primarily find technical use in non-destructive materials testing, where testing frequencies of several megahertz are used. Ultrasonic systems are outstanding in their suitability for investigating metallic materials on enclosed foreign particles or on material inhomogeneities. With the increased use of fibre composite materials in recent times, a predication can also be made about the internal structure of the materials with the aid of special ultrasonic testing methods [Pf 96a]. Apart from these test functions, ultrasonic measuring systems are particularly used for dimensional measuring functions in production metrology.

4.2.5.1 Fundamental Principles of Ultrasonic Sensors

Ultrasonic waves are always attached to matter and can therefore not spread in the vacuum. Apart from the material, the diffusion characteristics also depend to a considerable degree on the assigned ultrasonic frequency. While ultrasonic frequencies in the lower frequency range can be transmitted through the air, the ultrasonic frequencies used for materials testing between 1 MHz and 100 MHz require special coupling media. The reason is, on the one hand, the frequency-dependent transfer behaviour of air and, on the other hand, the fact that during the transfer of an acoustic wave from one medium to another, a proportion of the sound energy is always reflected back to the output medium. The proportion of the reflected sound is calculated from the so-called acoustic impedance (product of density and speed of sound) of the adjacent media. It can easily be shown that with the ultrasonic transmitters used for materials testing, a very high proportion of the sound is reflected to the air at the boundary surface, as the acoustic impedances from the air and the sensor strongly differ from one another. For this reason, a coupling medium should generally be chosen for the transfer of ultrasound to the workpiece, which has an acoustic impedance with a middle value between the sensor and the workpiece. By using a middle acoustic impedance, a sufficient linking of the ultrasound can generally be ensured on the part of the workpiece. Particularly suitable coupling media are liquids and gels.

Altogether, ultrasonic methods require a measuring chain which contains the following components:

- sound production (e.g. piezo transducers)
- coupling to the medium (liquids or gels)
- sound reception (with the pulse echo method, the transmission head serves as this at the same time)

- signal filtering (high and low-pass filter)
- evaluation and display (e.g. oscilloscope)

For the production and reception of ultrasonic waves, piezoelectric transducers are usually used.

Non-Destructive Materials Testing with Ultrasound

For materials testing, two particular procedures in particular have become established. With the so-called *transmission technique* errors are registered in the form of an intensity attenuation of the ultrasonic signal after passing through the workpiece, while with the *impulse-echo technique* a short ultrasonic impulse is produced and subsequently the reflection echoes of the impulse are received as a function of time. These reflection echoes can be recorded either with a separate ultrasonic transducer or, as is often the case with modern systems, with the same probe which is used for transmitting the signal **(Figure 4.2-15)**. A material error, such as the inclusion of a foreign particle, appears as an additional impulse in the temporal signal process between the so-called surface echo and the reflection echo from the rear of the workpiece **(Figure 4.2-15)**.

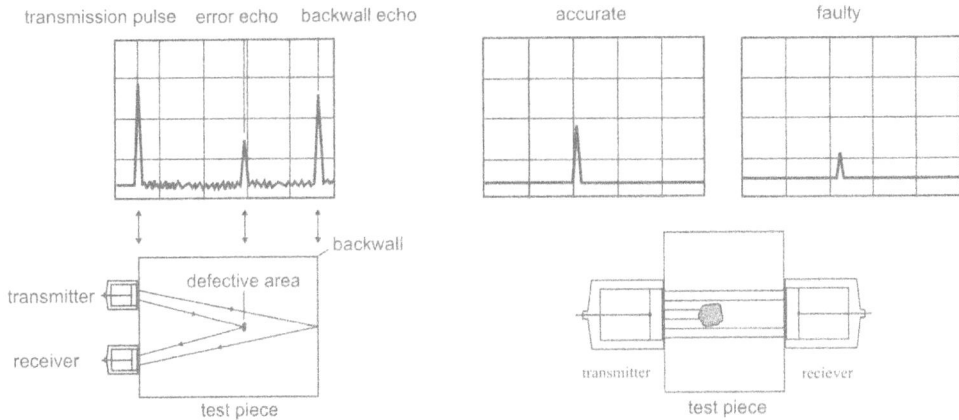

Figure 4.2-15: Non-destructive materials testing using ultrasound

The axial resolution (in the diffusion direction of the sound) for such systems is limited by the duration and form of the ultrasonic impulse. With higher ultrasonic frequencies, accordingly shorter sound impulses can be produced. Higher ultrasonic frequencies can also be used to produce narrower sound fields, whereby the transversal resolution (vertical to the diffusion direction of the sound) is increased. Increasing sound dispersion in the matter as well as higher sound absorption, which occurs with increasing ultrasonic frequency, has an unfavourable influence on the testing result, so that a compromise must always be made between resolution and acoustic attenuation.

Further details regarding the subject of ultrasonic materials testing can be found, e.g. in [Kr 96], [St 88].

Measuring with Ultrasound

With modern transient recorders, it is possible to resolve the temporal course very accurately, so that ultrasonic procedures can not only be used for non-destructive materials testing, but are also very well suited for dimensional measurement. Frequent measuring tasks are distance measurement, non-contact measurement of liquid levels, bulk materials or liquids, or wall thickness measurement.

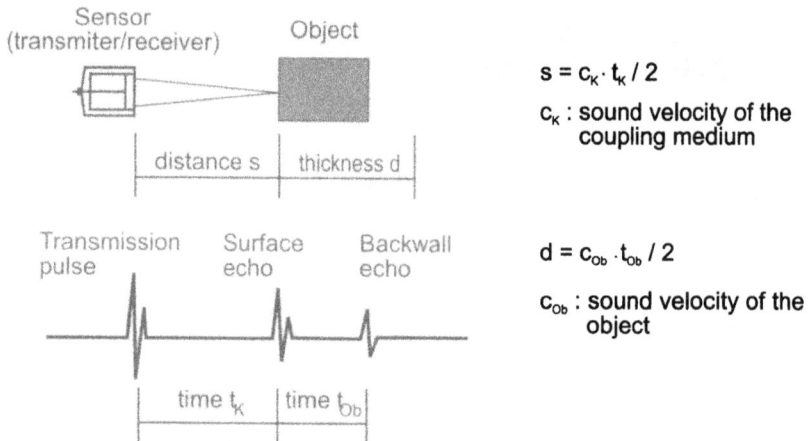

Sensor (transmiter/receiver) Object

distance s | thickness d

$$s = c_K \cdot t_K / 2$$

c_K : sound velocity of the coupling medium

Transmission pulse   Surface echo   Backwall echo

$$d = c_{Ob} \cdot t_{Ob} / 2$$

c_{Ob} : sound velocity of the object

time t_K | time t_{Ob}

Figure 4.2-16: Ultrasonic travel time procedure for distance or wall thickness measurement

For distance measurements, the run time of the sound between the transmission pulse and the echo reflected by a surface are measured with the aid of a master clock. With a known sound velocity, the distance is obtained from this **(Figure 4.2-16)**. The velocity of sound in air is dependent on the temperature. Therefore this influence must be compensated via an external temperature sensor or a standardised reference distance. In contrast, the influence of air pressure and moisture is small, so that it can usually be disregarded for technical applications. Depending on the frequency, ultrasonic distance sensors can be used for distances up to 30 m. With small distances, accuracies of below 0.1 mm are quite achievable.

If it is not the distance of the sensor from a surface, but the thickness of a workpiece, e.g. for wall thickness measurement which must be determined, the speed of sound of the material must be known. The thickness of the material can be determined from the temporal distance between the surface echo and the rear wall echo. It must also be considered here that solids display a temperature dependency at the speed of sound, which is, however, generally smaller than the temperature dependency of the speed of sound in air.

The accuracy of a distance or wall thickness measurement can be increased through the evaluation of repetition echoes (repeated sound cycle of the same measurement path).

4.2.6 Measuring Sensors with an Incremental Material Measure

In the previous sections, sensors have been described which supply a measurement value similar to a voltage level as an output quantity. As the measuring and evaluation process today often incorporates digital computers, particularly digital measuring systems have also attained widespread use in production metrology. Digital measurement means that the measurand is assigned an output quantity through the measuring system which is a quantified representation of the measurand with defined steps.

In digital measuring systems, sensors are often used which have an incremental material measures, e.g. in the form of a graduated scale. Most digital length or angle measuring devices are based on standards with a periodic structure. By using various physical principles, periodic - initially still analogue - electrical signals are derived based on the standards, from which the digital measurement value is formed.

The transformation of the transmitter signal into a digital measurement value is described in more detail in Section 4.2.6.4, using the example of a photoelectric incremental measuring sensor.

It is common to all incremental sensors, that the measurement value is only available after passing a reference position. This applies to the inductive incremental receivers described in the following section as well as the incremental sensors operating on a capacitive or photoelectric basis (Sections 4.2.5.2 and 4.2.5.3).

The following sections introduce different sensors with incremental material measures, which are based on different physical principles.

4.2.6.1 Inductosyn

One of the best known applications of the inductive measuring principle is the inductosyn. An inductosyn can be depicted as the winding of a resolver in a linear measuring system. The individual windings are not wound circularly, as with a coil, but placed along a line in a meandering path and applied to a non-conductive substrate (e.g. ceramic) **(Figure 4.2-17)**.

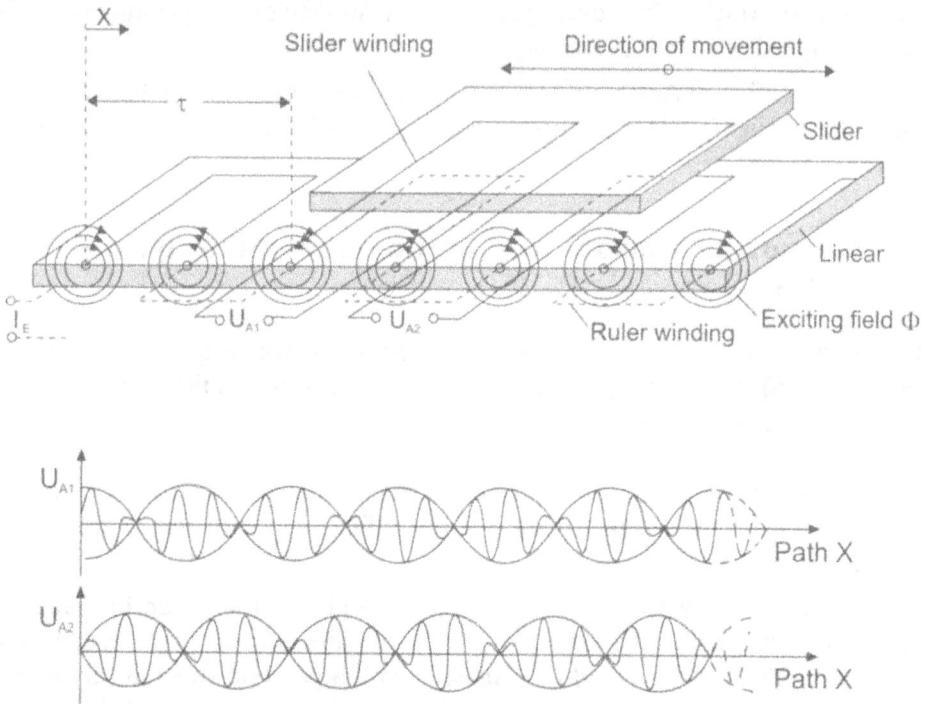

Figure 4.2-17: Principle representation of a linear inductosyn

On a scanning plate, which is guided across this meandering arrangement of windings, there is a likewise meandering conductive strip. If a high frequency alternating voltage between 1 kHz and 20 kHz is set on the fixed conductive strip, a voltage is induced between the ends of the conductive strip on the rider, according to the transformer principle. The amplitude of this induction voltage depends on the relative position of the two conductive strips to one another. With conductive strips directly opposite one another, the maximum voltage is induced. If the conductive strips of the rider are situated directly opposite a gap in the conductive strip of the scale, no signal is transmitted. A shift of the rider therefore periodically takes the same value after passing through the conductive circuit. Due to the cyclical repetition of the measuring signal process after each pole pitch, the inductosyn is also referred to as a "cyclical analogue measuring sensor".

In order to differentiate the direction of movement, a second meandering winding probe is attached to the probe plate of the rider, which is shifted by one quarter of a division period. Through the evaluation of both signals, the amount, as well as the direction of the shifted can be indicated.

4.2.6.2 Magnetic Incremental Measuring System

Magnetic incremental measuring systems have a similar structure to the previously described inductosyn. In this case, the function of the scale is taken on by a permanent magnetic bar, which has alternating opposite magnetised poles. This magnetic scale is probed with a coil system which is subjected to a high frequency carrier frequency. The electrical signal is produced by a double-levered ferromagnetic yoke as well as three coils.

The two energising coils on the levers of the ferromagnetic yoke are fed with an alternating current. In the receiving coil, an alternating current is thereby induced, for which the amplitude is dependent on the position of the yoke relative to the magnetic scale.

If the two levers of the yoke happen to be opposite a north or south pole of the scale **(Figure 4.2-18 right)**, an additional magnetic flow through both levers of the yoke is caused by the magnetic field of the bar. Each half wave of the alternating current now changes the magnetisation of both primary coils differently. In one lever of the yoke the magnetic flows accumulate, while in the other lever, the total magnetic flow is reduced through the opposing magnetic flow of the scale. In this way, magnetisation saturation is reached at different times in the two levers of the yoke. Every half wave therefore produces a flow modification by the receiver coil, through which an alternating current is induced with double the frequency of the field current.

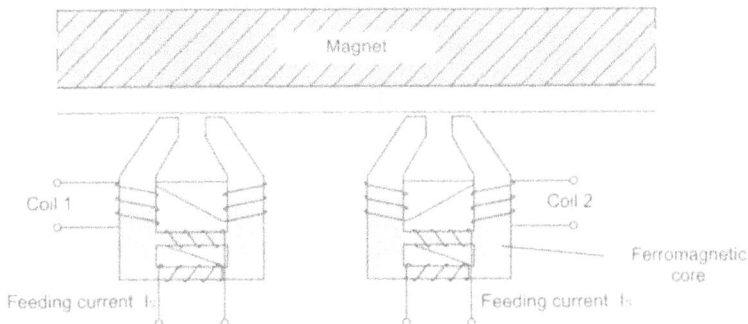

Figure 4.2-18: Magnetic incremental measuring system

When the pole pieces are positioned symmetrically, however, such an effect is not observed, as the bias point for both levers of the yoke lie in the point of symmetry of the nominal line of magnetisation. Magnetic flows are produced through the field coils in both levers of the yoke which short circuit each other, so that no signal develops in the receiving coil [NN 91].

As with inductosyn, two scanning units, which are offset against one another by a quarter period, are used to determine the direction of movement.

4.2.6.3 Capacitive Incremental Measuring Procedure

Capacitive incremental measuring systems are increasingly being used in operational measuring instruments such as callipers, micrometers, etc. They are based on the capacitive surface sensor principle described in Section 4.2.3. The material measures embodiment consists of a series of thin metal foils, which are applied to a non-conductive substrate and electrically connected to one another. Together with the scanning plate, these form individual condensers **(Figure 4.2-19)**. A shift of the scanning disk in relation to the material measures causes a change of the effective condenser surface and concomitantly, the capacity. The measuring signal is almost sinusoidal and can be subdivided up to 0.1 µm through interpolation. By connecting several sensor plates in parallel, the sensitivity can be increased and additionally recognise direction by using a second sensor which is offset by a quarter division period. This measuring procedure only requires a small amount of energy, so that it can even be operated with solar energy [Dut 96].

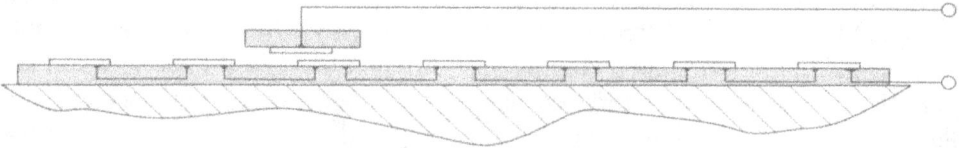

Figure 4.2-19: Capacitive incremental measuring system

4.2.6.4 Photoelectric Incremental Measuring Procedure

The vast majority of digital length or angle measuring devices use a photoelectric measuring principle.

So-called glass scales (Section 2.2) are used as a material measures in photoelectric incremental sensors. Optical scanning of glass scales is differentiated into the following execution forms:

- scanning in transmitted light

- scanning in incident light

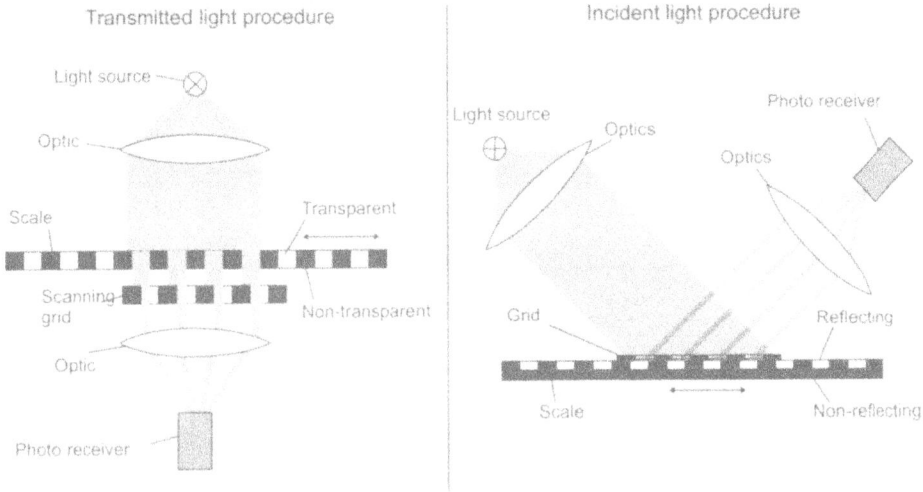

Figure 4.2-20: Photoelectric scanning of incremental scales using the transmitted light and incident light procedures

With the *transmitted light procedure* the light source and sensor are on opposite sides of the glass scale, so that the light which is transmitted from the sensor through the scale is evaluated.

In contrast, with the so-called *incident light procedure* the light source and sensor are arranged on the same side of a moveable glass scale with a reflective line graduation, so that the light reflected on the scale is received by the sensor.

With both procedures, the measurement segment is transferred through a shift of the scale against a second grid in the optical path in a series of light/dark transitions. If the scale is continuously moved, there is a relatively high change in light intensity at the sensor, if both grid divisions are exactly opposite one another with almost complete shading of the light when the lines of the scale are arranged opposite the gaps in the scanning grid. The relative shift of both glass scales against each other can be determined through the number of intensity maxima which have been passed through. If the scale has a grid constant D, with N intensity maxima passed through, the intensity maxima of the path Δx results in:

$$\Delta x = N \cdot D .$$

(4.2-13)

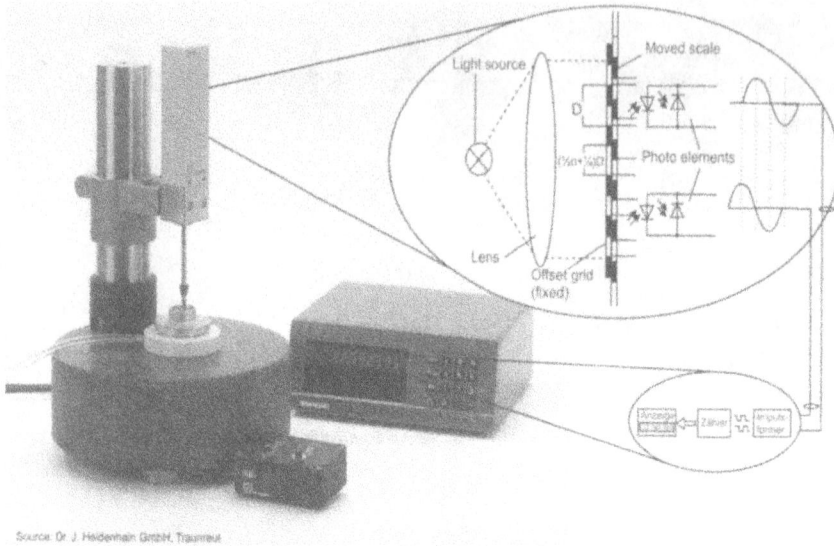

Source: Dr J Heidenhain GmbH, Traunreut

Figure 4.2-21: Photoelectric incremental measuring system

Glass scales can be produced to such high precision that, at the most, slight graduation errors occur. These, however, have almost no influence on the measurement result, as a large number of lines are always illuminated, so that these graduation errors are compensated through averaging.

The directional signal required to distinguish the direction of movement is usually obtained through a second sensor, which is arranged with its probing grid offset by one quarter division period to the first sensor.

With photoelectric incremental sensors, the achievable uncertainty of measurement is up to 1 μm/m. If the received signals are multiplied with the aid of electronic interpolation, the resolution can be increased up to 0.1 μm/m (with constant measurement uncertainty). Only the laser interferometers can achieve comparable targets with respect to resolution and measurement uncertainty (Section 4.3.3.6). An increase in resolution beyond the graduation of the scale can be further obtained by tilting the scale against the scanning grid, through evaluation of the resulting Moiré strips.

Interpolation and Transformation into Digital Measurement Signals

So far the transformation of an analogue measurement signal into a digital signal has not been dealt with in detail. This is made up for below, using the photoelectric measuring system as an example.

In order to generate a counting signal from the analogue signals of both photo elements which have a fixed phase relationship to one another, these signals must be converted into square wave impulses using comparators.

Usually, however, an interpolation procedure is used beforehand to increase the resolution beyond the scale graduation. A procedure which is often used is "interpolation with auxiliary phases". First, several phase signals shifted are produced from both of the photo receivers using an appropriate circuit **(Figure 4.2-22 B)**. These individual signals each exhibit a fixed phase shift against one another. **Figure 4.2-22 A**, for example, represents a 10 times subdivision with auxiliary phases, which are therefore phase offset against one another by 18°.

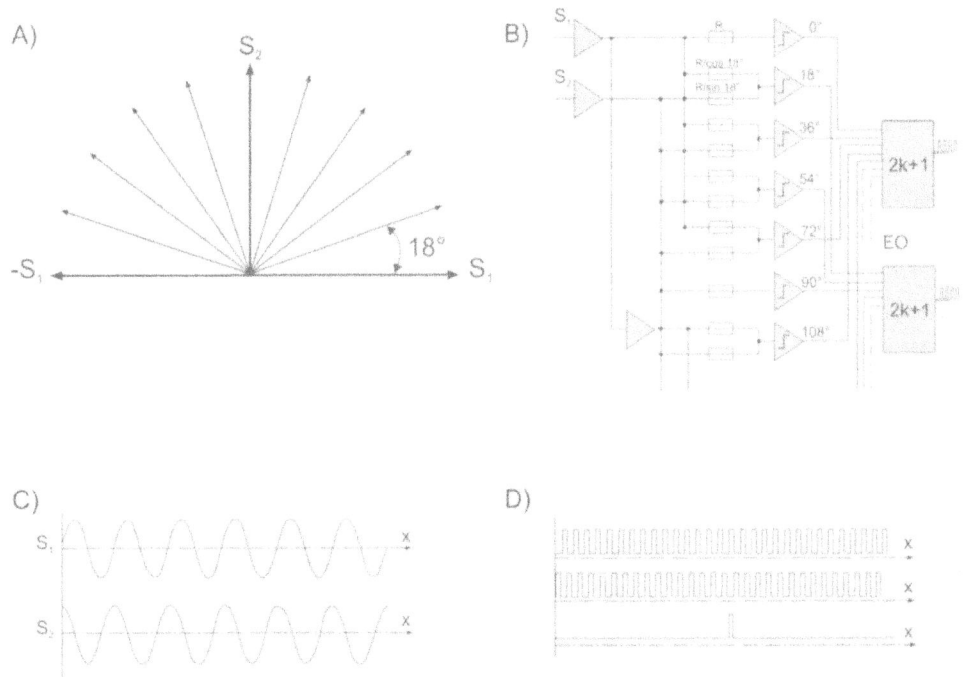

Quelle: Dr. J. Heidenhain GmbH, Traunreut

Figure 4.2-22: Interpolation with auxiliary phases

Each of these signals is subsequently transformed into a square wave signal with a comparator. These square wave signals are combined afterwards as exclusive-or-elements into two sequences of square wave signals. In this case with 10 auxiliary phases, the output signals have the fivefold frequency of the input signal, , and are out of phase against each other by a quarter period of the increased frequency. The difference between two sequential flanks corresponds to a measuring step, which in this case corresponds to the twentieth part of the division period on the scale (Figure 4.2-22 D). Digital interpolation procedures can be used for a much higher degree of partitioning [NN 91].

From these counting impulses, a discrete numerical value is now determine as a number of square wave impulses, usually using special counting devices or programmable insert cards in a computer and converted into a discrete measurand.

4.2.7 Sensors with Encoded Materia Measures

In contrast to the incremental measuring procedures described in the previous sections, the value for the current position is available immediately after switching on sensors with absolute encoded material measures. The measurement value is read off – without counters or reference labels – directly from the graduation. This is achieved by using a special code which absolutely represents the position. These codes are read off in the same way as with the incremental procedures, e.g. optical sensors, and converted directly into the absolute position value with the aid of suitable processing **(Figure 4.2-23)**.

In modern measuring devices with code measuring procedures, a binary code or Gray code are often used. The latter has the advantage of being a one-step code, where only one signal changes between measurement steps. Through plausibility queries, the probability of an incorrect position value can be clearly reduced using this code.

The operating principle of an absolute measuring digital photoelectric measurement system is represented in **Figure 4.2-23 top**. The binary code is, however, not particularly suitable for larger measuring lengths, as a large number of tracks are required for absolute encoding in this case. For this reason, other encoding processes which require fewer tracks are used for larger measuring lengths in photoelectric measuring systems. **Figure 4.2-23 bottom** represents, for example, a coding procedure by means of several incremental tracks with defined different division periods. The absolute position information is gained through simultaneous analysis of the phase angles of all incremental tracks. Systems are currently available on the market which measure absolutely with 7 incremental tracks on measuring lengths of up to 3 m with a resolution of 0.1 µm. This procedure, as well as the procedures based on serial codes (e.g. pseudo random code), are described in detail in [NN 97]. The advantage of the substantially smaller number of measuring tracks and the accordingly compact construction method, however, compare unfavourably with the increased amount of effort required to evaluate the measurement signals and the higher demands on signal quality.

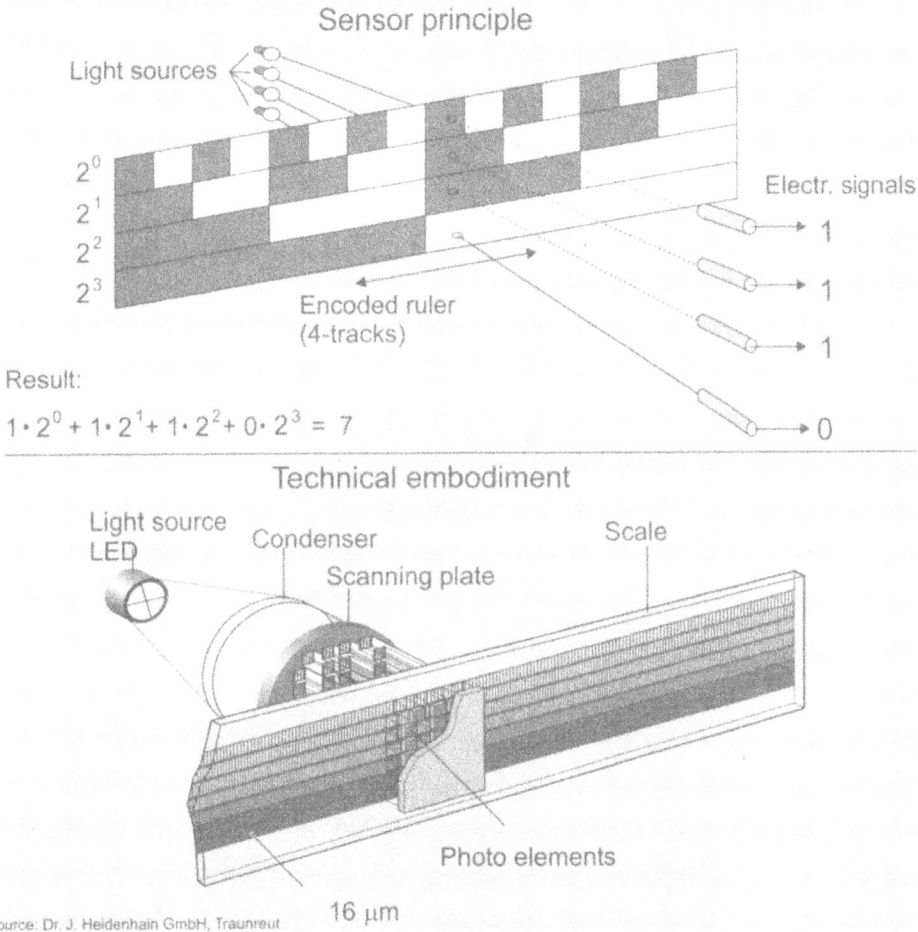

Figure 4.2-23: Operational principle and technical execution form of an absolute measuring digital photoelectric measuring system

4.3   Optical and Optoelectronic Testing Equipment

It seems clear that not only today but also in the near future will continue to grow the requirement for fast precision measurement technology for the highly exact geometrical measuring of the most varying workpieces. This demand can be directly derived from the fact that there is an increasing quality consciousness with a concurrent growing degree of automation and rising production rates. Generally, the determination of geometrical parameters such as distance, profile, form and surface microstructure belong to the most common measurement functions. This makes clear the meanings of process-integrated, highly precise optoelectronic geometry and micro geometry measurement.

Geometry measuring procedures by means of optoelectronic implement a distance proportional encoding of the measuring light to determine the distance between the measured object and the sensor. The subsequent procedure-specific decoding of the received measuring light permits a direct conclusion about the appropriate measure of distance modification between the sensor system and the measured object.

With today's state of technology, there are, in principle, five different possibilities for geometry encoding with light (Figure 4.3-1).

According to this it will be possible to differenciate, e.g. the time taken to travel along a measurement path at the speed of light, the distribution of light intensity on a detector, the location of the representation e.g. through triangulation, a phase shift between measuring and reference light, or determining the direction of polarisation to determine form.

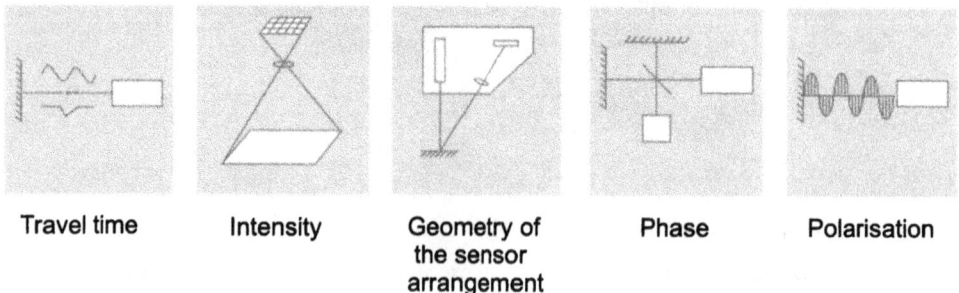

| Travel time | Intensity | Geometry of the sensor arrangement | Phase | Polarisation |

Figure 4.3-1:   Different possibilities for encoding geometrical features using light

Common to all procedures, on the transmission side, is one or more sources of light and, on the reception side, appropriate photo detectors which are selected according to the procedure-specific demands. Distinguishing features to be found are, e.g. spectrum, beam power density, coherency length, polarisation, etc. (light sources) or spectral sensitivity, plane dimensioning, dynamics, etc. (photo detectors).

The described principles of encoding with light have led to the development of a multitude of different opto-electric measuring procedures, so that one-, two- and three-dimensional measuring functions can be carried out with the optimum time and effort.

The following sections begin with a representation of the required sensor components and their system specific features. Subsequently, the important opto-electric measuring procedures will be introduced, beginning with the principles and then showing various examples of applications.

4.3.1 Optical and Optoelectronic Elements

4.3.1.1 Optical Elements

The following sections briefly describe the function of selected optical components. A detailed discussion about optical components as well as details of structural components, which can only be described using the laws of refraction, is contained in numerous textbooks on optics [Hec 99, Lip 97].

Beam Guidance
Mirrors are primarily used if a modification is required in the direction of light beam paths. These reflect the light, such that the angle of the reflected light corresponds to that of the incident light, for instance $\alpha_1=\alpha_2$ applies (Figure 4.3-2).

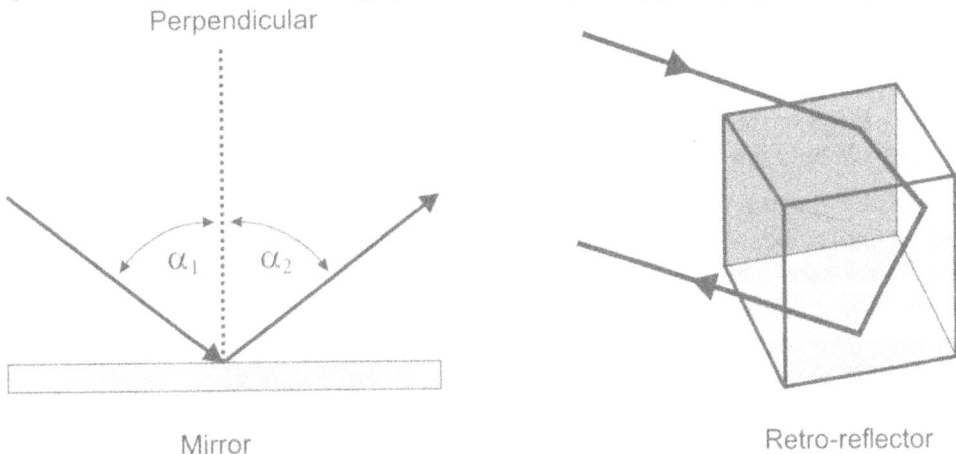

Figure 4.3-2: Schematic representation of a mirror and retro reflector

The retro reflectors shown in Figure 4.3-2 are mostly used in interferometers. These always reflect an incident light beam parallel to itself, irrespective of the tilting of a reflector, which, however, must not exceed a particular maximum angle. Retro reflectors are composed of three reflecting surfaces, which are arranged such that they form the corner of a cube. The surface of the reflector is perpendicular to its body diagonals.

In actual use, changes in the degree of reflection of mirrors occur due to modified directions of light polarisation. Likewise, a mirror can also change the status of the light. In order to avoid this, the function of mirrors is often realised through a detour via total reflection in prisms.

Focussing and Collimation
Lenses focus a parallel incident light beam in the distance of its focal length, which depends on the wavelength of the light used. For this reason, every wavelength in every beam of light is focussed on its own point. Therefore a coloured band, instead of a single point, is developed for white light. This effect, which is unfavourable for most applications, is corrected with achromatic lenses. It should be further noted, that the characteristics of a lens are calculated for beams which spread out in small angles to the optical axis. As they expand, the display characteristics of the lenses increasingly deteriorate.

Conversely, if a point-formed light source is brought into the focal point of the lens, a collimation of light takes place. If two lenses are combined such that their focal points come together, a cross section change of a light beam results, depending on the focal distance relationship of the individual lenses.

Polarisation Modification
Components which modify the polarisation status of the light are composed of double refractive materials [Hec 99]. Polarisers which transfer unpolarised light into linear polarised light can be used in combination with a second polariser to modify the light intensity.

$\lambda/4$-plates convert linear polarised light into circular polarised light, if the so-called optical axis of the double refractive materials encompasses a $45°$ angle with the polarisation direction of the light. Circular polarised light contains two partial beams which have a phase difference of $90°$ with respect to one another. For that reason, $\lambda/4$-plates are used, e.g. to generate two partial beams with an appropriate phase shift from one light beam.

Through $\lambda/2$-plates it is possible to rotate the polarisation direction of a light beam at will, whereby the orientation of the optical axis of the plate is modified relative to the polarisation direction.

Beam Splitting

The splitting of a light beam into two partial beams take place with semi-reflecting mirrors . The intensity relationship of the partial beams can be varied in wide areas through dielectric layers. Relatively thick glass plates serve as a substrate for these layers, which cause a usually unwanted beam misalignment. Thin diaphragms form a remedy as carriers of the layers, although they are mechanically quite unstable. Therefore cubic beam splitters are usually used, where the division layer lies between two prisms **(Figure 4.3-3)**. These beam splitter cubes are also available in a polarising version, which splits an incident beam into two polarised partial beams which are perpendicular to each other.

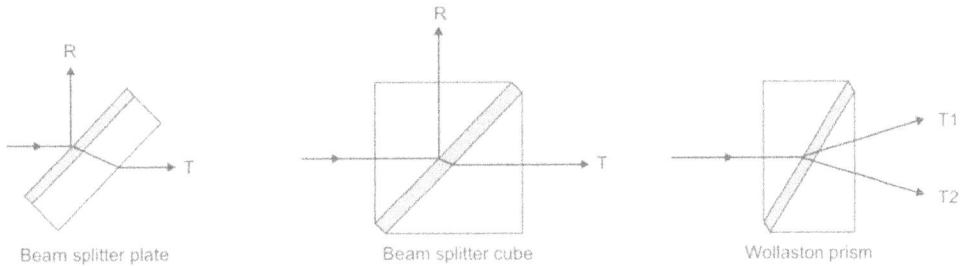

Beam splitter plate   Beam splitter cube   Wollaston prism

Figure 4.3-3:   Schematic diagram of different beam splitters

A Wollaston prism also splits an incident linear polarised beam into two partial beams which are perpendicular to each other. These components are of particular interest, as the partial beams encompass an angle which is adjusted through production of the prism. The degree of polarisation of the partial beams is also very high.

4.3.1.2 Optoelectronic Elements

The following sections deal with several fundamental optoelectronic elements, which are often applied in optical metrology for production purposes.

Photodiodes

The photodiode is a semi-conductor component. Their operational principle is based on the absorption of light in the p-n junction which leads to a generation of pairs of electrons (p-n photodiode). In the case of external short circuits or with adjacent reverse voltage, the electric field in the p-n junction splits electrons and holes from one another before a noticeable recombination can take place, thus increasing the cut-off current. This increase in cut-off current depends on the intensity of the incident light.

The efficiency can be improved if an undoped (intrinsic) intermediate layer (pin photodiode, Figure 4.3-4 a) is inserted between the 'p' and 'n' doped areas. This permits more light to be absorbed.

A further efficiency increase can be achieved by such a large pre-loading of the pin diode in the reverse bias that the optically produced electrons and holes cause impact ionisation on their way through the i-zone, thus releasing an avalanche of charge carriers (avalanche photodiode, Figure 4.3-4 b).

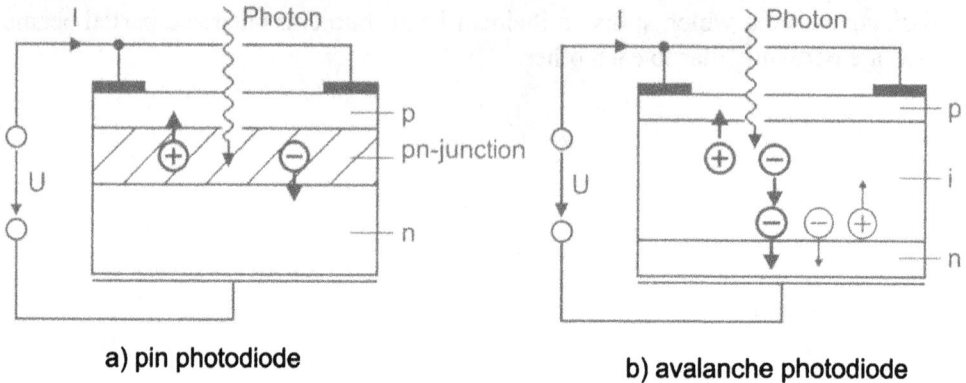

a) pin photodiode   b) avalanche photodiode

Figure 4.3-4: Types of photodiodes

Position-Sensitive Diodes
The position sensitive diode (PSD) is a wide pin diode which uses the so-called lateral photoelectric effect on the surface of the semiconductor. PSD's are available in the following models:

- differential photo receivers,
- quadrant photo receivers and
- homogeneous surface detectors.

Today mainly quadrant and surface detectors are used, whereby the former is particularly suitable for adjustment functions using the null method, while the latter is suitable for position measurement. The principle structure of the surface receiver is represented in Figure 4.3-5a. The operational principle is based on the fact that the charge carriers produced by the incident light distribute themselves differently according to the point of impact. The reason for it is that the charge carriers always seek out the path of least - here ohmic - resistance. The measurement of these currents then permits a conclusion to be drawn about the position of the light spot on the PSD. In order that a simple and fast evaluation of measured currents can take place, the p-conducting layer represented in Figure 4.3-4a should display the most constant surface resistance possible.

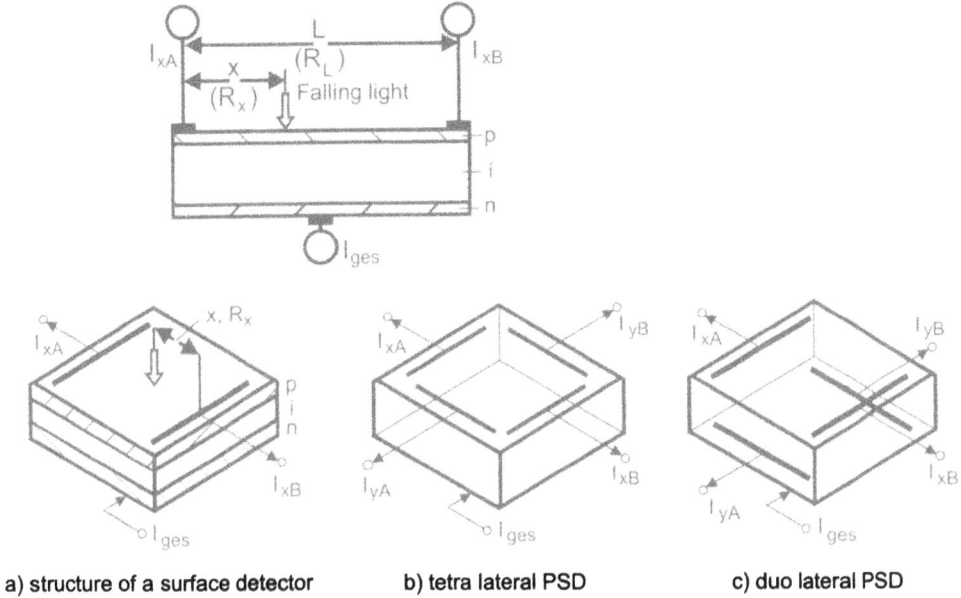

a) structure of a surface detector   b) tetra lateral PSD   c) duo lateral PSD

Figure 4.3-5:   Structure and types of surface detectors

The calculation for the position of the light spot on the semiconductor surface takes place according to the following relationship for the one-dimensional case [Sei 95]:

$$x = k_x \frac{I_{xA} - I_{xB}}{I_{xA} + I_{xB} + I_{yA} + I_{yB}}, y = k_y \frac{I_{yA} - I_{yB}}{I_{xA} + I_{xB} + I_{yA} + I_{yB}} \qquad (4.3\text{-}1)$$

Where k_x and k_y are calibration factors which must be uniquely determined for both coordinates (x and y)before applying the PSD calibration factors. There is a particular reference here to the denominators of the formulas for determining position. They represent a standardisation of the actual total value of current. Therefore, the position determination is independent of the light intensity (intensity compensation).

Surface detectors are available in two models (Figure 4.3-5b and c), whereby the duo lateral is preferable to the tetra lateral version. The reason is that the electrodes in x and y direction can influence the spatial distance less with the duo lateral model. This achieves greater position linearity [Sei 95].

CCD Image Sensors and CCD Cameras

CCD image sensors consist of a matrix arrangement of partially more than 10^6 MOS capacitors, of which each one represents an individual pixel. The abbreviation MOS stands for *metal oxide semiconductor* . An MOS capacitor (Figure 4.3-6) consists of a p-doped silicon layer, an insulator layer (e.g. SiO_2) and an electrode.

Light which meets the p-doped substrate produces pairs of electrons, the number of which depend on the strength and duration of exposure (integration time). As the initial potential of the electrode is maintained at approx. +10 V, a charge separation takes place. The positively charge holes are repelled by the electrode and flow off to the mass, while the negatively charged electrons are drawn to the electrode. They do not, however, reach the electrode, but are stopped by the insulator layer and accumulate under the electrode. Their number is a direct measure for the local exposure in the observed pixel.

In order to pick out the electrons from all of the individual pixels and receive the exposure information, the charge coupled transport system is applied, which owes its name to the CCD element. CCD stands for *"charge coupled device"*. Here the accumulated electrons are shifted from one electrode to the next through systematic setting of voltage on one or more electrodes until they reach the output level, where they are converted into a voltage signal which is used as a video signal .

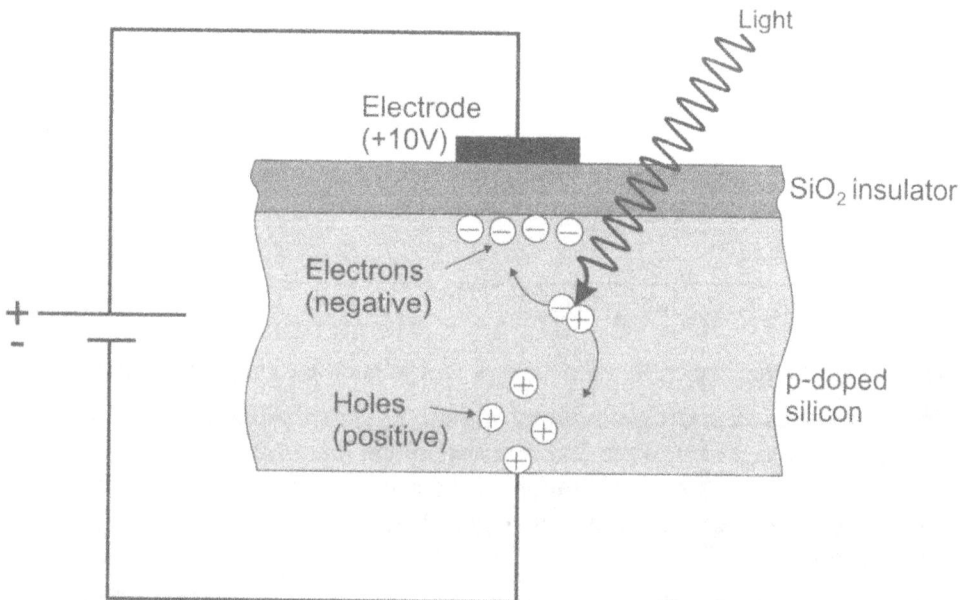

Figure 4.3-6: Structure of an MOS capacitor

The operation of the charge couple transport system is represented exemplarily in Figure 4.3-7. First a voltage is only set on Electrode A, under which several electrons have accumulated due to the influence of light. Now, the same voltage is set on the neighbouring electrode B, so that the electrons distribute themselves evenly under electrodes A and B. Subsequently, electrode A is placed on mass, through which all charges which were initially under electrode A, now move under electrode B. Thus a charge transfer of one electrode to a neighbouring one has taken place, which is repeated until all charges have been selected. The degree of efficiency of the charge transfer is between 99.99% and 99.9999%, so that the charge transfer taking place is essentially free of loss. The shifting of charges from one electrode to a neighbouring one takes approx. 60 ns, this means, for instance, that every 60 ns a pixel is selected at the video output. If the pixels are arranged in lines or in matrix form, the result is a light-sensitive CCD chip, which represents the basic element of a CCD line or CCD matrix camera. The light sensitive pixels arranged in a uniform grid have a typical size of $8 - 13$ μm [Pf 92a].

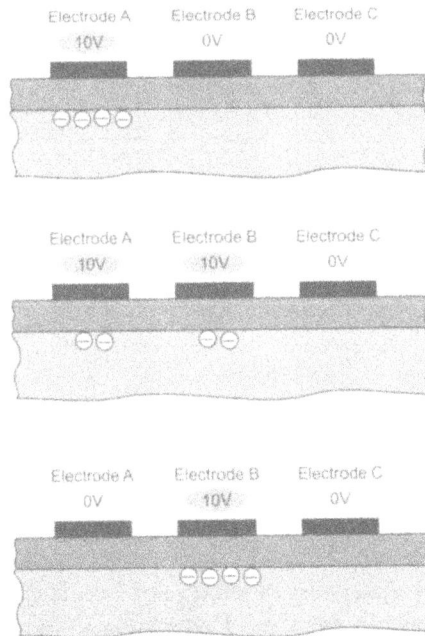

Figure 4.3-7: The charge coupled transport system

Figure 4.3-8 shows the structure of a flat CCD camera chip, with which grey tone or colour images can be produced electronically. In a CCD chip, however, an additional darkened column is responsible for the charge transfer of every light-sensitive pixel column, which, in principle, represents an analogue shift register and is also called a "CCD bucket chain". According to the external voltage cycle, the charges of the light-sensitive pixels are first shifted over to the analogue shift register after the integration period and temporarily stored there. This shifting of charges takes place through the initially described direct creation of an external voltage on the individual potential wells. In a subsequent step, the columns are selected in the same manner by creating an additional voltage, but with a lower carrier frequency.

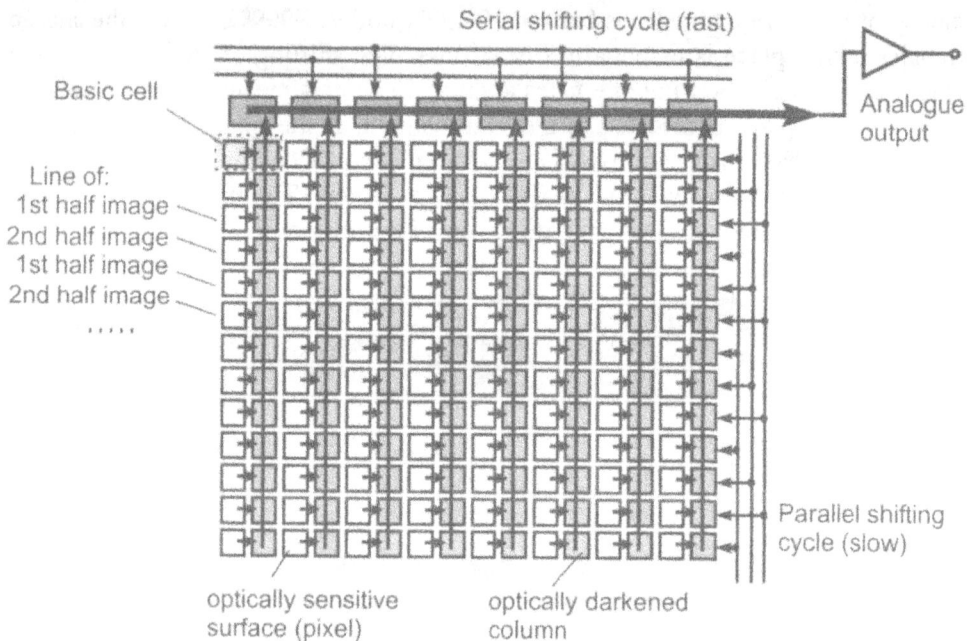

Figure 4.3-8:   Structure of a CCD camera chip (here: according to the interline principle)

Through cyclical repetition of this voltage cycle, the charges shift successively, as in a bucket chain from potential well to potential well toward the output. There the individual charge bundles are changed into an output voltage by a selection diode and the signal, ultimately consisting of voltage pulses, is changed into a continuous signal through a Sample-Hold member. This analogue signal is passed on via a co-axial cable to an interface card (Frame-Grabber) in an evaluation computer, where the analogue signal is digitised and stored in the memory.

The selection of an image does not usually take place at once, but through the transfer of two half images, which are then reassembled to the overall view. While the half image is selected from all odd lines, the second half image results at the same time through exposure of the even lines. In the next cycles the roles of the image lines are exchanged.

With the alternative selection procedure according to the frame transfer principle, an equally sized darkened storage zone is located under the CCD-Chip in which the generated charge of the individual light-sensitive pixels are temporarily stored after exposure. Therefore there are no gaps between the light-sensitive pixels, as there are with the interline procedure. The image (frame) is transferred as a whole within a time <1 ms into this storage area and can be selected as a whole according to the bucket chain procedure (frame transfer), while exposure again takes place in the upper layer. CCD chips need the double chip surface according to the frame transfer procedure, but they are much easier to produce than chips in accordance with the interline transfer procedure.

CCD's are sensitive in the wavelength range of 0.4 µm to 1.1 µm and also span from visible light to the near infrared range. By using more specialised components, the latter also enables applications in the area of thermography, e.g. for technical diagnosis or process monitoring.

With colour cameras, three images of the object are made simultaneously, whereby each is only sensitive for one of the spectral regions *Red*, *Green* and *Blue* (RGB). The entire colour image can then be configured from these three colour image components. Every colour image level thereby always has the same structure as an individual grey tone image, which is composed of the uniform arrangement of the intensity values encoded in the pixels.

Technically, there are two solutions available for the simultaneous recording of the individual colour image levels. The more expensive, but precise version consists of three separate CCD chips (3 chip colour cameras), which are each sensitised to one of the above mentioned spectral regions through upstream colour-sensitive deflection filters. These deflection filters divide the incident light and thereby the image into the three spectral regions. With single chip cameras, three pixels each lie next to one another, which differ in their spectral sensitivity. The resolution is thus smaller, due to the neighbouring arrangement of three pixels per image point.

4.3.2 Camera Metrology

In recent years, camera metrology has gained in importance more than many other measuring procedures. This is justified, on the one hand, by the high level of flexibility and processing speed of the systems which has taken the rapid development of computer technology in its wake. On the other hand, measurement or inspection using a camera image represents a technology which strongly accommodates human perception and therefore makes it suitable for many industrial tasks.

In many areas of production technology today CCD cameras are linked to a measurement system with subsequent automated image processing. Therefore the necessary device-related fundamentals of camera metrology will first be described in section 4.3.2.1. Subsequently, the fundamentals of digital image processing which are required in production metrology will be described in Section 4.3.2.2.

In the concluding sections, typical fields of application for camera metrology will be introduced for industrial 2D image processing and 3D form acquisition of workpieces using photogrammetry and stripe projection procedures (Sections 4.3.2.3-4.3.2.5).

4.3.2.1 Device-Related Fundamentals of Camera Metrology

This section will first describe how an electronic image is made in the first place and stored in an image processing computer and which components and standards are available for this. Only when the image has been stored in the memory of image processing computers, can numerical procedures for image processing be applied, for example, to automate certain measuring and inspection tasks.

Image Processing Steps
In order to better arrange the individual image processing steps within the larger context, they are represented using the block diagram in **Figure 4.3-9**.

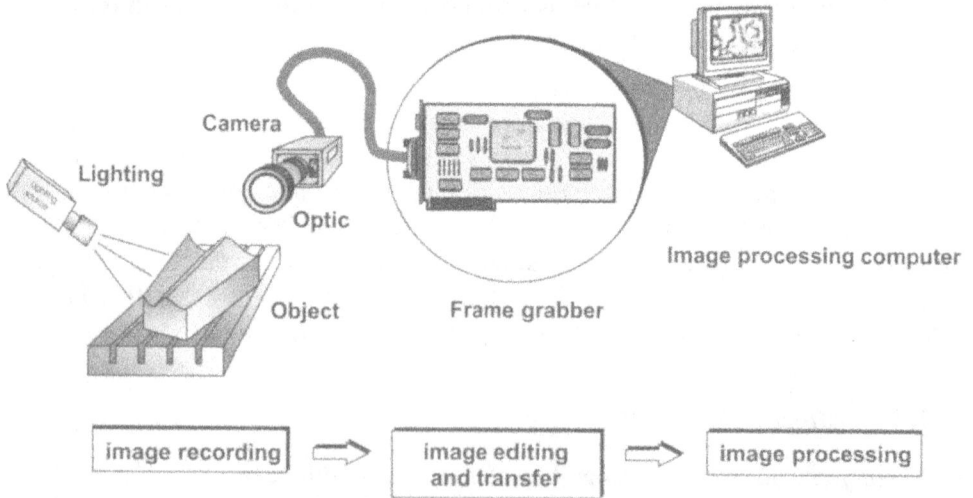

Figure 4.3-9: Image processing steps and the required components

First an image is made of the scene of interest with an electronic CCD camera and digitised in an interface card in the computer. The image is then available in the memory of the computer, for the computer CPU to access for image processing purposes. These encompass numerical procedures for simple image improvement such as noise reduction or increasing contrast, as well as more complex procedures for automatic feature extraction or pattern recognition (Section 4.3.2.2). Depending on the requirements, the image processing computer can be a PC, a workstation or parallel computer architecture.

Electronic CCD Cameras and Frame Grabbers

The most common image recording medium today is a so-called CCD camera. The abbreviation CCD stands for "charge coupled device" . With the "coupled charge transfer" procedure – also called the "bucket chain procedure" – the individual light sensitive image sensor elements of the camera chip – the so-called pixel (name of the picture element) – and the information contained within them in the form of electrical charge are selected through the local image brightness of this image point.

The structure and operating methodology of light sensitive CCD chips are explained in Section 4.3.1. There, it is described in which way the scene displayed on the CCD chip by means of objectives is converted from a brightness distribution into a discrete distribution of electrical charge bundles, with their size being a measure of incident local light intensity. After selecting the charges out of the CCD chip, an analogue voltage signal is generated and passed on to an interface card in the image processing computer, which edits the image into a form appropriate for the computer.

In image processing, the interface card is commonly described as a "frame grabber" , which digitises the analogue image signal of the camera. This analogue image signal essentially consists, apart from synchronisation impulses for lines and image wrap, of a sequence of brightness information on the individual pixels, the matrix arrangement of which forms the total image. The brightness values are present as electrical voltage levels and are usually digitised to 8 bits and can thus be digitally stored in a computer. The 8-bit digitisation corresponds to 256 grey tone levels, which range from white (max. brightness value 255) to black (grey value 0). In comparison, the human eye can only distinguish between approx. 64 grey tone levels.

It should be noted that, during image recording, the image is discretely scanned and digitised through the discrete and matrix-formed arrangement of pixel sensors on the chip, whereby the entire brightness is integrated within a pixel sensor. This can be described with communications technology methods and shows that a CCD camera possesses a type of low-pass effect. The resolution of a CCD camera is limited and extremely fine structures can thus not have a better resolution [Lez 90], [Brü 96]. Therefore, the finest line grid which can be resolved by a pixel matrix has a spatial frequency of a half pixel thickness (Nyquist limit), this means, with a line grid, a bright line falls on one pixel column and the dark line falls on the neighbouring pixel column [NN 95a]. For example, the spatial capacity with a

wide, high resolution CCD chip is 4096 × 4096 pixels on a surface of 28 × 28 mm²
with over 70 pairs of lines per millimeter.

In certain applications, the resolution limit can be artificially increased beyond the
Nyquist limit by applying numerical interpolation procedures to the light dark
transformation in the stored image of an edge (sub-pixeling) [Pf 92a], [Pf 90],
[Chi 95]. Finally, it must still be considered that when representing fine stripe
structures, the matrix form of the pixel arrangement can lead to pixel Moiré effects
which overlay the image as disturbing, low-frequency intensity variations.

Special forms of CCD surface chips found in line cameras and colour cameras are
used just as frequently in image processing systems and are based on the same
technical principle.

With line cameras, it must be noted that it is not a matrix of light sensitive pixels
which is present, but only a single image row. In order to generate a planar overall
picture of an object, the object and the camera must be moved towards one another, so that a line by line scanning of the object takes place. In order to avoid distortion effects, the relative velocity of the camera and the object must be coordinated with the image recording frequency of the CCD line. A typical field of application for line cameras is surface testing on line assembly material [Kör 95],
[Swa 97].

Standards and Formats
Within an image processing system, several components are linked together and
exchange signals with each other which are synchronised and must be coordinated
with regard to formats and voltage levels. In this way, the Sample & Hold member
must be synchronised with the selection cycles when choosing a CCD chip; the
image digitisation which takes place in the frame grabber must occur in a pixel
synchronised cycle, etc. A relatively free Plug & Play problem is only possible by
guaranteeing the required standards

For selecting an image in the CCD camera and transferring it to the frame grabber
card, the CCIR/PAL television standard exists in Europe (in the USA: RS-
170/NTSC television standard), under which the integration time – i.e. the exposure time – amounts to half images of 20ms or 50 Hz. According to this norm, 25
frames are taken per second and passed on to the frame grabber as an analogue
signal.

The CCIR standard thereby defines the substantial characteristics of the so-called
VBS signal (video, blanking and sync signal), which encompasses substantial synchronisation pulses for horizontal and vertical diversion and scanning, transfer frequencies, voltage levels, etc., which are important for the representation of a black
and white image on a screen.

Beyond that, it is also possible to encode this signal with colour information. The BAS signal is now expanded to a colour CBVS signal (CVBS), in which the colour information is integrated in an additional colour carrier (composite coding). The carrier frequencies of the colour carriers and the type of colour encoding in these colour signals distinguish the western European PAL standard from the US American NTSC standard. Alternatively, the individual RGB colour channels can be transferred separately through three channels, like black and white images.

The camera and frame grabber components are normally coordinated with one another in such a way, with the following standards typically resulting for Europe:

An image consists of 572x768 active pixels with a half-image repetition frequency of 50 Hz. A black and white image with 8 bits/pixel therefore require 429 Kbytes of memory. In the US American NTSC format there is a half-image repetition frequency of 60 Hz with a resolution of 640x480 active pixels.

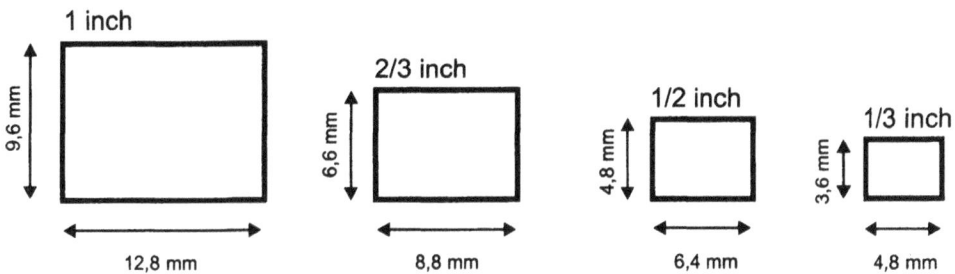

Figure 4.3-10: Standards for image recording formats of CCD chips for surface cameras

In addition to the number of elements of the image matrix which is determined by the signal frequencies, standards also exist for the size of CCD chips. The size of these chips is indicated in the form of chip diagonals in inches. The size of the imagable object field varies, depending on the chip size.

Typical values have meanwhile also become generally accepted with line cameras. CCD lines usually consist of 1024, 2048 or 4096 pixels and thus have typically more pixels than the CCD line of a surface camera. If the application allows it, higher resolutions can be reached with line cameras than with surface cameras, also with greater magnification, relatively large object fields can still be displayed. However, the object must be scanned in order to obtain a flat image. Scanning can also be advantageous for highly precise applications. If the incrementation is effectively smaller than the pixel distance (micro scanning), an extremely high resolution image originates in the scanning direction, which is superior to the image from a surface camera .

Standards also exist regarding objectives for CCD cameras, which are occasionally also still referred to as CCTV objectives (closed circuit television), which enable the simple exchangeability of objectives. For linking objectives which typically have focal lengths within the wide angle range from 8 mm - 50 mm, there are standardised threads which are called c-mounts. Thus, objectives from different manufacturers can easily be screwed onto the camera.

For accurate measuring functions, there are special corrected distortionless optics available nowadays, with accuracies to 1:10,000 and distortions below one per thousand [Jäh 96]. Their representation quality reaches near to the natural diffraction boundary of the optical representation, so that an ideal object point can be represented on a pixel of only a few micrometers. However, the dissolution-limiting effects mentioned at the outset must still be considered with image recording.

For highly accurate measuring functions, so-called *telecentric objectives* are used. These objectives are characterised by the fact that the scale of representation does not change with the slight defocusing of an image [Srö 90], [NN 95a],[NN 95b].

Concretely, this means that a scale line, for example, can have a somewhat unsharp display with slight defocusing, but, contrary to conventional objectives, does not move. This is an important characteristic for accurate measuring functions. Furthermore, telecentric objectives are characterised through the representation of the object in parallel projection, as shown in **Figure 4.3-11**.

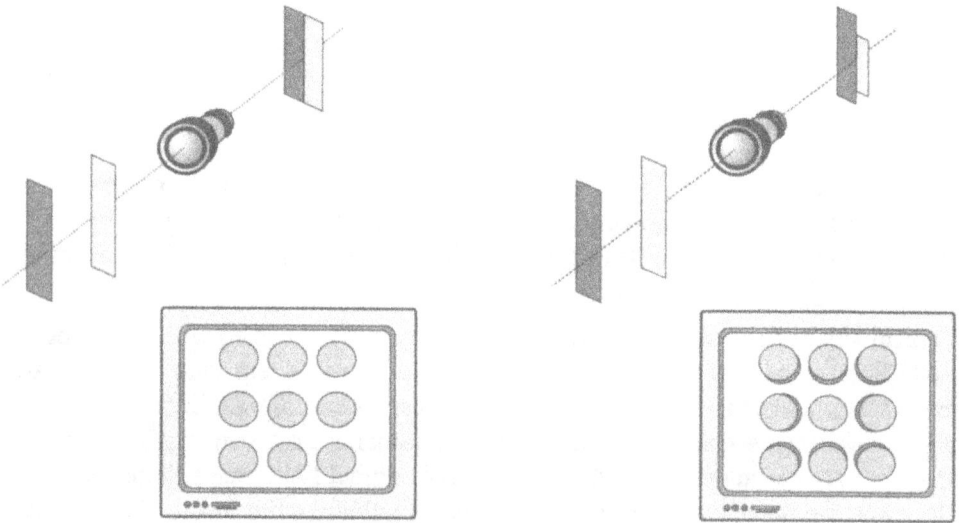

Figure 4.3-11: Representation characteristics of telecentric objectives (left) and conventional objectives (right), bottom: control of nine cylinders

Telecentric objectives have fixed focal lengths and a fixed representation scale. For highly accurate measuring functions with a resolution of approx. 3-8 μm, the object field shown is typically quite small at a few mm² and the sharpness depth smaller at less than 1 mm [NN 95a, Sör 90].

With so-called *spacers* the distance can be increased between the objective and the camera chip, which is necessary for close-ups with great enlargement. It must, however, be considered that the telecentric characteristic is lost through the application of spacers with telecentric objectives!

With regard to the components in image processing computers, standards also exist for frame grabber cards. The older generation of frame grabbers not only digitised the incoming analogue camera signal, but already performed arithmetic operations on the images on the frame grabber cards with the aid of special arithmetic chips. The problem was that ISA Bus device drivers in the computers were too slow to make the relatively large quantity of image data available in the fast time cycle of the computer's CPU, or RAM. The frame grabber cards were therefore relatively complex and expensive. In the meantime, however, the PCI Bus had established itself as standard and enables the fast transfer of images to the CPU, where the actual image processing can then take place on the basis of high level languages such as C/C++, independently from the frame grabber hardware. An advantage of this development is that the exchangeability is increased between the frame grabber and the image evaluation software, as these components are decoupled.

Additional Options and Deviations from Standards
For more specialised applications, there are systems which deviate from the above mentioned standards. The costs of such systems rise rapidly, as it concerns individual cases which do not serve a large market.

For certain measuring functions, high resolution surface cameras with large format CCD chips are required in order to still be able to record entire scenery, especially with higher magnifications.

For applications where highly dynamic processes such as crash tests are examined, special high speed cameras are available on the market. Further specialisations also represent cameras which are particularly appropriate for infrared or x-ray areas. With such cameras, for example, thermographic investigations and monitoring functions can be carried out, which are important for many technical processes.

For applications in areas strongly affected by EMI, digital cameras are available which do not pass an analogue image signal to the frame grabber, but a signal which has already been digitised in the camera itself. Disturbance sensitivity during the transfer of the image signal, for example in a strongly affected EMI machine environment, is thereby reduced. These cameras have additional advantages for highly precise measuring functions. The analogue synchronisation of camera and image memory causes problems with respect to the maximum achievable accuracy (pixel jitter) and is omitted with digital cameras. These cameras are also advantageous for precise measuring functions on fast moving sceneries, as the somewhat time-delayed half-image transfer is omitted [Jäh 96].

4.3.2.2 Fundamentals of Digital Image Processing

The high demands on modern quality management result in the necessity for flexible tools to be made available for quality assurance in production companies. Image processing systems represent such a tool in quality testing areas, as well as in automated measuring functions and in the area of technical visual inspection.

The fact that development towards automated image processing systems in production today can in no way be viewed as completed, is clarified by the estimate that so far, in the area of visual technical inspection, only approx. 10-15% of the applicable test functions are automated.

Important advantages of image processing systems are the high measuring rate which can be achieved, and the high level of objectivity which particularly distinguishes the systems from human inspectors. For a human inspector, e.g. a high manufacturing cycle, concentration weaknesses or just features which are difficult to distinguish from one another can cause problems which, in principle, can be solved with an automatic image processing system.

Because a technology which strongly accommodates human perception is behind image processing, the possibility of local interaction by the user is also given. For example, a worker can detect simple errors quickly and independently with the aid of a camera image. Even for less qualified workers, suitable tools in control software can clarify how the optimum status of the system should look e.g. using stored and commented sample images. Additionally, lighting failure or a dirty objective can be detected with the many conceivable internal safety and plausibility queries in image evaluation software. The capability of self diagnosis is only given with very few measuring systems at all.

Through the high abundance of information in the camera image, it is also possible to combine several characteristics, through which some evaluation functions only become possible. For this, it is however necessary that the features of a workpiece to be measured or evaluated can solely be derived from its optical appearance and a representation succeeds based on suitable characteristic values. This representation of the optical appearance based on characteristic values, however, often represents one of the greatest problems in practice.

Natural and Artificial Sight

How can a computer with a CCD camera connected to it measure the diameter of a drill hole or determine whether a groove is correctly situated on a component?

An inspector frequently detects "at sight" whether a drill hole exists on a component. Neither lighting fluctuation nor positional misalignment, tilting or twists of the component cause difficulties here. Object forms, brightness and colours are intuitively detected and assigned to these terms. It must not, however, be forgotten, that the visual system of a human represents a highly developed "image processing system, which is trained for decades and falls back on a huge foundation of knowledge.

Analogue to this, an image processing computer must first derive and learn from a mass of pixels which are transmitted by the camera, through which criteria individual objects can be detected and differentiated from other objects or the image background. Such criteria can, in the simplest case, be differences in brightness between object and background, or in one of the most difficult cases, different *textural features*. Image evaluation therefore strongly depends on the features which the image scene makes available. If the areas can already be separated from one another through differing grey tones, image evaluation is substantially less complicated, than in the case where objects a only differ from the background through surface texture.

Neurophysiological studies have concluded that for human sight, the recognition of object edges is of great importance. Likewise – although often for other reasons – the recognition of object edges also plays an important role for technical applications of automated image processing.

For the automatic processing of camera images, a great multitude of procedures and techniques have been developed in order to extract the relevant image information from the camera image. In the following sections, the most important procedures for image analysis will be briefly introduced, with the emphasis being on the recognition of object contours.

Colour Images, Grey Tone Imagesand Binary Images
As previously described, CCD cameras generally supply image information in the form of grey tone or colour images, whereby the images usually have 256 grey tones or 256^3 different colour levels. This is equivalent to a digitisation depth of 8 bits for grey tone images or 24 bits for colour images. The latter are composed of e.g. 3x8 bits for the colour components *red*, *green* and *blue* (*rgb*). For special applications, cameras with larger digitisation depths of 10 or 12 bits are available. A colour image can also be represented by many representations other than the colours red, green and blue. This corresponds to a representation in another coordinate system in the so-called *colour space* [Pra 91]. Special cameras therefore supply image information encoded in other colour space coordinates, which are transferred into the conventional *rgb* representation through a colour space transformation.

The function of image processing is to separate the relevant image information from the rest and e.g. automatically extract objects from an image. In production metrology this means recognising measurement objects and separating them from other image contents in order to subsequently perform measurements on the image. For separating the objects from the background, it is in many cases, e.g. through suitable lighting, already possible to differentiate measurement objects alone through differences in brightness from the image background. This technique is applied, e.g. with so-called *transmitted light illumination* where the object is illuminated from the rear, so that the camera records a shadow projection of the object. In this case it is possible, with the aid of a so-called *threshold value procedure* to differentiate between object and background. For subsequent processing, the pixels are divided into two categories, depending on whether a fixed value – the *threshold value* – was exceeded or not. All pixels which have a higher grey tone than the threshold value are set to "white" (grey tone 255) and all pixels which are less than or equal to the threshold value are set to "black" (grey tone 0).

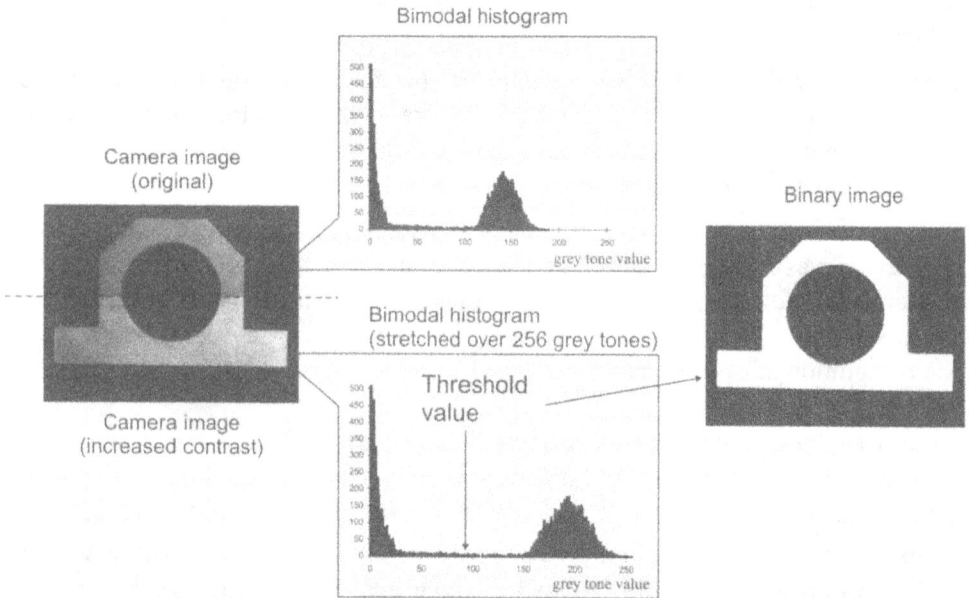

Figure 4.3-12: Contrast increase and subsequent binarisation according to the threshold value procedure

The binary image, which is very important for image processing, is developed in this manner. The optimum threshold value is determined using the so-called *histogram*. This is a representation of the respective number of grey tones occurring in an image (with colour images, histograms can similarly be produced for each of the colour levels). With a grey tone image, where light and dark areas are clearly defined, e.g. with the transmitted light procedure, the histogram has two pronounced maxima within the area of low, or high, grey tones. In this case, one refers to a *bimodal histogram* **(Figure 4.3-12)**. With this type of histogram, the traditional threshold value formation uses the grey tone as the threshold value, which lies at the minimum position between the two maxima. A separation according to the simple features is only possible in production metrology if the image scene is relatively simply structured and is already well structured from the outset.

Point Operations

When converting a colour or grey tone image into a binary image, an image processing operator is used, which makes an allocation based on the same principle for all pixels and only takes into account the colour or grey tone of the pixel itself. Such operators are called "point operators", indicating that the neighbourhood relations in no way enter into the law of formation of the filtered value. Such point operations are applied for segmentation, as with binarisation, or in the case of increasing contrast and are used for pre-processing the images [Pf 92a]. With grey tone images, the contrast increase takes place through a linear stretching of the grey tone histogram over the full range of 256 grey tones **(Figure 4.3-12)**.

Local Two-Dimensional Filter Operations

Apart from the point operators, local operators are also responsible for image processing, whereby the neighbourhood environment, which is taken into account when filtering the image, is of great importance. Often a usually square operator window with an odd number of columns or lines is shifted over the image.

The calculation of the filtered new grey tone takes place by multiplying the unfiltered surrounding values with the coefficients of the filter mask and subsequently adding them up. The result is written back to the current image position as a filtered value [Zam 91]. This corresponds to a folding of the image with a square operator window.

Within the context of this book, a selection of the most frequently used filter methods, out of the multitude of mask filters which have been developed, will be described in more detail. An example of an important function in image filtering is the emphasis of edges, which present themselves in the form of grey tone jumps. As particularly high spatial frequencies occur in the spectrum of a grey tone jump, which do not exist in places with a relatively smooth grey tone path, edges can be emphasised with the aid of a *high-pass filter*. Conversely, a *low-pass filter* smoothes the image. Due to computing time required for technical applications, frequently only immediate neighbourhood relations can be considered for image filtering through the use of 3x3 filter masks. The most important realisations of high and low-pass filters are briefly described in more detail below.

In practice, the most frequently used operator for emphasising edges is the so-called *Sobel filter* grey tones **(Figure 4.3-13)**. This operator emphasises edges in x or y direction of the image. If the geometric mean is formed from the filtered images in x or y direction, one obtains satisfactory object edges in almost all directions. A further improvement can be achieved with the aid of the so-called compass filter (e.g. Kirsch operator), which combines the results of several individual filterings with various direction-dependent filters [Zam 91]. Both filter masks of the Sobel filters have the following structure:

$$Sobel_x = \begin{bmatrix} 1 & 0 & -1 \\ 2 & 0 & -2 \\ 1 & 0 & -1 \end{bmatrix} \text{ and } Sobel_y = \begin{bmatrix} 1 & 2 & 1 \\ 0 & 0 & 0 \\ -1 & -2 & -1 \end{bmatrix} \qquad (4.3\text{-}2)$$

While the Sobel filter emphasises the object edges in one direction by differential formation of the grey tones, a weighted average of the grey tones is determined in the other direction. Edges which run diagonally can also be emphasised well through the formation of the gradient amount. It is also possible to calculate the edge direction, with the aid of the above mentioned mask filters [Pf 92a].

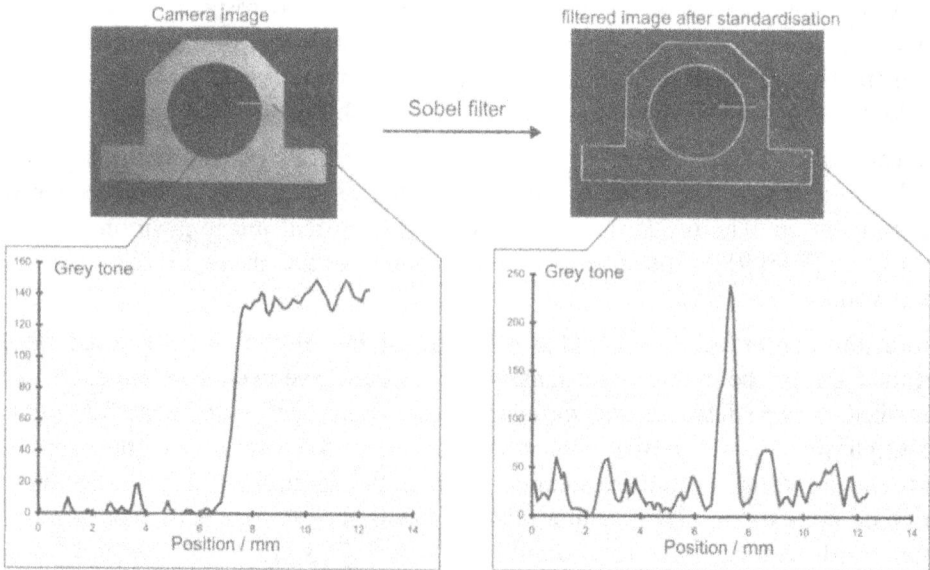

Figure 4.3-13: Effect of the Sobel filter

In contrast to the Sobel filter, the "Laplacian filter" represents a direction-independent high-pass filter. This filter can be regarded as a numerical approximation of the second derivation of the image. The effect of this filter on a grey tone image is represented in **Figure 4.3-14**.

The Laplacian filter possesses the following filter mask:

$$LaplacianFilter = \begin{bmatrix} -1 & -1 & -1 \\ -1 & 8 & -1 \\ -1 & -1 & -1 \end{bmatrix} \tag{4.3-3}$$

However, through the numerical representation of the second derivation of an image, the filter becomes extremely noise sensitive, so that there are only very few cases where it can be applied for image processing functions in the area of production metrology.

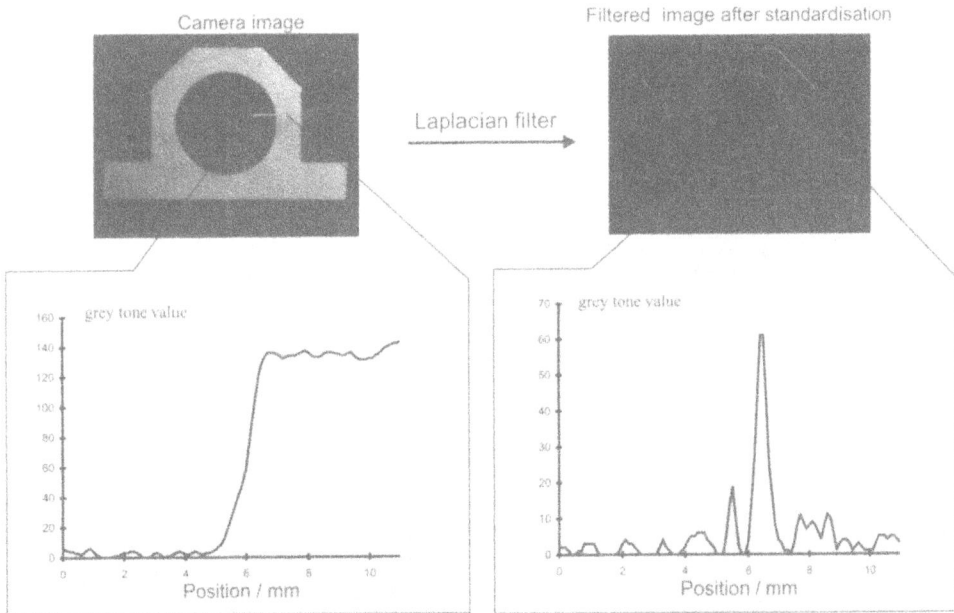

Figure 4.3-14: Effect of the Laplacian filter

A substantial improvement in the characteristics of the Laplacian filter is achieved by combining it with a noise suppressing low-pass filter, e.g. the Gaussian filter [Mar 80]. In its simplest form, it possesses the following structure:

$$GaussianFilter = \begin{bmatrix} 1 & 2 & 1 \\ 2 & 4 & 2 \\ 1 & 2 & 1 \end{bmatrix}$$ (4.3-4)

With a Gaussian filter, a weighted averaging of the grey tones is undertaken from the environment of the pixel, whereby the weightings decrease, depending on the distance from the current pixel, as with a Gaussian curve.

The mask filters introduced so far correspond to a linear transformation of the image. Some characteristics of an image can, however, be better extracted or even eliminated from an image with non-linear operators. A typical representative of non-linear operators is the *median filter*, often used to eliminate point-form noise. In contrast to the Gaussian filter, there is no undesired smoothing of edges with the median filter.

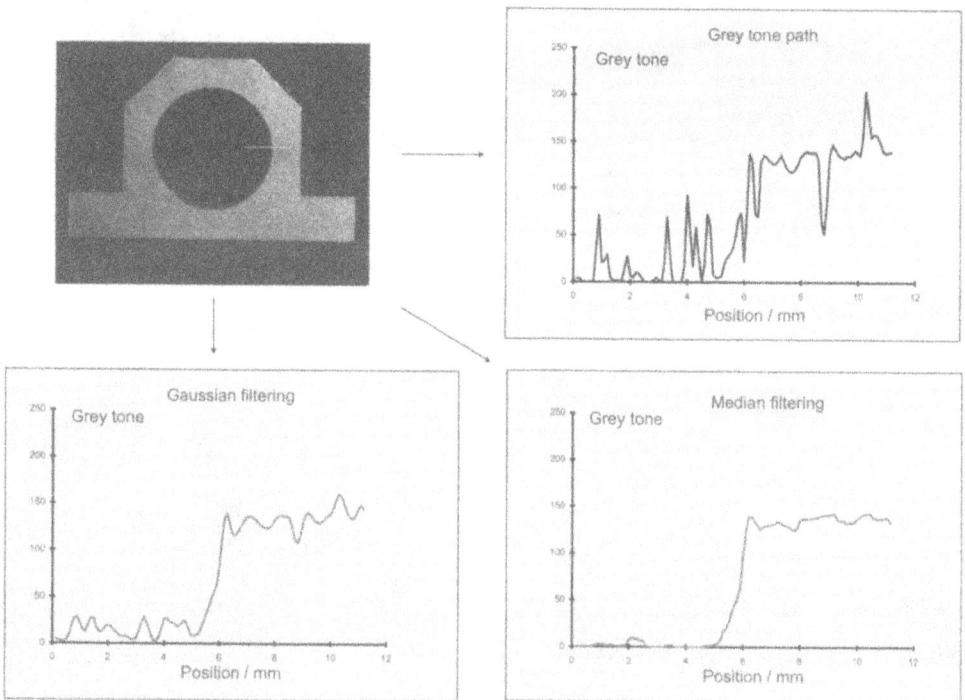

Figure 4.3-15: Comparison of Gaussian and median filtering

The median filter belongs to the class of the rank filters. The grey tones are written into a list sorted by size within a neighbourhood environment around a central pixel. Subsequently, the value which is arranged in the middle of this list is written back as a filtered value in place of the central pixel. As a single point-form outlier never lies in the middle of this listing, such disturbances are eliminated by median filters. Edge profiles, however are preserved without being smoothed, as is the case with linear low-pass filters.

Morphological Image Processing

Morphological operators rank among the most important digital image processing tools. Their main application is in the detection of small defects or cracks during the industrial visual inspection. The morphological edge detector is characterised by a high insensitivity to noise. The two elementary operations of mathematical morphology are *erosion* and *dilatation*. These process the image by means of a structural element, which could be an additional object or a mask.

With binary images, an *erosion* causes the quantity of all pixels of an object, with which the reference point of the mask can be harmonised through shifting in such a way that the object is fully contained within the object, to be set to the value 255. This corresponds to an AND connection of the binary mask with the binary object. The edges of the object are thereby eroded, while inside holes will be enlarged.

Conversely, the result of *dilatation* is when the quantity of all reference points for which a shifting of the mask over the object causes one point of the mask to coincide with one point of the object. This corresponds to an OR connection of the binary mask with the binary object. The edges of the object are enlarged, while the inside holes are reduced in size. The gradual execution of erosion and dilatation is often referred to as "opening" and conversely, the gradual execution of dilatation and erosion as "closing". The two functions represent the most important compound functions of morphology. By "opening" an image, smaller objects such as points and fine structures in the image are deleted, while large objects are preserved. The "closing" operation, however, fills out small gaps and cracks. An extension of dilatation and erosion functions on grey tone images is possible, whereby the logical AND and OR connections are replaced by the minimum and maximum determination of the object grey tones within the structuring mask. Significant geometrical changes can, however usually only be observed with repeated gradual application of the procedures, whereby computational complexity increases, which prevents a wider use of these procedures within production metrology.

Contour Point Chaining and Line Thinning

The binary image represents the ideal result of edge filtering, where all true contours are complete without gaps, with a width of 1-2 pixels and no additional points remain which have been produced by the noise.

In order to achieve this goal, additional steps in image analysis are usually required following emphasis of the edges using a high-pass filter. In this way, for example, the edge image can be subsequently transferred into an appropriate binary image with the aid of a fixed or also a local adaptive threshold value. The resulting image then usually contains edges which are represented through lines varying in brightness and width. Additionally, however, gaps of the true contours and noise-induced pseudo edges appear after such filtering.

For this reason, a binarisation of the edge image is usually followed by line thinning and contour point chaining.

Initially, widened lines are reduced to the width of individual pixels with special algorithms. Numerous procedures for line thinning are described in literature. With the simplest procedure, the intensity profile is evaluated vertically to the detected edge and the grey tones which are less than the local maximum are suppressed. In this way, a line will ideally remain, which is exactly the width of one pixel. Should the edges, however, not only indicate a maximum value, but a plateau of several equal grey tones, further procedures are necessary [Bäs 89].

The chaining of points generated in this manner is an important, if difficult, procedural step, as the actual detection of contours takes place through this procedure. There is a differentiation between local and global procedures for contour point chaining. While local procedures search for continuation elements in a neighbourhood near to the central pixel, global procedures use information from the entire contour chain found so far. In addition to the Hough transformation described in the following section, the "heuristic search" and "dynamic programming" have achieved wider use. A detailed description of this procedure is in [Bäs 89].

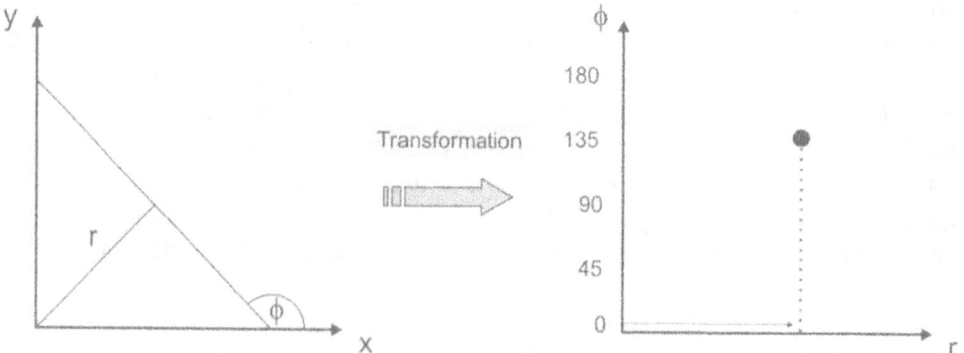

Figure 4.3-16: Parametric representation of a straight line in polar coordinates

Hough Transformation

The Hough transformation is a technique for recognising global patterns of the image space, such as segments, curves or closed forms from the ideal point-form image in a suitable parameter space. It goes back to P.V.C. Hough, who laid down this transformation in a patent specification in 1962 [Hou 62].

The greatest selectivity is obtained if the Hough transformation is applied to an already segmented edge image. The computing processes are represented using the example of a straight line, which, for practical reasons, is described in Hesse standard format:

$$x \cdot \sin(\varphi) + y \cdot \cos(\varphi) = -r \qquad (4.3\text{-}5)$$

A straight line in the image is illustrated as a single point in a parameter space, with the Cartesian axes r and φ. In order to now find all straight lined edges in an image, the edge image can undergo a Hough transformation. With this transformation, a segment of straight lines is suspended in every possible edge point and thus, in each point, which indicates the grey tone 255 after binarisation and increases a counter store by the value of one in the appropriate place in the parameter space. At the end of this procedure, the accumulated values are then summed in the counter store (**Figure 4.3-17**).

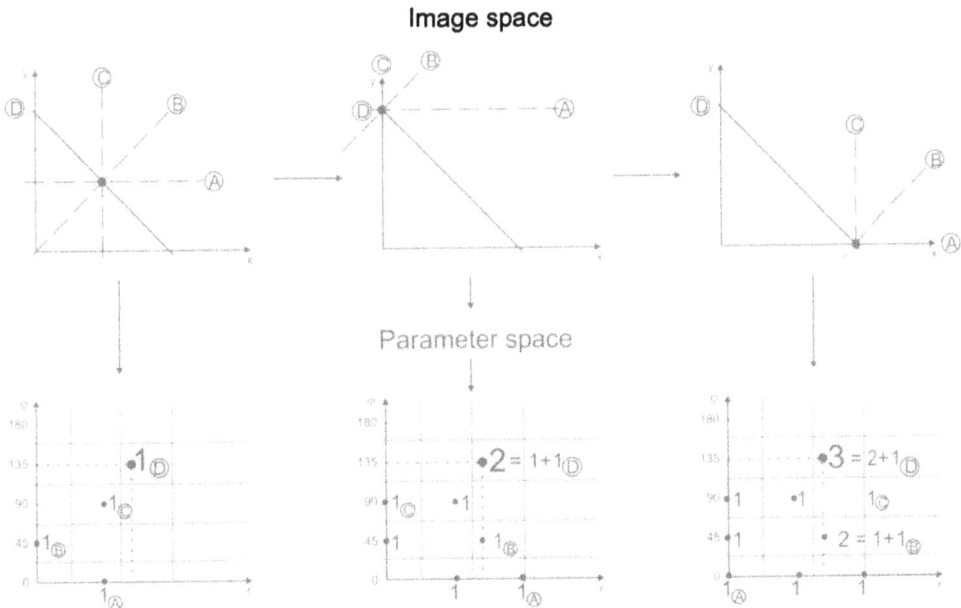

Figure 4.3-17: Principle of the Hough transformation

For points which remain in the edge image due to noise, this has the consequence that the memory counters are increased in various places within the parameter space. However, for contour points lying on a straight line, amongst other things, the appropriate counter reading always increases in exactly the same place in the parameter space which belongs to the actual object edge. This leads to significant maxima in the parameter space, which represent the actual object contours. In addition to recognising straight lines, circles in the image can also be detected with the Hough transformation [Lv 91].

Fourier transformation
The previous section introduced a representation of image information in an appropriate parameter space. In addition to direct filtering of images in the local space, image processing in the so-called local frequency space also represents an important variation. In a similar manner to time signals, with a one-dimensional Fourier transformation whereby the frequency contents of the signal can be extracted, a transformation of image information into harmonious sine or cosine portions is often useful. As information from the entire image contributes to this transformation, the Fourier transformation is a so-called *global image processing operator*..

For image processing, a two-dimensional extension of this Fourier transformation is usually used which takes advantage of special symmetries and is therefore limited to square windows with a side length of 2^n. For this special Fourier transformation, the name FFT (Fast Fourier Transformation) has established itself [Jäh 97].

Through selective treatment of individual Fourier coefficients, direct image improvements can be achieved relatively simply and made visible after an appropriate inverse transformation. A further field of application is texture determination, as some surface structures in the local frequency range are easier to separate than directly within the local range. A typical application is the technical visual inspection, where a periodic grinding structure on components can be detected and assigned a characteristic value with the aid of the Fourier transformation.

Correlation Analysis
Correlation analysis is traditionally used in the area of signal processing, for the investigation of time signals. With correlation analysis, a predication can be made about the degree of similarity (correlation) between two signals. In this case, one refers to *cross correlation* between two functions. A predication can also be made about the similarity of a signal with itself, using the *autocorrelation function*, whereby periodic portions in a signal can particularly be detected, as they present themselves in characteristic maxima of the correlation function.

Within industrial image processing, correlation analysis is predominantly used to identify given patterns in an image of which the position and orientation are presumed to be unknown. With this procedure, known as *template matching*, a point-by-point search is undertaken for the existence of a particular, defined pattern (*template*). By comparing the image information with the ideal template, a similarity measure is obtained, which can be interpreted as a probability that the desired template corresponds with the environment of the point. As the points within the environment of every pixel are compared with the template through cross correlation, the calculation complexity of such procedures rises quickly with an increasing window size, so that a multi-level procedure must be used in these cases. Template matching is used, e.g. when checking the automatic assembly of electronic components.

Feature Extraction and Classification Procedures

For production metrology, it is crucial that a statement can be made with the acquired image information, e.g. about the dimensional accuracy of a workpiece, the size of a drill hole or the existence of a groove. For this purpose, the components must be classified according to a set procedure. Classification means determining whether or not a particular feature lies within set tolerances

Depending on the task, however, substantially more complex classification methods can become necessary. For the case describe above, the feature space is one-dimensional (1 feature) with one class (good or bad), within a framework of tolerance limits.

In general, however, several independent features of a workpiece must be inspected at the same time and it can possibly be required to accept and distinguish between components of differing quality. In this case, the feature space can be high-dimensional and consist of several classes (e.g. 1st choice, 2nd choice, reject). Examples of class features can be, e.g. contour length or the angle difference between two corner points. The difficulty now lies in the fact that, in industrial reality, the features are also subject to measurement value dispersion, even with objects that are actually identical. In addition to that, slight deviations in components can already lead to deviations in feature values with set lighting.

The task of classification now lies in determining to which of the previously determined reference templates an image pattern belongs.

Apart from the features "diameter of a drill hole" or "existence of a groove", more abstract quantities are also described as a feature in image processing, such as the local variance of grey tones or the energy content of an image. These features are often called *textural features*, as they are very important in texture analysis. These textural features are of great importance in the area of technical visual inspection, e.g. if a scratch is to be detected and automatically measured on a component which is otherwise similar in appearance. Important textural features such as *local grey tone variance, inclination, energy* or *entropy* are based on statistical methods of image evaluation [Zam 91]. In addition to these procedures based on statistics of the first order, procedures based on higher order statistics, such as *co-occurrence matrices* have only obtained little importance in production metrology, due to the usually high level computing time required, although they possess great potential from the point of view of image processing.

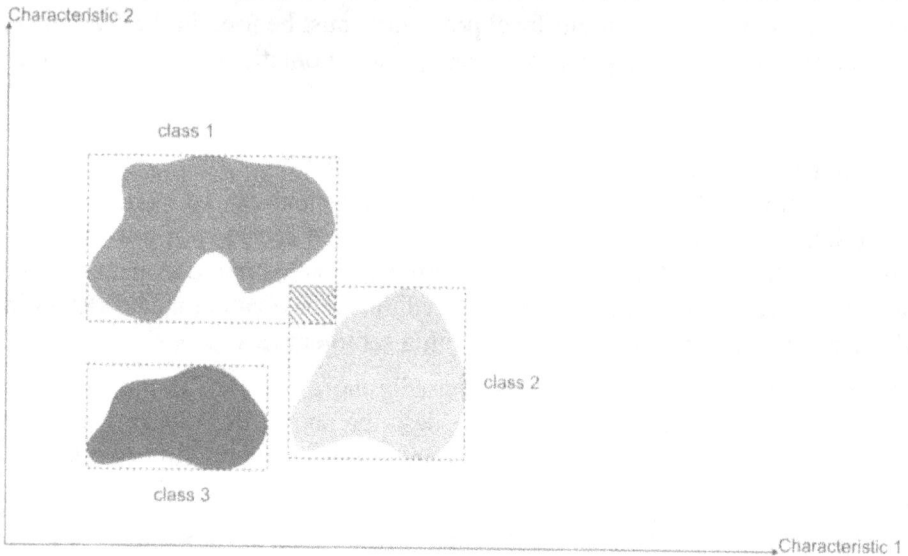

Figure 4.3-18: Classification according to the cube classifier

Following the calculation or detection of features, a classification generally takes place according to these features. In order to do this, different classification procedures can be used. Known standard classifiers are:

- The *minimum distance classifier* (minimum distance method):

  With reference to the centre of a sample class the Euclidean distance is used as a distance measure. The geometrical analogue of this classifier is a high-dimensional ball in the feature space. All points inside of this ball belong to the sample class; while all points outside do not belong to the sample class.

- The *maximum likelihood classifier* (maximum likelihood method):

It is based on a statistical foundation where the sample classes are described by n-dimensional distribution or density functions.

- *Geometrical classifiers*: The cube classifier, as the simplest representative of geometrical classifiers, has the advantages of easy implementation and short computing times, as the testing of whether a feature lies within an n-dimensional cube only consists of comparison operations.

A prerequisite for a unique classification is the overlap flexibility of the individual classes in the feature space **(Figure 4.3-18)**. If an overlap of two classes occurs, either further features must be introduced or another classification procedure must be selected.

4.3.2.3 Industrial 2D Image Processing

In this section, several example applications of industrial image processing are explained from different branches of industry. The selected examples demonstrate the wide spectrum of applications which are covered by industrial image processing, from technical visual inspection of metallic surfaces for the monitoring and control of handling procedures to workpiece and product testing using the example of engine blocks.

Technical Visual Inspection of Metallic Surfaces

As efficient as image processing systems are, the required engineering complexity for their implementation is often underestimated. This is because the fast and unreflected comparison with our efficient human vision leads to the fact that other particularly important boundary conditions for image recording are overlooked. Particularly in the area of technical visual inspection, one often comes across polished or metallic surfaces, where reflections or shadows can lead to problems and therefore require special attention when laying out image recording parameters such as the arrangement of lighting or the image definition. These questions must be considered in advance of a system solution, in order to ensure reliability of the measuring and inspection systems. As an example, **Figure 4.3-19** shows two indexable inserts in top illumination, which are lit up from the side with directed light.

Their surfaces have engravings, which only stand out in contrast with the remaining surface under certain directions of illumination, furthermore recesses in a countersunk hole, metallic reflections and much more. It is clearly recognisable that much of the image information in the edge image is lost due to this poor illumination arrangement, partly because of the brightness saturation in reflective areas or darkening in shadows. It would no longer be meaningful to carry out many conceivable measuring or testing functions here. This example shows that image processing is much more than just the application of different numerical procedures to stored images. It already begins with the actual recording of the image. Whatever is not contained within the original grey tone image can not be evaluated later.

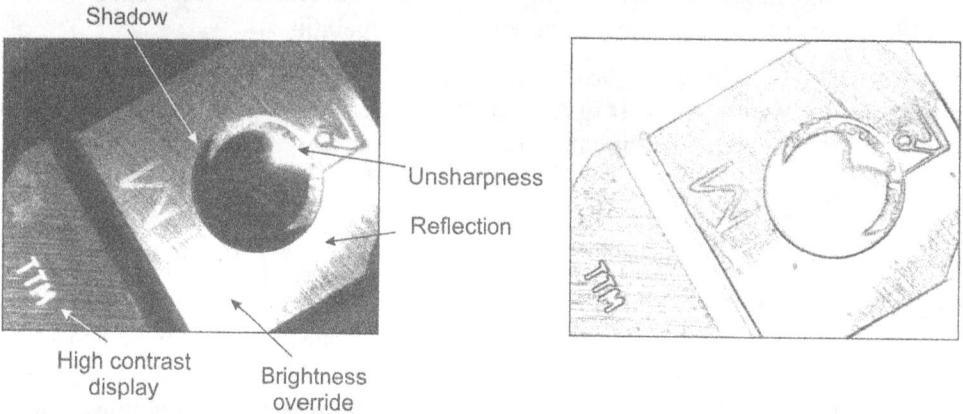

Figure 4.3-19: Example of reflective surfaces with poor illumination, left: grey tone image, right: edge image after Sobel filtering

In contrast, **Figure 4.3-20** shows the same scenery after optimising the illumination arrangement and the edge image generated in the computer. Now the described contours of the features of interest in the image can be evaluated with certainty, as all contours are complete and further disturbing effects will not occur in the image. The target contours are represented as a comparison.

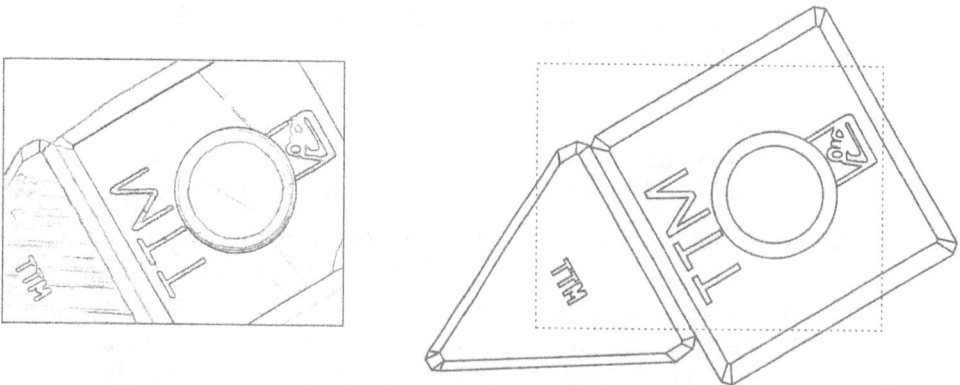

Figure 4.3-20: Edge image after optimising the image recording parameters (left); representation of the target contours as a reference (right) for an example from a CAD database

Test functions here would be, for example, the identification of the existing index-able inserts on the basis of the engraved letter combination and the testing of the correct diameter and position of the drill holes as well as the investigation of surface defects such as scratches, which can be carried out as a function of the detected type. Such measurement and testing functions can only be automated through optoelectronic measuring procedures such as image processing.

Even if continuous production requires a high clock rate, image processing systems can keep up here, as after the actual image recording, which occurs very quickly, the image evaluation in the computer can take place in the background, while a new insertion takes place in parallel via a handling system.

An application which is only enabled through optimum illumination is the measurement of tool wear with milling tools **(Figure 4.3-21)**. The recording of tool status prior to processing provides valuable information, not only for process monitoring, but also possibly for process control, as well as for useful life of tools. The causes of wear, mechanical abrasion, diffusion processes, built up edge formation and scaling, lead to geometrical modifications on the face and flank of the tool, which can be recorded using an incident light inspection system.

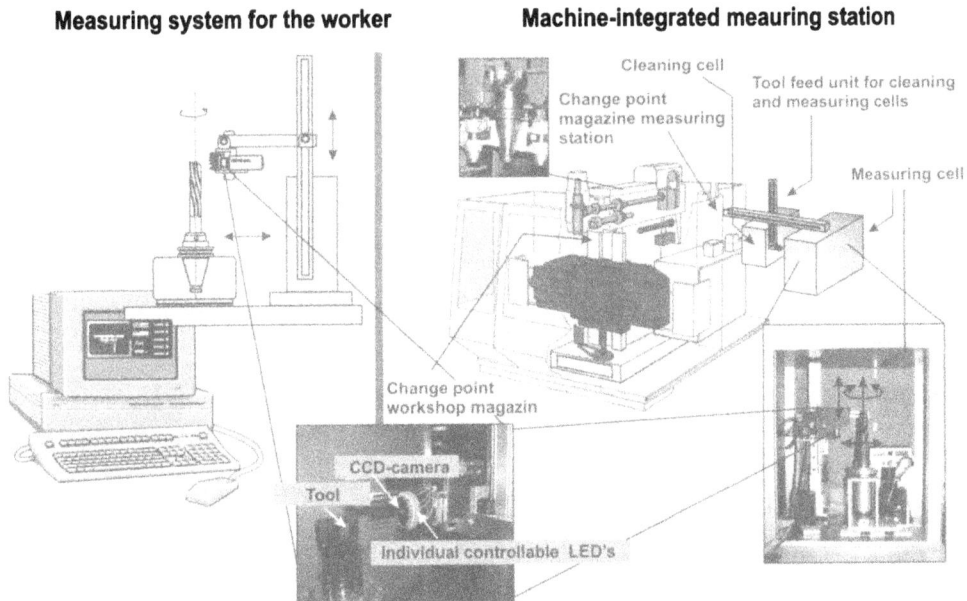

Figure 4.3-21: Tool wear measurement

Various scenarios are conceivable for the application of an automated wear measuring station in production. A manually fed wear measuring system is used in tool preparation, in order to determine service life optimising specifications for regrinding the tools, or it is made available to the workers as a measuring island in order to undertake an automatic and thus repeatable check of the tools, in parallel with processing. A fully automatic wear measuring system can, however, be applied directly at the tool magazine of a tool machine, as long as it is protected from dust, chippings and oil mist. Through a control command, a tool is submitted to automated tool wear measurement in specific intervals after being processed and reinserted in the tool magazine.

Automatic Monitoring and Control of Handling Processes

For the production of numerous products it is necessary to position two parts exactly in relation to one another (e.g. two workpieces or a tool and a workpiece) prior to a subsequent processing step. Often this requires first recording the position of at least one workpiece metrologically. This frequently takes place with intelligent image processing systems, the results of which are used for further process control or the control of handling systems.

While camera controlled handling is already widely spread within the automobile industry (e.g. bonding work with camera controlled industrial robots), the use of camera metrology for the processing control in other branches of industry is much less common, although the potential exists.

For this reason, the application example introduced here is selected from a branch of industry which has so far been characterised by manual or semi-manual manufacturing methods.

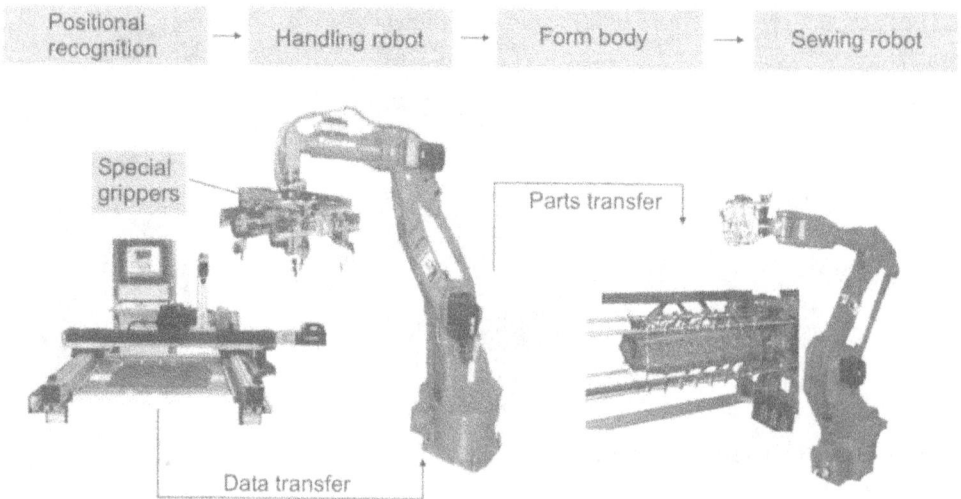

Figure 4.3-22: Image processing for automatic position detection of pre-cut textile parts

In the area of textile industry, there have hitherto been a high proportion of manual processing steps required, especially in the subsequent processing of pre-cut parts to final products. In the early 1990's, this industry already recognised the necessity to automate in order to ensure competitiveness and therefore developed new manufacturing concepts. Such innovative manufacturing methods, based on three dimensional production of textiles with the aid of industrial robots is currently in the development phase. With this manufacturing concept, the production of a textile product takes place in several successive production steps, beginning with automatic cutting to the finished sewn end product. In particular, prior to joining several pre-cut parts with the aid of an industrial robot, they must be exactly positioned. This requires metrologically recording the position in order to subsequently bring the pre-cut parts into the desired target position with the aid of a handling system. Image processing systems are particularly suitable for this task. By combining the image processing system with a positioning unit, the position of the textile part to be handled can be recorded and transferred to the handling system control (e.g. industrial robots) **(Figure 4.3-22)**.

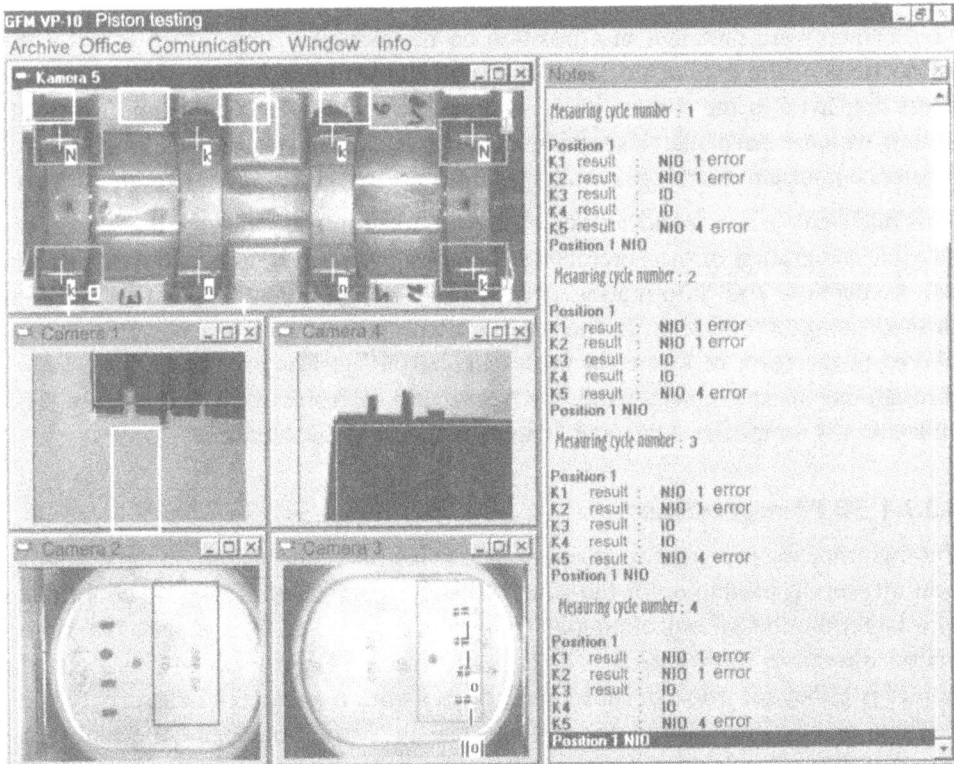

Source: Gesellschaft für Meßtechnik (GFM); application: Mercedes Benz, Projekt NVM

Figure 4.3-23: Automated visual inspection of piston rings in assembly

In this way transfer of the pre-cut part into the final position for finishing (sewing process) can take place through the position of defined grab points.

As a large spectrum of different textiles will be used, the image processing system must be in a position to ensure equally certain recognition of different materials, colours and even patterns. Such functions can generally not be realised through simple operations such as high-pass filtering alone, but usually require the simultaneous evaluation of several independent textural features in order to ensure certain object recognition.

Automated Visual Inspection of Engine Blocks
Figure 4.3-23 shows a screen display of the control surface of an image processing system, where various inspection functions are carried out on engine blocks. This concerns such differing functions as inspecting the assembly of piston rings and the appropriate position and correctness of impressed type markings such as the existence of further grooves and markings, which are important for subsequent processing and assembly processes (recognising the position of the piston).

Five cameras are mounted in a position on the assembly line, which monitor the subsections of the engine block which are of interest in each case. The rectangular areas displayed in the camera images define the so-called region of interest (ROI) within which a particular inspection function is carried out, as well as additional reference lines and markings which visualise fast inspection results for the worker.

This highlights a special advantage of image processing systems for the worker accessible integration of measurement technology, which lies in the similarity of visual perception and information processing through human vision as well as through image processing. In the case of disturbances, they can not only be displayed in the form of text error messages, but are just as immediately visualised through the camera image, which the worker can interpret easily and quickly. This substantially simplifies a fast and adapted reaction to disturbances.

4.3.2.4 3D Photogrammetry

Photogrammetry is an optical imaging measuring procedure based on the triangulation measuring principle, for the simultaneous recording of 3D spatial coordinates of a larger number of object points. The object is recorded from at least two observation directions with analogue or digital cameras **(Figure 4.3-24)**, whereby the object is either recorded in subsequent frames with one camera from different directions or with several cameras at the same time, using one camera per observation direction. The 3D coordinates of the individual object points can be calculated absolutely in the space, by bringing the appropriate display beams defined by 2D pixels to the section. By including further observation directions, redundancy is

increased during measurement recording, so that measurement uncertainty is reduced and measuring problems due to shadows and undercuts are minimised.

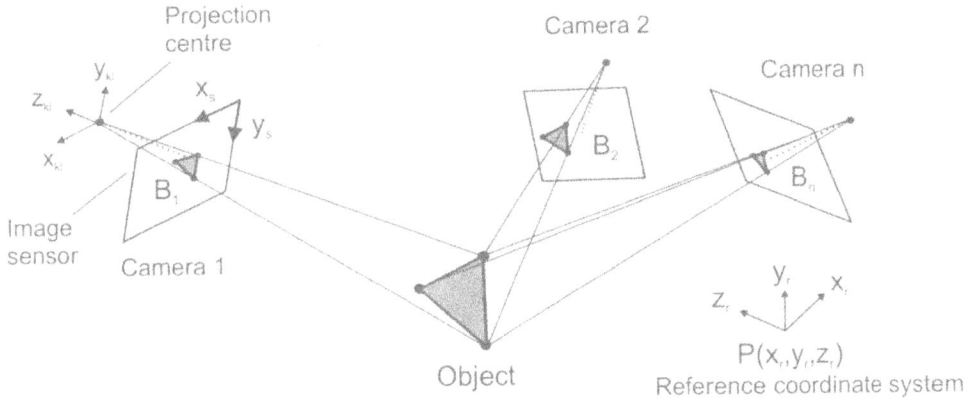

Figure 4.3-24: Principle of photogrammetry

Homogeneous white-light can be used for lighting purposes. Within the context of
2D images, by using digital image processing algorithms, it is possible to measure
either displayed object features (such as edges or drill hole center points) or markings applied and shown on the object (such as circles or lines) as in the majority of
applications. Alternatively, the object can be measured with structured white light,
such as cross section grids or stochastic patterns, so that the time consuming application (bonding, etching) of markings is omitted. Further advantages of illumination with cross section or point grids and the use of LCD or mirror array projectors
are the flexible adaptation of the template to the individual measuring function
with regard to measuring points and measurement point density, a well as understanding the projector as an 'inverse camera' and including it in the measurement
as an additional observation direction.

A prerequisite for determining the 3D coordinates with a small measurement uncertainty is an illustrative model, which describes the measuring system as accurately as possible, as well as a calibration strategy which is simple to carry out, in
order to determine the model parameters. With the aid of the central projection
model and the camera constants as representative parameters, the display beams
which are defined by the individual 2D image coordinates and the appropriate projection centres, are described mathematically. In order to reduce the measurement
uncertainty, systematic display errors are also modelled using additional parameters. Such display errors are, e.g. main point shift, radial symmetrical, radial
asymmetrical and tangential distortions of the objective, as well as affinity errors.

In the context of calibration, these parameters, as well as the parameters describing the position and orientation of the individual recording viewpoint, must be determined with respect to a reference system. If the measuring system consists of one camera per recording direction which is structured with sufficient mechanical stability (**Figure 4.3-25 left**), calibration takes place once after structuring or modifying the system. For this, a three dimensional reference body is positioned in different positions in the measuring volume (**Figure 4.3-25 right**). From the 2D image coordinates measured on the circles applied on the body, all parameters are automatically calculated within the context of bundle block compensation.

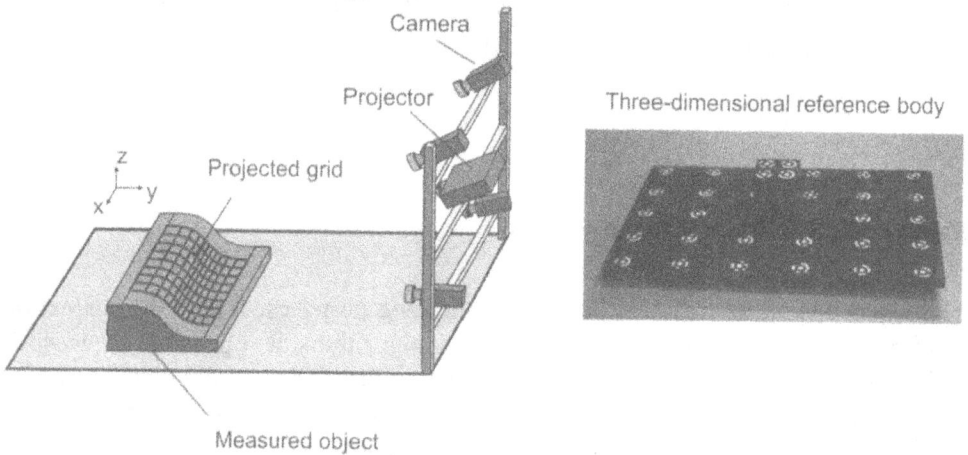

Figure 4.3-25 Photogrammetric multi-camera system with the appropriate 3D calibration body

Should, however the recording of the object take place with only one camera, successively from different directions, the parameters are automatically determined when calculating 3D object coordinates using 2D measurement values in the context of a simultaneous calibration. With both procedures, the measuring link results from the additional recording of calibrated paths.

The measurement uncertainty of photogrammetric measuring procedures is determined by the representation scales of the camera (relation of camera constant and object distance), the position and number of recording viewpoints and the uncertainty with which the two dimensional image coordinates can be measured. A prerequisite for precise measurements are high-contrast and well defined displays of the object features, so that the highest accuracies can be achieved when measuring adhesive markings. As with all optical procedures, when using structured illumination, the measurement uncertainty is also influenced by the reflectivity of the object surface, whereby diffuse surfaces are most favourable.

Due to the low relative measurement uncertainty of minimum approx. 1:70,000 with digital cameras and 1:200,000 with analogue cameras (standard deviation in reference to the length of the measuring range diagonals) [Bre 93], the short recording times of minimum one image interval, as well as flexible applications, photogrammetric measuring systems have increasingly gained in importance in production metrology. Particularly the progress regarding the resolution of modern CCD cameras (micro and macro scanning, 4Kx4K chip) and the efficiency of evaluation computers enables economical automated systems with short measuring times.

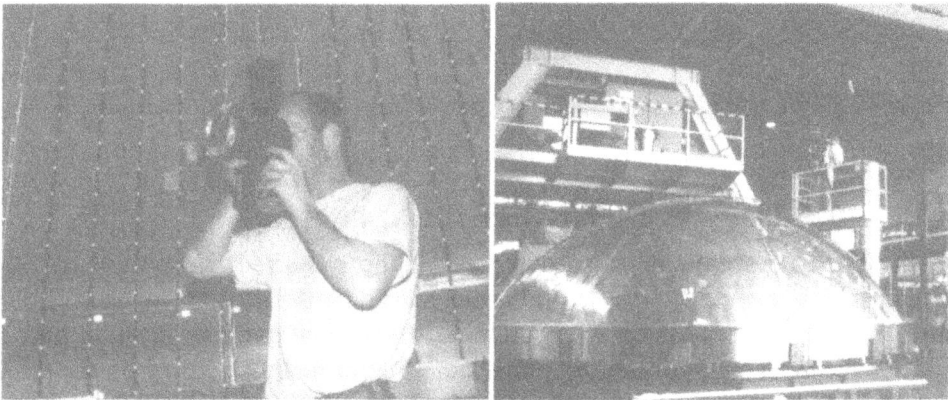

Quelle: GDV Ingenieurgesellschaft

Figure 4.3-26 3D measurement of the tank lid and base of the Ariane 4

A main field of application within the context of quality assurance, is the measurement of large objects, especially on the areas of space travel, airplane-, ship-, and plant-manufacturing. Photogrammetric measuring systems are used as mobile optical coordinate measuring machines, whereby surface points with markings and drill holes are signalled with marking adaptors. The object to be measured is recorded with one camera successively from different directions, whereby generally short exposure times are used by applying flash lighting **(Figure 4.3-26)**. With such systems, measuring times (bonding, recording, evaluation) of a few hours are achievable.

A further field of application is the form acquisition of smaller objects ($< 1 m^3$) with relative measurement uncertainties of approx. 1:15,000 within a few seconds. Stationary multi-camera systems with structured illumination are used for this purpose, with possible recording times of a few milliseconds [Sha 97].

These can be used for positional determination and identification in handling processes and well as for quality assurance within the manufacturing cycle time **(Fig-**

ure 4.3-27). Additional fields of application are the digitising of shapes and models, as well as the acquisition of object deformations.

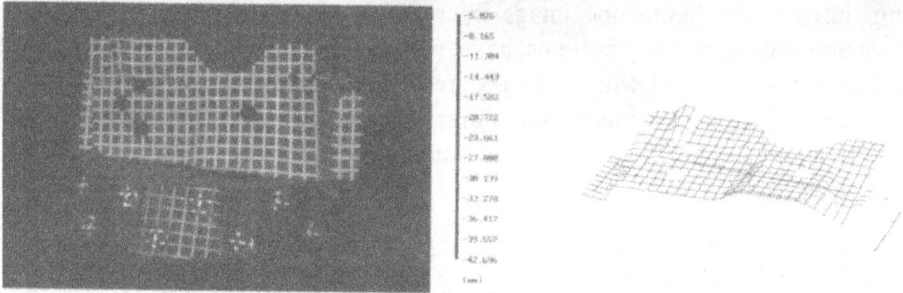

Figure 4.3-27 3D measurement of a sheet metal part for quality assurance (grey tone image, 3D plot)

4.3.2.5 3D Stripe Projection Procedure / Moiré Procedure

With 3D stripe projection, also described as the projected fringe technique or topometric procedure, a periodic grid is projected onto the object from a projection direction and recorded extensively from a different observation direction with a CCD camera. The equidistant stripe pattern is deformed, depending on the three dimensional form of the object **(Figure 4.3-28)**, so a three dimensional coordinate can be calculated for each pixel by using triangulation methods. As with photogrammetry, a prerequisite is knowledge of the exact position and orientation of the projector and the camera, as well as further representational parameters. These can be determined, for example, with a photogrammetric calibration using a 3D reference body, as shown in **Figure 4.3-25 right,** or through repeated measuring of an individually defined shifted level.

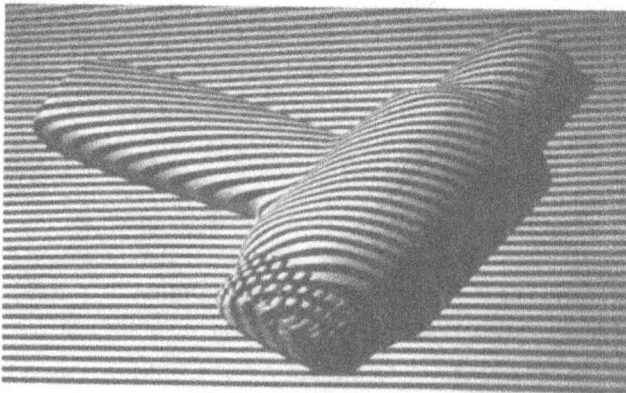

Figure 4.3-28 Deformed equidistant stripe pattern on a housing component

For the quantitative evaluation of a stripe image, similar evaluation procedures to interferometry are used in order to determine the phase position of the respective stripes and the stripe number [Bre 93]. The phase position can, for example, be determined using the carrier frequency or the Fourier transformation procedures, whereby phase accuracies of only approx. 1/20 of the stripe width are achievable.It is to mention that, only one stripe image must be recorded and evaluated (static stripe projection), so that moved objects can also be measured extensively.

An increase in the phase accuracy of up to 1/100 of the stripe width is achieved by applying a dynamic procedure, whereby the phase shift method is most often used. The stripe pattern is shifted gradually within a stripe period and the phase position determined from the individual recorded intensity values for each pixel. Three stripe images are usually required for this, so that only resting objects can be measured. The evaluation procedures for calculating the phase position always requires a sinusoidal intensity distribution of the stripe pattern. With fine line structures, this can be approximated by more economical rectangular line patterns due to the low-pass behaviour of projection optics, although the phase noise increases.

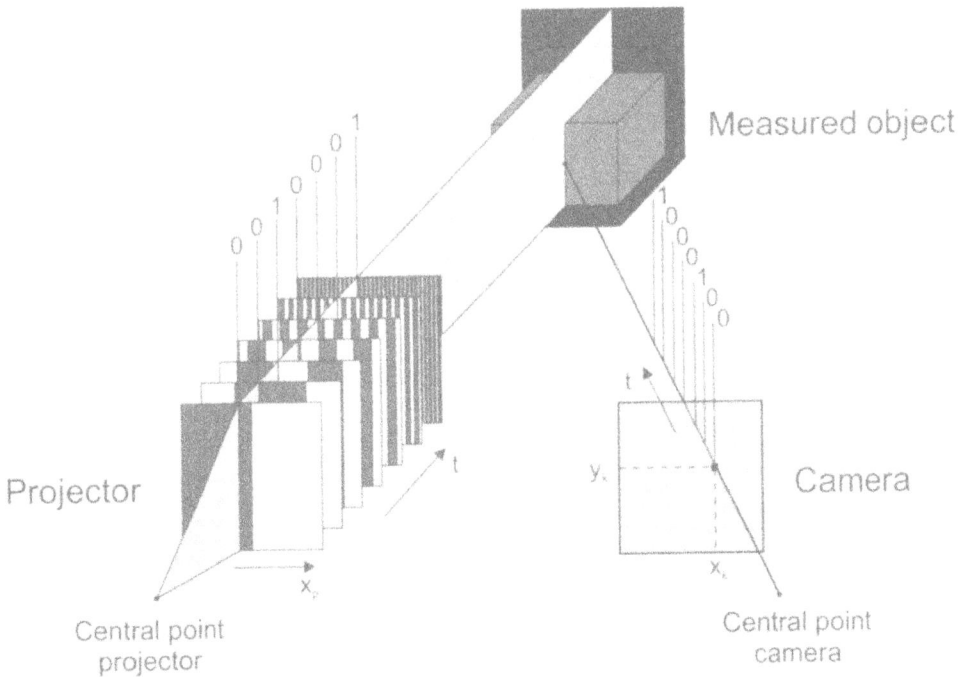

Figure 4.3-29 Principle of the encoded laser cut procedure

Determining the stripe numbers (counting) is only possible relative to each other, due to the absence of an absolute reference point, so that only connected continuous surface areas can be measured using the appropriate demodulation procedures. As however, shadows, undercuts and discontinuities occur in the surface, when measuring three dimensional objects using the stripe projection procedures described above, many applications are expanded to an absolute measuring system by combining them with an encoded laser cut procedure. The object is illuminated with several different line patterns and the course of the measured intensity is recorded for every pixel of the camera. This is used to form a code word, from which the stripe number of the projector can be uniquely decoded. A binary grey code coding is most frequently used **(Figure 4.3-29)**.

Different stripe patterns are projected with the aid of LCD display projectors or with slide projectors, whereby the patterns are applied to a common glass substrate and successively projected by shifting the glass substrate.

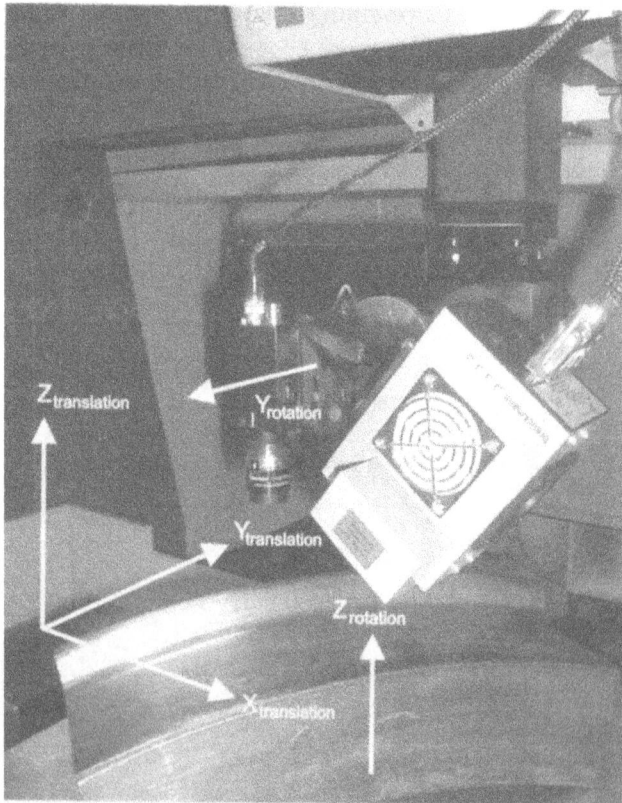

Figure 4.3-30: Stationary stripe projection system at the axis portal

Due to the complexity of the three dimensional form, the variable reflectivity and size of the measured objects often necessitates the linkage of different object views for recording object topography. Stationary system linear modules are used for this. Through the positional data of the axles and angle gauges, the individual measuring ranges of different object views can be transformed into a well ordinated coordinate system. In this way it can be ensured that even complex workpiece geometries are completely recorded **(Figure 4.3-30)**. A further advantage of this procedure is the possibility of using a relatively small measuring range of the stripe projection system. Depending on the principle, the measurement uncertainty of stripe projection systems scales down linearly with smaller measuring ranges. Thus, large workpieces with a small measurement uncertainty can be recorded optically by applying linear modules and their measuring link in comparison with the measuring range of the optical system.

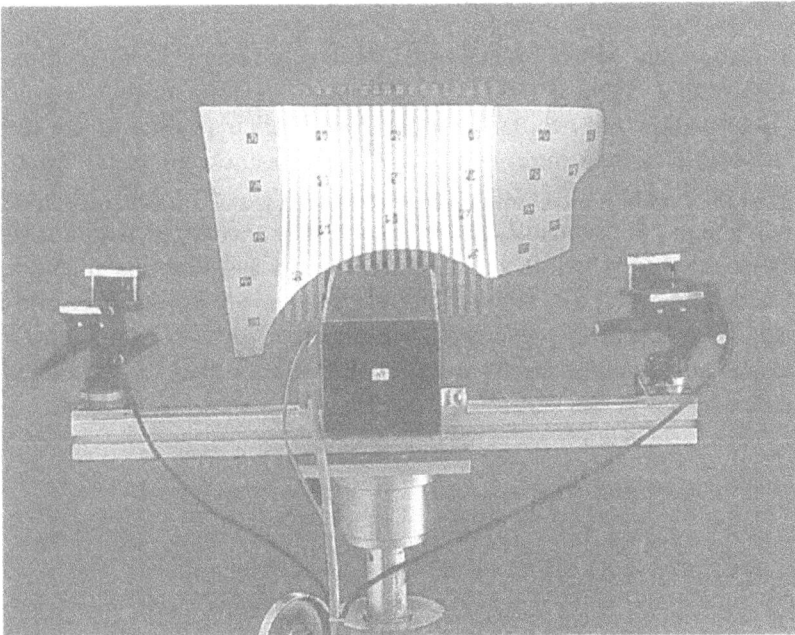

Figure 4.3-31: Mobile stripe projection system

With mobile systems, which are characterised by lower costs and higher flexibility, the total object can be recorded by measuring additional code markings applied at or near the object and the subsequent photogrammetric linking of the partial views **(Figure 4.3-31)** [Sha 97].

The measurement uncertainty of stripe projection systems is larger when compared with photogrammetric systems and is similarly determined through the geometry of the recording arrangement (triangulation angle), the representation scale of the camera and the reflectivity of the surface, whereby diffuse surfaces are most favourable. Areas which are too low in stripe contrast, or too dark or bright can therefore only be measuredwith a smaller degree of accuracy, are automatically masked out as being invalid.

Measurement uncertainty can always be reduced by minimising the stripe width. The stripes must, however, have a minimum width due to the limited resolution power of the camera. However, with the aid of the Moiré technique, it is possible to over come this limit and improve the stripe width, and with it the resolution, by factor 10 [Bre 93]. The stripe pattern on the object is not observed directly, but through a reference grid (projection Moiré). If the stripe densities of the projected grid and the reference grid are coordinated, a greater width results from the optical overlay of the Moiré lines, which can then be resolved by the camera again **(Figure 4.3-32)**.

Stripe projection and Moiré procedures extensively record the three dimensional surface topography. Typical fields of application are form testing (e.g. evenness testing), deformation and oscillation metrology. Particularly the Moiré technique, thereby enables a simple interpretation of the measurement result when testing form deviation. A grid is projected on a master part and recorded with a camera as a reference image . Subsequently, the master part is exchanged with the part to be tested, the position of which is adjusted accordingly and illuminated and recorded with the same grid. With optical overlay (reference image as reference grid in front of the camera) or electronic overlay (multiplication of both images) Moiré lines are developed at the site of the form deviations, later on they can be directly visualised and evaluated in this way.

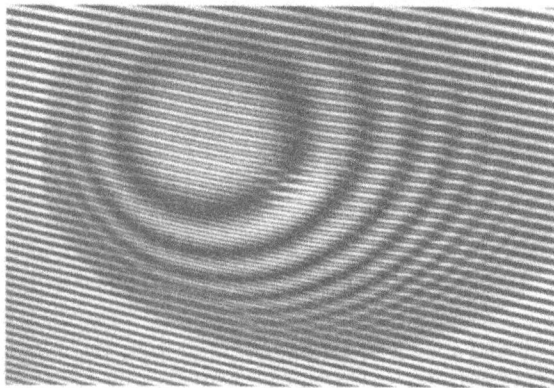

Figure 4.3-32: Moiré pattern of a ball

A further emphasis of the stripe projection procedure is the digitisation of free format surfaces within the context of model and mould making and rapid prototyping **(Figure 4.3-33)**. In subsequent processing steps, with the aid of appropriate surface feedback programs, CAD data is generated from the measured point clouds and from this, NC programs are generated for processing machinery.

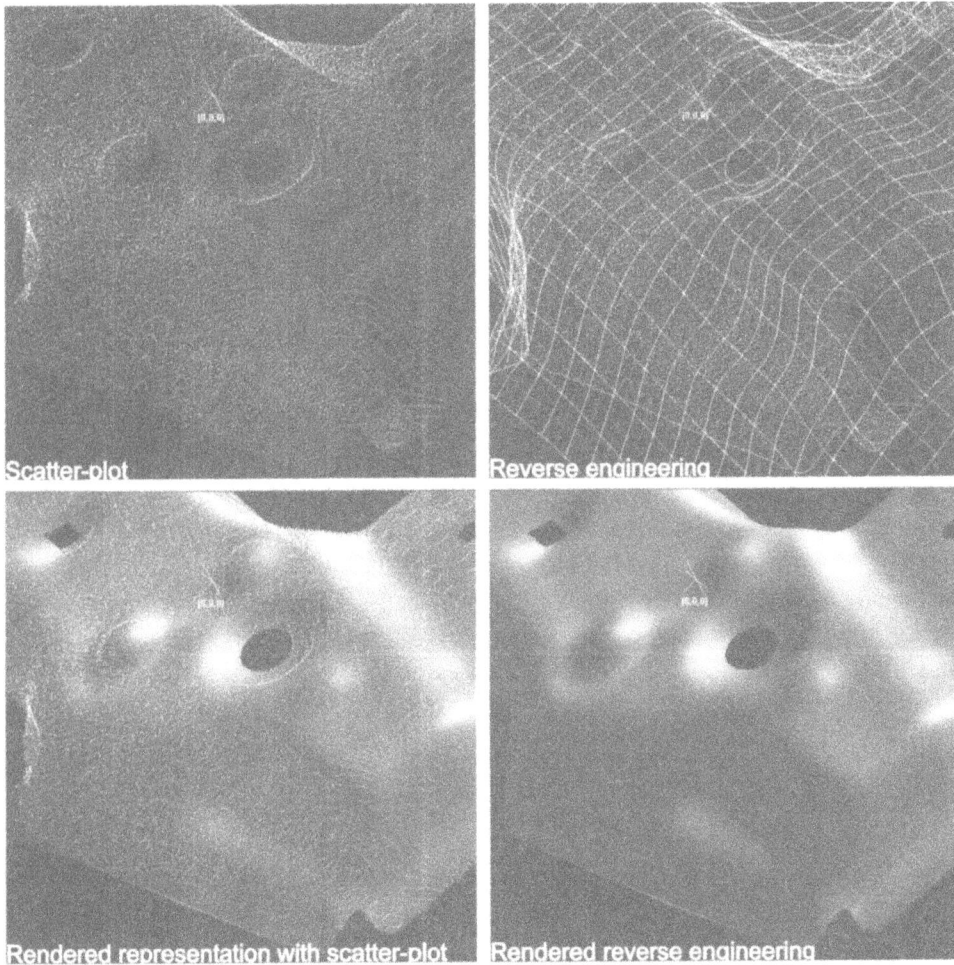

Figure 4.3-33: Digitisation and surface feedback for three dimensional object geometries

4.3.3 Laser Metrology

4.3.3.1 Triangulation Procedure

Distance Measuring Laser Triangulation
Triangulation sensors which are based on the laser triangulation principle today belong to the most common optoelectronic distance measuring systems.

Although technical literature has been describing triangulation sensors for more than 10 years, they have only achieved a strengthened industrial breakthrough during the past few years, as their technical advancement is closely connected with the progress of semi-conductor technology.

The rapid development of laser diodes, detectors, analogue signal processing and digital signal processing enabled the conception of compact, fast and disturbance-insensitive sensor systems. In terms of signal technology, electrotechnical sensor hardware has achieved a mature stage and found a continually growing acceptance in many assembly and packaging lines through mass production.

The triangulation (lat. triangulum = triangle) measuring principle is based on the determining of a side of a triangle by determining two triangular angles while knowing the length of the triangle side included by them **(Figure 4.3-34 left)**. If the length of the segment AB is known, the distance AC can be determined by measuring the angles α and γ. With the technical realisation of a laser triangulation sensor, the light from an appropriate laser source is focused on the workpiece surface through beam forming optics and the resulting point of light is displayed on a position detector.

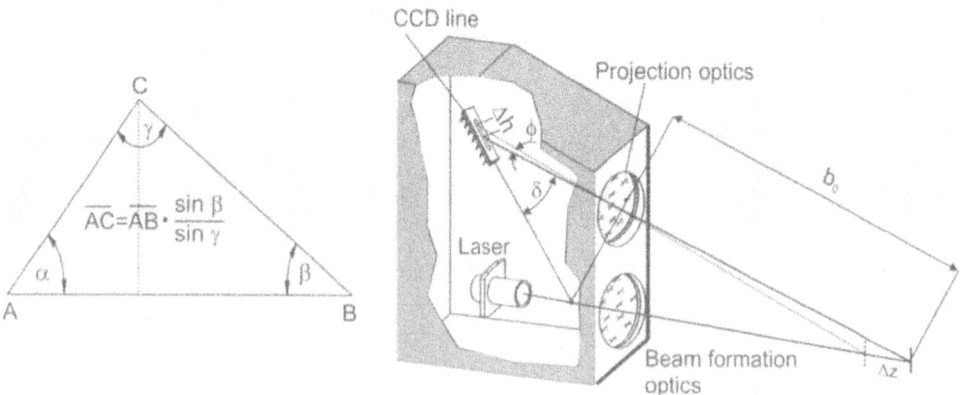

Figure 4.3-34: The triangulation principle and the structure of a triangulation sensor

The image beam path is thereby arranged under the triangulation angle γ to the light axis **(Figure 4.3-34 right)**. If the distance of the measured workpiece surface changes, the display position of the projected light point also changes on the detector, the length of which limits the measuring range. In order to avoid an unfocused display of the measuring light point on the detector, which is caused by the change in the representation ratio within the measuring range, the line detector must be arranged according to the Scheimpflug principle.

The detector is inclined at the angle δ, so that the beam axis of the light source, the level of the representative optics and the detector level intersect at one point [Moh 90].

A Δz shift of the measured object from the base distance along the light beam axis leads to a shift of distance Δh by the measuring mark displayed on the detector.

$$\Delta h = b_0 \cdot \frac{\sin \phi}{\sin(\delta - \phi)} \tag{4.3-6}$$

Due to their size, performance, and beam quality, diode lasers are usually used as sources of light.

Depending on the required resolution, measuring range size and application, either position sensitive diodes (PSD) or CCD lines (charge coupled device) are used as line detectors (Section 4.3.1). While the advantages of the CCD line are evident in an application-specific digital evaluation of line information, substantially higher resolutions of small measuring ranges and a higher measuring frequency can be achieved with position-sensitive diodes.

With laser triangulation, the attainable measurement uncertainty depends on the surface texture of the sample. With an ideally diffusely dispersed surface, a Gaussian distribution of light intensity results on the line detector. The better the intensity distribution on the line detector (correlated with the ideal Gaussian distribution) the lower the influence of the surface on the measurement uncertainty. With surfaces which indicate diffuse reflective characteristics such as paper or ceramics, a very low measurement uncertainty can be attained. With strongly reflective surfaces, here metallic surfaces with very little roughness can be included and surfaces which the laser beam penetrates, such as various plastics or glass, the surfaces substantially influence measurement uncertainty. Furthermore, attention must be given that the workpiece does not creating shadowing of the transmission and reception beams.

The sensors which are used in production technology have a typical measuring range of 1 mm to 100 mm with a working distance of 10 mm to 100 mm. Accordingly, the attainable resolution is 0,5 µm to 50 µm. The achievable measurement uncertainty, as described above, depends on the type of surface. With the above mentioned measuring ranges and a ceramic surface it is approx. 1.5 μm to 150 μm. The light spot diameter varies from 20 μm to 250 μm. It is a standard procedure to carry out a laser power adjustment to adapt the light performance to the available surface.

Numerous industrial applications evidence the high place value of laser triangulation. For example, with an arrangement as shown in **Figure 4.3-35**, the thickness of a paper tape can be tested on-line in the production process. With an arrangement as represented in **Figure 4-3-35**, for example, radial run-out measurement is used to check the roundness of a rotating shaft. Laser triangulation is also applied for functions which require recognition of a specific component or component feature **(Figure 4.3-35)**.

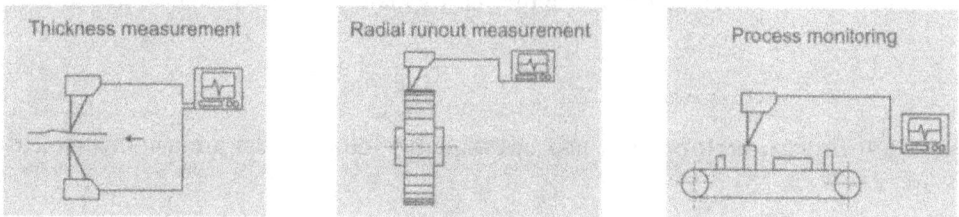

Figure 4.3-35: Practical examples of laser triangulation

4.3.3.2 Travel Time Procedures

The travel time procedure is suitable for distance measurement, particularly for large measurement segments. What is measured here is the time which an amplitude modulated luminous laser beam requires to pass through a measurement segment and back. From this time interval, the travelled path can be calculated with a known speed of light. As very short time periods must be measured due to the high speed of light, the phase measuring procedure is often used, apart from the quite complex direct time measurement.

Figure 4.3-36 schematically shows the typical beam path of a travel time measuring system. The high frequency amplitude modulated laser beam is projected on the workpiece surface through two movable deflecting mirrors and reflected back diffusely. The reflected luminous beam is then focused with a collecting lens and recorded by a measuring detector. This luminous beam is then compared with a signal, which has been directly uncoupled from the laser beam, in order to determine the phase change.

The measuring range of the travel time procedure reaches from 0.2m to 1 km and permits a resolution of approximately 1 mm. It is, however, at times heavily impaired by environmental influences (such as air pollution or object surface). Typical fields of application are positional, distance or fluid level measurements, but 3D surface recording systems can also be realised through the application of movable reflecting mirrors.

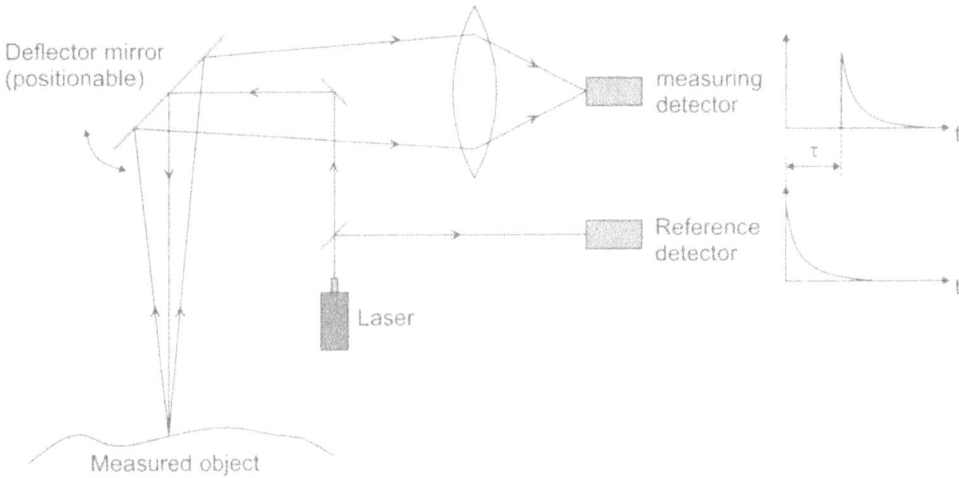

Figure 4.3-36: Travel time procedure

4.3.3.3 Autofocus Procedure

The term autofocus procedure refers to measuring procedures which carry out the distance measurement with calibrated object-sided focal length due to the sharpness definition of the image. Two procedures can be differentiated: the video autofocus procedure and the laser autofocus procedure.

With the *video autofocus procedure,* the distance measurement is carried out by reflecting the measure object on a CCD surface sensor, determining the degree of sharpness of the refflection by determining the contrast and reaching a conclusion about the measuring distance from the required refocusing. This measuring procedure is applied, for example, with CNC controlled optical three coordinate measuring machines. The video autofocus procedure will not be detailed further in the following section, since the sole emphasis will be on the laser autofocus procedure.

Principle of the Laser Autofocus Procedure

The operational principle of the *laser autofocus procedure* described and applied as below, has found widespread use through the compact disk technique.

The coherent light of a laser diode is divided into its main vertical swing directions. The proportions of light swinging in the direction of polarisation can pass the diagonal direction of the dielectric multiple layers generated by the beam splitter cube, while swinging light is reflected in the reverse bias.

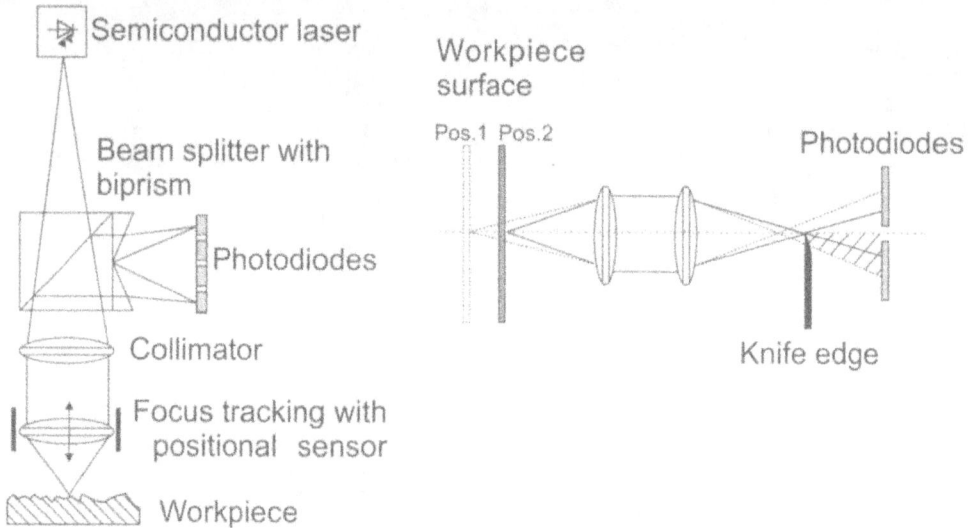

Figure 4.3-37: a) Sensor principle of the autofocus procedure b) Foucault knife-edge principle

At the output of the beam splitter almost linear polarised light is available, which is parallelised by collimator optics and subsequently transferred from a $\lambda/4$ disk into circular polarised light. The measuring objective is stored in a moving coil and focuses the laser light on the measured object. Measuring-light reflected by the workpiece surface passes again through the measuring objective and the $\lambda/4$ disc. The resulting polarisation vector is now rotated 90° around the beam axis, so that the measuring light is reflected at the beam splitters towards the focus detectors. **Figure 4.3-37 a** clarifies the sensor principle of the autofocus procedure.

Various procedures can be used to influence beams in the direction of the detector. With the *"knife edge procedure"* or *"Foucault knife-edge principle"* **(Figure 4.3-37 b)** one half of the optical beam path is shaded with the aid of an optical cutter. Depending on whether the beam meets the measured object intrafocally or extrafocally, the left or right part of the reflected measurement beam is displayed on a double photodiode.

In place of a cutter, a wedge prism is often used in conjunction with two pairs of photodiodes. It is also possible to use a cylindrical lens in conjunction with a pair of photodiodes.

As the principle-dependent measurement error development is independent of the beam influence through the different optical elements, signal evaluation will be described below, representing the most often used sensor configuration with a double wedge prism (biprism). The advantage of the biprisms is that no intensity loss develops from the fading out of half of the backscattered light bundle and leads to a simplification of the sensor adjustment.

In order to regulate the movable measuring objective on a constant workpiece distance, a focus error signal FE is determined from the four photodiode halves:

$$FE = \frac{(I_{D1} + I_{D4}) - (I_{D2} + I_{D3})}{I_{D1} + I_{D2} + I_{D3} + I_{D4} + C}$$

Accurate focusing on the workpiece surface results in a focus error signal of zero (principle of optical balance). With a shift of the measured object in the direction of the optical axis from the focus, the asymmetrical illumination of the respective pair of photodiodes leads to an electrical error signal, which contains the directional information of the distance change as well as the amount.

A control loop is realised, which refocuses the movable suspended measuring objective through an electromechanical drive on the workpiece surface. The objective shift is measured with an appropriate measuring system (e.g. optical or inductive) and emitted as a measurement value.

The distance measurement generally takes place independent of the reflected light intensity; however a minimum reflectivity is required of approx. 2% of the measured surface.

Regulation of the laser output performance can be carried out for adjustment to different reflective conditions with the aid of the diode composite signal $R = I_{D1} + I_{D2} + I_{D3} + I_{D4}$.

Fundamentals of Sensor Dimensioning
The focus diameter of the measuring light beam is approx. 1- 5 μm, depending on the realisation. According to the reflectivity of the measuring surface, the required output power of the laser diode is between 100 and 300 μW.

In order to realise a high sensor sensitivity, the sharpness depth range of the optical display

$$a_h - a_v = 2u'k\frac{\beta'-1}{\beta'^2}$$

must be minimised. Here a_v describes the front (near to sensor) and a_h the rear (away from sensor) sharpness boundary, k the f-number, β' the representation scale and u' the admissible unfocused circle diameter. According to the common prefix system, the numerical values of a_v and a_h are negative with real objects.

For the difference $a_h - a_v$ to be as close as possible to zero, the f-number k must be minimised, or the numerical aperture maximised to $A = \frac{1}{2k}$. The largest possible representation ratio must be chosen.

Apart from a high longitudinal resolution (e.g. resolution in the measurement direction), a high lateral resolution (local resolution) should also be aimed at. Therefore the smallest possible measuring light point is required. For imaging systems with diffraction limitations, the Airy disk is applied as the radius:

$$\rho_{1Min} = 1.22\frac{\lambda}{D}f'$$

In order to realise a small measuring light point, a large diameter D of the entry pupil is implemented with a short object-lateral focal point. The measuring light wavelength results from the available light sources. For autofocus metrology, laser diodes with a wavelength λ of 630 nm or 780 nm are usually used.

In order to meet the described basic physical conditions and the demands for a compact and manageable sensor configuration at the same time, autofocus sensors generally have a relatively small free working distance of 1 mm to max. 15 mm, according to the type of sensor. With the most common devices, the measuring ranges can be selected according to the resolution. Depending on the make and choice of options, typical measuring ranges are 6 μm, 60 μm, 100 μm, 600 μm and 1000 μm. The dynamics of the measuring systems are determined by the mechanical components for refocusing. Depending on the objective masses and the height differences to be controlled, measuring frequencies of up to 1200 Hz are realised.

4.3.3.4 Laser Based Straightness Measurement

The laser beam procedure (LBP) is - like the alignment telescope - used to measure *transverse* deviations from a reference line.

With the LBP an expanded and collimated laser beam is used as a reference. The principle structure of a laser beam measuring arrangement is represented in **Figure 4.3-38**. The laser beam is expanded and collimated by an optical system after its production in a laser. The laser beam is thereby available as a reference beam with the most constant as possible diameter and a constant beam profile. A semiconductor photodiode is used as a detector, which provides the point of impact of the laser beam on the diode in the form of beams as x and y coordinates. The detector is mounted on a carriage, which is moved across the surface to be inspected.

On marked places on the surface or while the carriage is moving, measurements are triggered and, for example, processed further and stored on a PC.

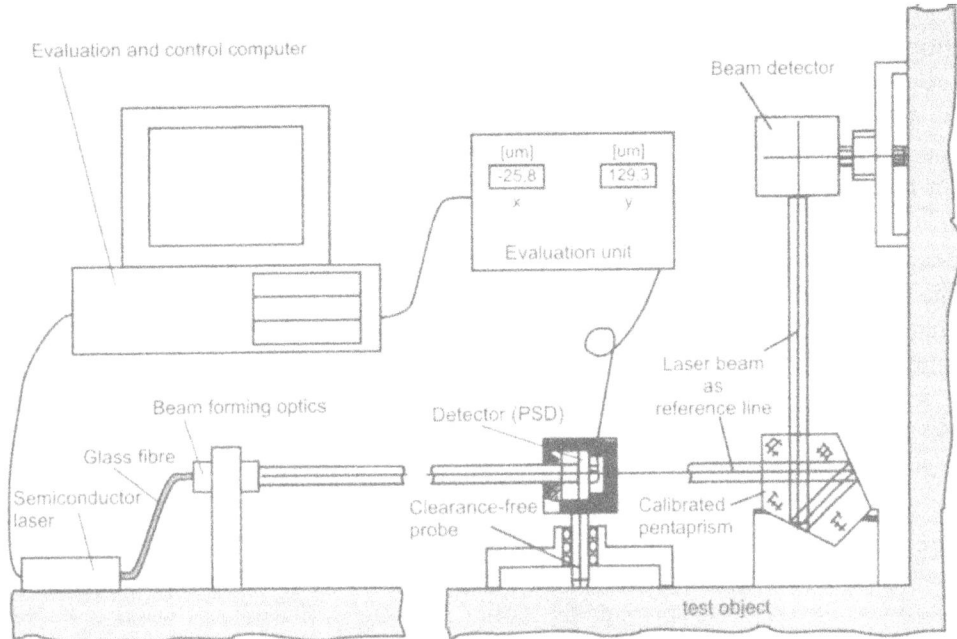

Figure 4.3-38: Arrangement of a measuring system according to the laser based straightness measurement

The following sections introduce and discuss individual components of the measuring arrangement and its influences on measurement uncertainty.

The Laser

For cost reasons and due to their compact design, semiconductor laser diodes are almost exclusively used as sources of beams. Semiconductor laser diodes have the characteristic that, due to thermal effects during operation, the spatial position of the emitted light beam is not temporally constant. After formation of the beam, this leads to the reference direction during a measurement also not being constant, thereby making the measurement unusable.

This can be remedied on the one hand, through using two additional detectors in a fixed position in the space, whereby the current position of the reference beam is also measured during a running measurement. However, the disadvantage is that additional optical components, such as beam splitters (Section 4.3.1) must be placed into the path of the beam, which lowers the beam intensity and could change the beam profile.

On the other hand, the laser beam can be linked into a glass fibre with a suitable device after it is produced. At the other end of the glass fibre, a spatially and temporally stable laser beam is available, which subsequently meets beam formation optics. Positional instabilities of the laser beam from the glass fibre coupling thus only become apparent as intensity fluctuations, which can be eliminated through the appropriate processing of detector signals. Furthermore, the laser can be set up in a protected position through *fibre optic coupling*. Only the beam formation optics must be firmly fixed in the space.

Beam Formation
The light emitted from the glass fibre is collimated and expanded to an appropriate beam diameter. The size of the beam diameter depends considerably on the maximum distance of the detector, as a collimated laser beam in fact always has a more or less diverging beam path. The stability of the beam diameter is meaningful, in that the transversal measuring range of the detector must remain constant over the distance to be measured.

The Laser Beam in Air
Collimated laser light only spreads in a straight line in air if the air temperature and humidity are constant over the entire measuring distance, in relation to the propagation direction transversal gradient of air pressure [Pf 72]. As this is only rarely given, even under laboratory conditions, this condition can be accommodated by bringing, *for example, directed axial atmospheric turbulences* into the beam path. The momentary position of the beam is then distributed over a certain time period in such a way, that the exact beam position can be determined by *averaging*.

The Detector
PSD's (position sensitive diodes, Section 4.3.1) in the most diverse designs are usually used as detectors. They are characterised by the fact that they output the centroid of the light point directly on the detector surface in the form of two streams per coordinate direction (x and y). A complicated calculation of this position, which would, for example, be required when using a CCD camera as a detector, is not required. Therefore very high measurement frequencies of up to 30 kHz are possible.

Foreign Light Influences
Light-sensitive planar measuring sensors not only receive the usable light (laser beam), but also the surrounding light. A PSD, however, provides the centroid from all of the light arriving at it. Therefore incorrect measurement values can be ex-

pected. There are two possibilities for remedying this problem. On the one hand, a colour filter can be positioned in the the beam path of the laser light, which filters out the interfering light. On the other hand, the interfering light can be suppressed through an amplitude modulation of the laser light and a corresponding demodulation when evaluating the detector signal (*"lock in" procedure*). The advantage of this method is that no additional optical components, which could also be additional sources of error, need to be positioned in the beam path of the laser.

Measurement Uncertainty
The measurement uncertainty of laser beams depends on the following factors:

- stability of the device used for fixing the beam formation optics

- environmental influences of the surrounding air

- resolution of the detector

- quantisation noise, such as the A/D converter card for downloading the measurement data into a PC

Under ideal conditions, uncertainties of under 1 µm are attainable on a measuring distance of 10 m. Realistically, however, uncertainties are 2-10 µm/m measuring distance under production conditions in a manufacturing hall.

Examples of Applications
With high-precision turning, the even guidance of the turning tool along the rotary axis is crucial for the accuracy of manufactured workpieces, as deviations in the given relative movement between the tool and the workpiece inevitably lead to geometrical dimensional deviations in the manufactured component. For highly precise checking of the guide axles during the turning process, laser beam systems can be used, which are directly integrated into the lathe. For this, the detector (PSD) is fastened to the lathe tool and the laser beam is aligned parallel to the rotational axis **(Figure 4.3-39)**. If, during the turning processing, the workpiece shifts from the target position due to highly dynamic process forces, the laser shifts equally, relative to the aligned laser beam. The evaluation unit of the laser beam system registers the movements of the sensors and passes the measurement data to the NC control via an external data interface. In this way it is possible to permanently determine the straightness deviation of the tool and immediately compensate with the aid of the NC control. The effect of this compensation is clearly shown in **Figure 4.3-40**, where the positional deviations of a tool are represented which briefly comes under influence of a strength of 200 N transverse to the drive direction.

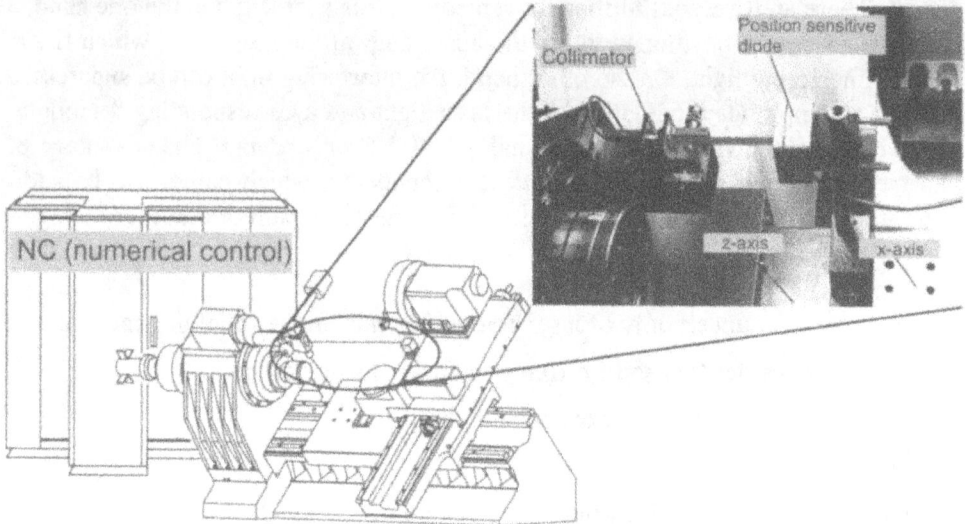

Figure 4.3-39: Integration of a laser beam system into a high precision lathe

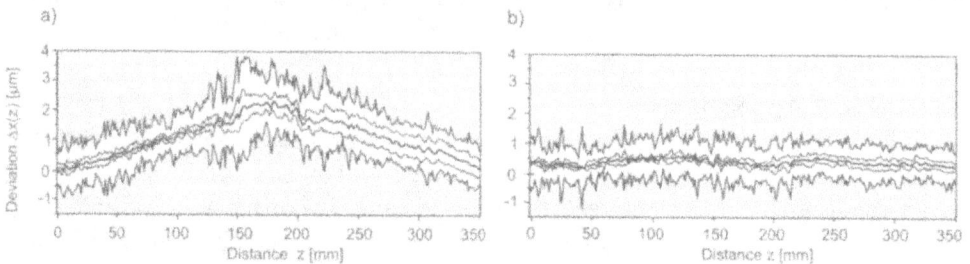

Figure 4.3-40: Deviations of a tool from the target position. a) without compensation, b) with compensation

4.3.3.5 2D Laser Light Cutting Procedure

The light cutting procedure is based on the triangulation principle. A light curtain is produced by expanding the laser beam with a special cylinder lens through holographic projection or an oscillating mirror. With the light cutting procedure, the line detector is replaced with a CCD matrix camera. With the aid of image processing algorithms, the position of the light line on the camera is determined. After calibrating the sensor technology, the elevation profile is obtained along the laser line in fractions of a second. In contrast to point triangulation systems, as described in Section 4.3.3.1, no additional mechanical axle is required to realise a 2D measurement.

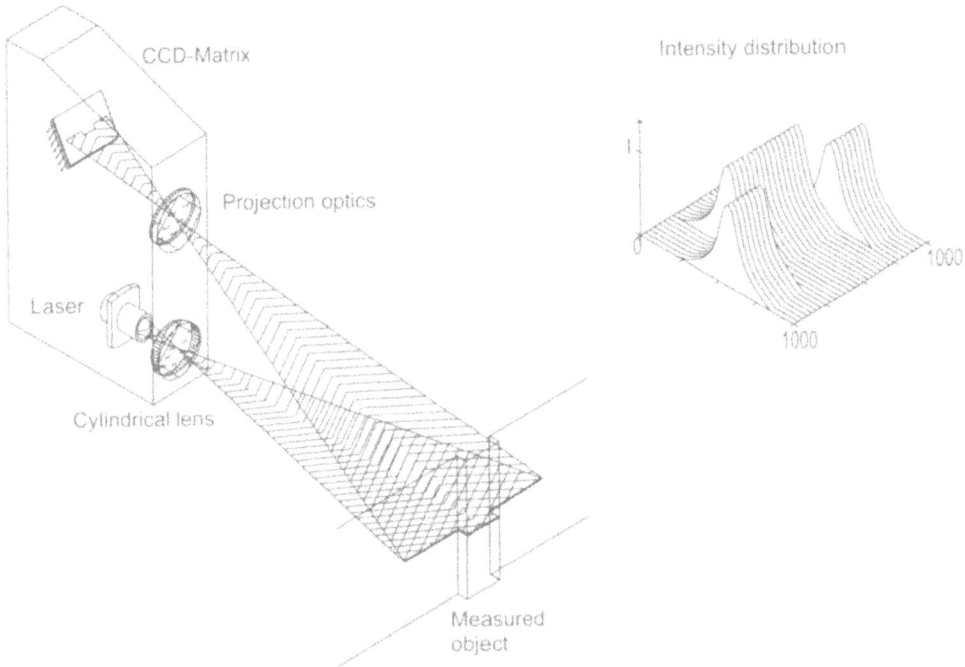

Figure 4.3-41: Structure of a light cutting sensor

The resolution of such a system depends on the measuring range and the resolution of the camera used. In production metrology, the width of the projected laser line varies from approx. 2 mm – 15 mm. If a CCD matrix camera with a resolution of 1000 x 1000 pixels is used, it is possible to obtain a lateral and transversal resolution of about 2 μm - 15 μm. The resolution can still be increased by detecting a displayed laser line in the subpixel region. The measuring frequency depends on the resolution of the CCD matrix camera used and the efficiency of the image processing system. Thus the measuring frequencies vary from approx. 2 Hz to 50 Hz. The attainable measurement uncertainty depends on the surface texture of the sample, as with point triangulation systems.

4.3.3.6 Interferometric Procedures

The basis of interferometric procedures is the coherent overlay of lightwaves, which have travelled different paths. Due to scaling with the wavelength of the light used, very high resolution procedures are involved, for example, length measuring interferometers enable the determination of shifts with a resolution of a few nanometers. In this section, the basic functions of an interferometer are followed by a discussion of the criteria which make them important for practical application. This will be followed by a representation of length measurement with interferometers. A further section introduces the extension to form measuring techniques.

Structure and Operation of Laser Interferometers

The basis components of all interferometers is an optical element for splitting an electromagnetic wave (beam splitter) as well as one with the function of reunifying the partial waves after passing through different paths. Both components can be identical, as in the case of the Michelson interferometer. The prerequisite for an interference signal which can be evaluated is a constant phase reference of the two partial waves. This characteristic described as "coherence of two waves" is fulfilled when the path difference of the partial beams is less than the coherency length of the light source used. Lasers are primarily used in interferometry due to their good coherency features.

The description of the function of interferometers is based on the Michelson structure. The result can be easily expandable to other structures as well as to two dimensions.

The structure of a Michelson interferometer is represented in **Figure 4.3-42**. The outgoing light from the laser is divided up by the beam splitter. The resulting partial waves are reflected by mirrors after passing through the paths s1 or s2. After both partial beams have passed through the respective paths again, they are overlaid behind the point of interference. Depending on the relative phase position of the partial waves, an intensity is registered between a maximum value and almost complete cancellation.

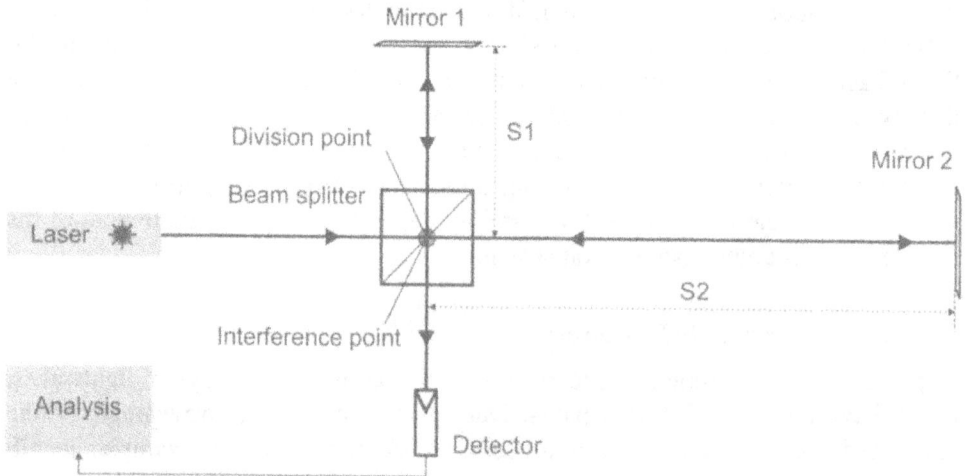

Figure 4.3-42: Structure of a Michelson interferometer

For the mathematical description of an interference signal, the complex notation style is used for the fields.

The field strength emitted by the laser is described by equation 4.3-7.

$$\vec{E}_a(x,t) = \vec{E}_0 \cdot \exp\left[i \cdot 2\pi \cdot \left(f \cdot t - \frac{x}{\lambda}\right)\right]$$
(4.3-7)

f is the frequency and λ the wavelength of the light used. The quantity t describes the temporal behaviour of the wave and x the path travelled by it.

The beam splitter, assumed to be ideal, splits the wave into two identical waves with the amplitude $E_0/2$. With their overlay in the interference point, the waves have additionally travelled the paths $2 \cdot s1$ or $2 \cdot s2$. Behind this point, the field strength is given by equation 4.3-8.

$$\vec{E}_{sum}(x,t) = \frac{\vec{E}_0}{2} \cdot \exp\left[i \cdot 2\pi \cdot \left(f \cdot t - \frac{x_g + 2s_1}{\lambda}\right)\right]$$
$$+ \frac{\vec{E}_0}{2} \cdot \exp\left[i \cdot 2\pi \cdot \left(f \cdot t - \frac{x_g + 2s_2}{\lambda}\right)\right]$$
(4.3-8)

The path which both beams travel together is described with x_g.

The intensity impacting on the detector I is here from calculated to:

$$I = |\vec{E}_{sum}|^2 = \frac{|\vec{E}_0|^2}{4} \cdot (1 + \cos\phi) \quad \text{mit} \quad \phi = 2\pi \cdot \frac{2 \cdot (s_2 - s_1)}{\lambda}$$
(4.3-9)

The derivation of this equation is based on the following prerequisites:

- ideal optical components exist
- the refractive index of air can be disregarded

Without these assumptions, which are not realistic when using interferometers in the production environment outside of optical laboratories, equation 4.3-9 changes into equation 4.3-10.

$$I = I_0 \cdot (1 + \gamma \cdot \cos\phi) \quad \text{with} \quad \phi = 2\pi \cdot \frac{2 \cdot (s_2 - s_1) \cdot n_L}{\lambda}$$
(4.3-10)

γ is defined as the degree of modulation of the interference signal and I_0 is a constant offset. In the phase term, which contains the desired information, the geometrical path difference has been replaced with the optical path difference. In practice, these modifications have extensive consequences, which are discussed below.

First of all, equation 4.3-10 contains three unknown quantities. In order to determine the interference phase, two further equations of these values must therefore
be determined. In practice, the phase shifting procedure, use of quadrature signals
and heterodynamics have proven to function here (Section 4.3.3.2).

First it must be noted, that the spatial information is contained within the argument
of the cosine. Due to its periodicity, a registered intensity value repeats itself if the
optical path length of one of the partial beams s_1 or s_2 is modified by an integral
multiple of $\lambda/(2 \cdot n_L)$. For the wavelength of HeNe-Lasers (633nm) this means a signal repetition interval of only 316.5 nm. Therefore the range of unambiguity of an
individual intensity measurement is limited to this value. The removal of the mentioned restriction is the subject of numerous research projects. Interferometric
length measurement is expanded to an absolute measuring system by using a wavelength continuum [Thi 93]. In form measurement, several discrete wavelengths enable extension of the range of unambiguity.

Furthermore, it must be observed that, in place of the geometrical path difference,
the optical path difference is measured, which follows from the first mentioned
through multiplication with the refractive index n_L. The refractive index of air depends on the temperature, air pressure, humidity and CO_2 content, as well as other
parameters. In terms of order of magnitude, n_L changes by $1 \cdot 10^{-6}$, if the temperature
in the measurement beam path changes by 1°K or the air pressure changes by 3.7
hPa. This shows that the refractive index for precision measurements with accuracies in a range less than μm must be determined very precisely. This can either be
done using the parameter method, e.g. measurement of the air parameters and calculation of the refractive index [Edl 66] or through refractive index measurement
by means of refractometers. The latter technique is the more accurate of the two
procedures.

As the final influence quantity, the laser wavelength must be mentioned. For
highly precise measurements, this must be stabilised at relative accuracies around
$1 \cdot 10^{-8}$. Sufficiently stabilised HeNe lasers are commercially available due to the
meanwhile widespread use of interferometers.

Length Measuring Interferometry
Due to adjustment requirements and insensitivity to mechanical disturbances, interferometers are, in practice, not constructed with flat mirrors, but with the retro
reflectors described in Section 4.3.1. If the measured path difference should be
uniquely allocated to a shift distance, the position of one of the two interferometer
mirrors must be recorded. For this reason, the mirror in the reference arm is usually directly connected with the beam splitter.

Nearly all commercially available interferometers operate according to the heterodyne procedure. Here, the shifting rate of the measuring mirror is determined via
the frequency shift of the laser light through the Doppler Effect. The mirror position results through integration of the shifting rate over time. The fundamental
structure of a heterodyne interferometer is shown in **Figure 4.3-43**.

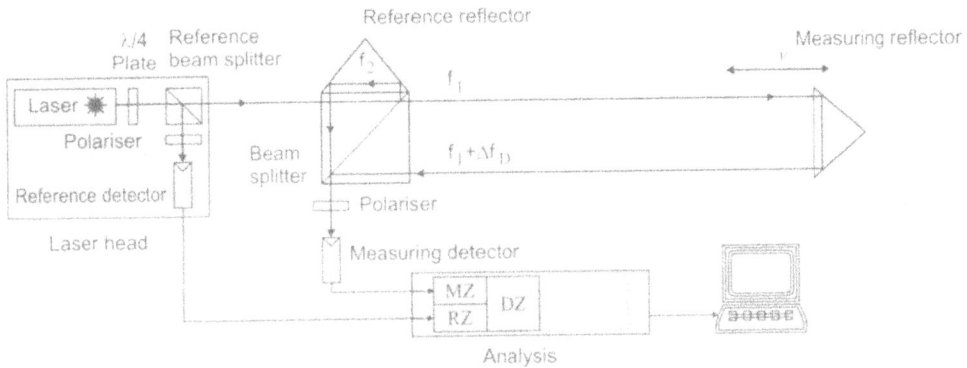

Figure 4.3-43: Schematic structure of a heterodyne interferometer

A laser is used which emits two frequencies which are typically around 2 MHz
apart and circularly polarised opposite to one another. A λ/4 disk converts these
into two polarised components which are perpendicular to one another. A small
proportion of the light from both frequencies is decoupled with a reference beam
splitter. The photo detector, which is positioned behind a polariser, registers as a
reference signal the differential frequency:

$$f_{\text{Ref}} = f_2 - f_1 \tag{4.3-11}$$

In the represented Michelson interferometer, the laser beam is split by a polarising
beam splitter in such a way that the partial beam with the frequency f_1 passes
through the measuring arm and that with f_2 passes through the reference arm. If the
measuring mirror is resting, the differential frequency f_{Ref} is registered analogue to
the reference signal. With a movement of the measuring reflector, the frequency f_1
is shifted by Δf_D. Due to this, the detector now displays the frequency:

$$f_{\text{Mea}} = f_2 - (f_1 + \Delta f_D) \quad \text{with} \quad \Delta f_D = -2 \cdot f_1 \cdot \frac{v}{c} \tag{4.3-12}$$

Here v is the mirror speed and c the speed of light. With downstream electronics,
the differential frequency f_v is determined.

$$f_v = f_{\text{Mea}} - f_{\text{Ref}} = 2 \cdot f_1 \cdot \frac{v}{c} \tag{4.3-13}$$

The shift distance of the measuring mirror results here from to:

$$L = \int_{t_1}^{t_2} v(t)\, dt = \frac{\lambda_1}{2} \int_{t_1}^{t_2} f_v\, dt \qquad\qquad\qquad (4.3\text{-}14)$$

The wavelength λ_1 linked with the frequency f_1 through $c = \lambda \cdot f$. In addition to the advantage that the differential frequency f_v directly contains the direction of the mirror shift through the sign of the mirror rate, the heterodyne procedure has the advantage of less adjustment sensitivity and thus disturbance sensitivity. More complex electronics, as well as a shifting rate limited by f_{Ref} are disadvantageous.

Application Examples of Laser Interferometers
In addition to the described length measurements, highly precise angle, straightness and trueness measurements are carried out with interferometers. For this purpose, the arrangement of interferometer components must be modified in such a way, that the interference phase in equation 4.3-10 is uniquely linked with the desired geometrical quantity [Don 93]. For this, corresponding options are available for commercial interferometers.

Laser interferometers are more widely used in machine tool manufacture, particularly in ultra precision and fine mechanics. Machines in which the drive is controlled with laser interferometers, serve the production of complex components with production tolerances of a few nanometers.

Figure 4.3-44 shows a triple axis ultra precision processing machine, in which the movement of all three axles are controlled with laser interferometers. On the right side of the illustration, the ray path of the interferometer is represented schematically. The beam splitters ST1 and ST2 divide the incident laser beam into 3 partial beams of approximately the same intensity. After passing through several deflective optics, these partial beams are available in the interferometers at the appropriate axes. The beam splitters of the X and Y interferometer with forced reference reflector are connected with the bed of the machine and the retro reflectors R_X and R_y are connected with the carriages of the appropriate axes. In contrast, the interferometer for the Z axis is mounted on the X carriage, as well as the Z carriage, as these must be moved with the X carriage during the processing procedure.

Beyond that, laser interferometry today is an indispensable aid to accuracy testing and calibration of tool machines, coordinate measuring machines and robots. For the most often built-in incremental measuring systems, correction templates can be produced and the accuracy optimised or tested.

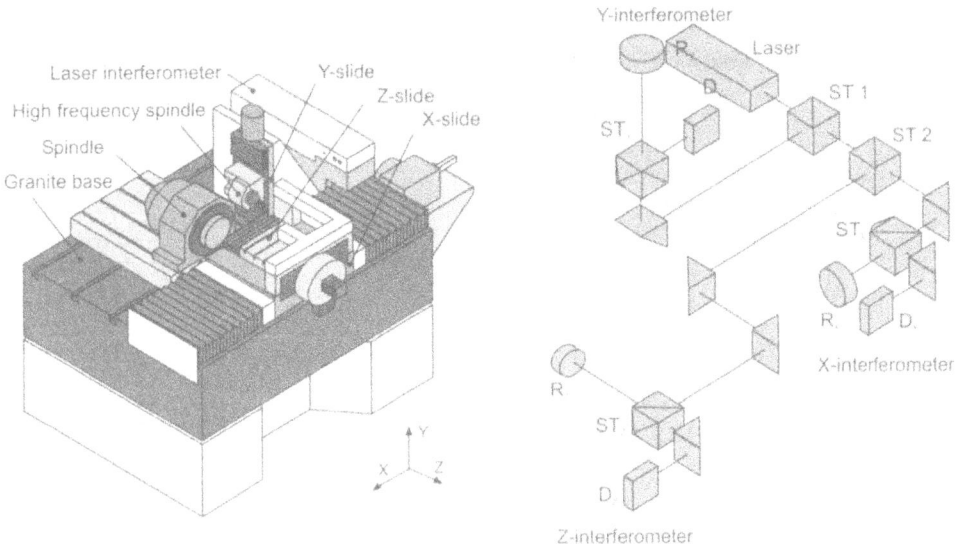

Figure 4.3-44: Application of a laser interferometer in an ultra precision processing machine

3D Form Testing Interferometry

In the previous section, the effect of interference is used in order to obtain distance information. In this section, this will be supplemented by describing form testing interferometry as a further method of production metrology.

The principle of form testing interferometry is the optical comparison of the tested surface with a flat or spherical reference surface of an exactly known form. In order to attain the required measurement uncertainties in the sub micrometer range, the wavelength of a helium/neon laser of 632.8 nm can be used as a scale. However, for purposes of form testing, a wave front of known high form accuracy must be produced.

As **Figure 4.3-45** shows, the even wave front of a small diameter emitted by the laser is released from high frequency disturbances by a spatial filter and transformed into a spherical wave. This impacts the collimator, specifically laid out high quality optics, which transform the spherical wave into an even wave. Deviation of this wave front from the planarity in the range of only approx. 30 nm ($\lambda/20$) can be achieved.

According to the same principle as in the previous section, two waves, the measuring wave and the reference wave, are then overlaid and interference occurs. In form testing interferometry, however, it is made certain that the waves are expanded to wave fronts of high form quality with diameters up to 150 mm, as they are crucial for the quality of this measurement technology.

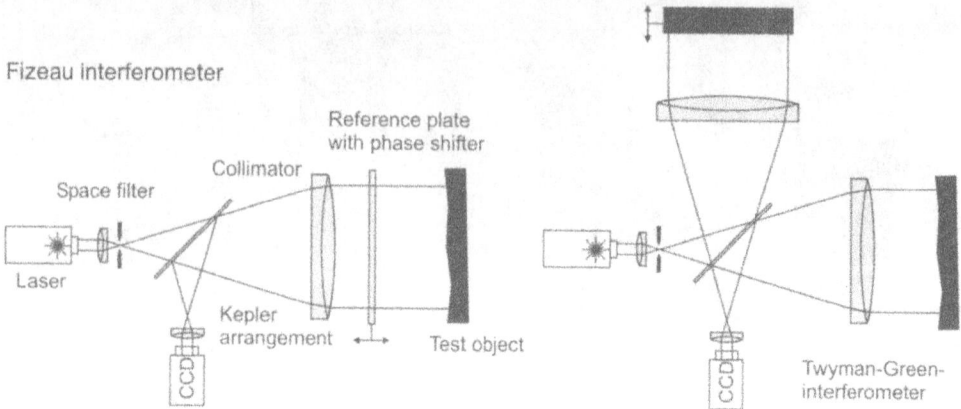

Figure 4.3-45: Principle of the form testing interferometer according to Fizeau and Twyman-Green

In the case of the Twyman-Green interferometers, represented in **Figure 4.3-45** on the right, the reference and measuring arms are spatially separated from one another. In order to achieve this, a beam splitter is used which divides the wave front in the intensity ratio 50:50. Both partial wave fronts are collimated in each case with optics. While the reference wave front is reflected back in itself without modifying the phase by a precision flat mirror functioning as a reference, the measuring wave front impacts the test piece. With the reflection at the test piece, the profile of measuring wave front is now, as it were, impressed to the measuring wave front. A modification of the originally constant phase ϕ of the measuring wave front into a distribution $\phi(x,y)$ is the consequence. The information about the height profile $z(x,y)$ of the test piece is thus contained in this phase distribution. The distorted measuring wave front formed in such a way interferes with the reflected reference wave front after their overlay at the beam splitter and an interferogram is developed.

With the Fizeau interferometer represented in **Figure 4.3-45** on the left, the second frequently used execution form in form testing interferometry, the reference arm is, as it were, turned inwards, making a second collimator unnecessary. In place of a reference mirror, a glass plate, usually made of BK7, is used as a reference, which reflects the reference wave with approx. 4% of the intensity, while the measuring wave is transmitted. However, in order to adapt the intensities of both waves and optimise the contrast attenuating filters must then be used.

$$\gamma(x, y) = \frac{I_{max} - I_{min}}{I_{max} + I_{min}}$$ (4.3-15)

The interference pattern therefore represents the form deviation of the test sample surface relative to the reference surface used. In the case of a flat reference, this form deviation corresponds to the height profile of the test sample.

The collimator and further optics, which together represent a Kepler arrangement, form the interferogram in a reduced form on the camera in the evaluation arm. It is detected by a CCD sensor and supplied to the evaluation software with a frame grabber card.

The mathematical description of the occurring interference pattern almost corresponds to that in equation 4.3-10. It must, however, be considered that the phase of the wave front is not constant, but represents a phase distribution expanded over the surface. Therefore the interferometer equation

$$I(x, y) = I_0(1 + \gamma(x, y) \cos(\phi(x, y) + \alpha))$$ (4.3-16)

applies, whereby it was assumed that the constant phase of the reference wave front is 0. Again it applies that the distribution $\phi(x,y)$ is extracted from an equation with three unknowns, as it contains the information. This can, for example, take place with the phase shifting technique, which will now be briefly described.

If, in the case of the Fizeau structure, the reference surface is shifted in a defined manner, the optical path lengths change in the interferometer. This leads to a shift of the stripes in the interferogram and thus to an additive phase term α in equation 4.3-16. By executing n shifts around angle $n \cdot \alpha$ and in each case the intensities of the interferogram $I_n(x,y)$ are measured, a solvable set of equations can be developed [Rob 93][Mal 92] and the phase distribution $\phi(x,y)$ of the measuring wave front can be calculated.

In the technical implementation, the reference surface is shifted over piezo motion elements by defined fractions of the wavelength and an interferogram is taken up in each case by the CCD camera. The determined grey tones of the camera pixels are interpreted as intensity values $I_1(x,y)$,.., $I_n(x,y)$ with $n \geq 3$ in relation to equation 4.3-16. For e.g., n=5 [Rob 93], this results in:

$$\phi = \arctan \frac{2(I_2 - I_4)}{2I_3 - I_5 - I_1}, \phi := \phi(x, y), I_n := I_n(x, y)$$ (4.3-17)

However, due to the arc tangent function arising with this calculation, the resulting phase distribution is ambiguous. All phase values are calculated modulo 2π. Therefore an additional multiple of 2π must first be added to every individual value in order to establish the unambiguity („unwrapping"). The criteria for determining this multiple results directly from the scanning theorem (Section 4.5.1).

Very many inceptions and procedures exist for unwrapping [Rob 93]; however an ideal solution can not be indicated. It generally depends on the quality of the measurement result, whether complex or simple methods find application. Finally, by scaling with the assigned wavelength, the quantitative height profile is reconstructed:

$$z(x, y) = \frac{\lambda}{2 \cdot 2\pi} \phi(x, y) \qquad\qquad (4.3\text{-}18)$$

The height profile of the test sample is thus determined.

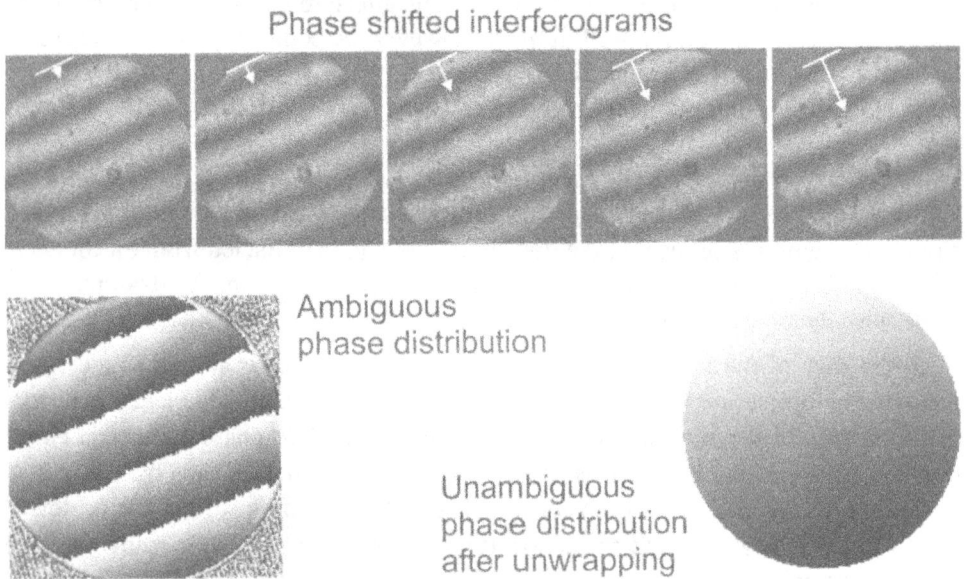

Figure 4.3-46: Form testing interferometry with phase shifting

In **Figure 4.3-46**, the phase shifting procedure is illustrated. At the top, the 5 phase shifted interferograms are represented. With the 5 image procedure [Mal 92] used here, the phase is shifted in five steps by a total of 2π, so that the fifth interferogram must correspond to the first one. This condition can be drawn on in order to calibrate the piezo motion element of the phase shifter. After calculation by means of equation 4.3-17 the representation in **Figure 4.3-46** on bottom left-hand corner results. It can be recognised that the phase information is received, but is even more ambiguous due to the arc tangent. Finally, in **Figure 4.3-46** bottom left-hand corner, the result is shown after the unwrapping procedure and scaling according to equation 4.3-18. This example concerns the measurement of a tipped flat mirror.

In addition to form testing interferometry using the phase shifting technique, other methods of phase evaluation [Rob 93] exist, however, they will be left beyond the context of this text.

4.3.3.7 Laser Scanners

The laser scanner is a measuring system to determine workpiece dimensions, which operates according to the light barrier principle, i.e. the optical path between the source of the laser and the photo detector is interrupted by workpiece under examination. The in principle structure of the laser scanner is shown in **Figure 4.3-47**

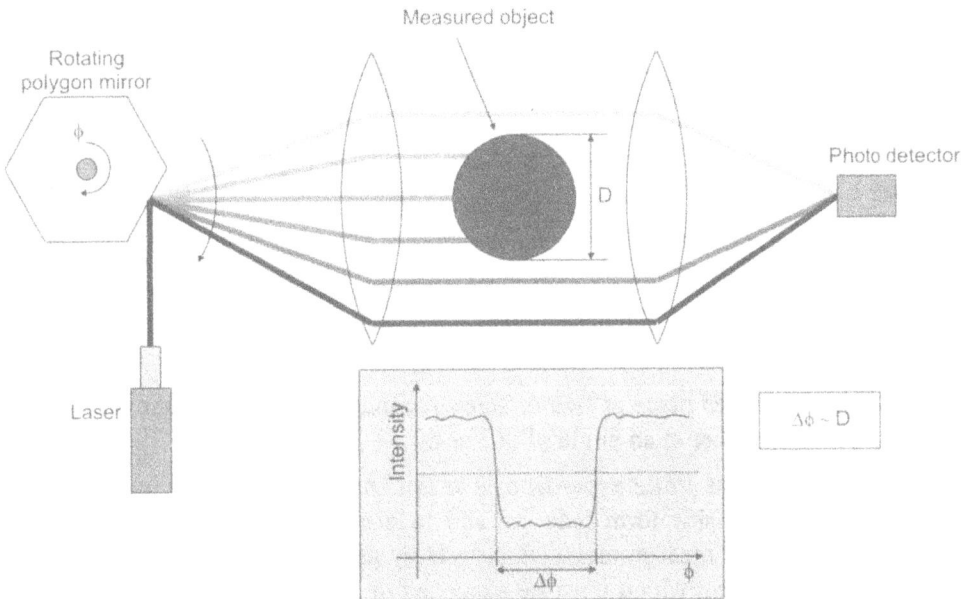

Figure 4.3-47: Schematic structure of the laser scanner

A laser beam is periodically deflected by a rotating polygon mirror, whereby the scanning range is determined by the number of corners of the mirror (**Figure 4.3-48**). The reflected laser beam then meets a lens, the focal point of which the laser beam has been reflected into and thus, for each angle position ϕ of the polygon mirror, the laser beam is deflected in the same direction, so that the beam behind the lens is continuously shifted in parallel.

Subsequently, the beam meets as second lens with a photo detector in its focal point. Between the two lenses, the measured object is attached, which shades the detector from laser light at certain angles. With the appropriate angular positions, the detector indicates an intensity decrease, the duration of which permits conclusions to be made about component geometry. The diameter D of a test piece is proportional to the angle $\Delta\phi$ where the optical path is blocked by the measured object.

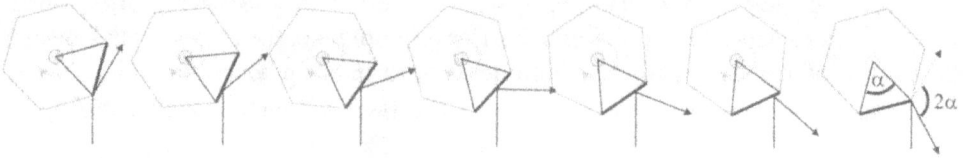

Figure 4.3-48: Operation of the polygon mirror

The momentary angular position of the polygon mirror is measured e.g. through incremental transmitters. With this arrangement, a single laser and a single photo detector are sufficient to scan an area which is clearly larger than the diameter of the laser beam itself.

There are basically two methodologies available for measuring bevelled objects:

- The measured object is scanned several times, whereby after each measurement, a defined turn of the object takes place. The course of the shadow dimension provides information about the profile geometry of the test piece.

- Measurement takes place at two or three scanners as the same time, which are offset to one another at an angle of 90° or 60°.

As the laser beam has finite expansion, it is not shadowed abruptly by the object edges, but finds a cosine form fade out and fade in. Nevertheless in order to acquire the component measurements as accurately as possible, a threshold value is determined in practice, which must be exceeded or fallen below in order to detect an interruption of the ray path. The resolution which can be achieved with a laser scanner lies below 1 µm with a measuring range of 2000 µm. This corresponds to a relative resolution of $5 \cdot 10^{-4}$. The measuring rate is approx. 400 scans / second.

4.3.4 Optical Measuring Devices

4.3.4.1 Alignment Telescope

The alignment telescope operates according to the principle of the measurement of *transversal* deviation of consecutively situated aligning marks. The optical axis of the telescope, which is defined by a crossline mark in the eyepiece of the telescope and an external reference crossline mark in the space, serves as a reference line, which also called an alignment line.

Following the exact alignment of the telescope with the external cross mark, the telescope remains rigid in the space. Between the external reference mark and the telescope, there are additional crossline marks (measuring marks) which are individually aimed at by focussing the telescope on the measuring marks. The following difficulties occur:

- The scale changes with the distance.

- The telescope must be refocused with the distance. The position of the telescope or the telescope target axis must remain stable in the space.

This problem will be discussed below.

Modification of the Scale
The scale becomes ever smaller with increasing distance of the measuring mark. On the one hand, this has the consequence that the measurement uncertainty becomes accordingly smaller. On the other hand, the eyepiece scale would have to adapt accordingly.

As a solution to these problems, a scale could be applied to the measuring mark. However, through the variable scale, this external reference scale would be too coarse in proximity and too fine in the distance.

Furthermore, the telescope could be shifted laterally in parallel until both of the crossline marks are aligned (*null method*). The deviation could be determined by measuring the lateral shift. This, however, requires highly accurate and sensitive linear modules on which the telescope must be mounted and highly precisely aligned.

Finally, there only remains the possibility of carrying out the lateral shift vertically. This is realised by the inclinable plane-parallel plates in front of the telescopic objective (**Figure 4.3-49**). For small tilts of these plates, the angle of rotation ϑ is directly proportional to the lateral offset Δy. The relationship

$$\Delta y = \vartheta \cdot d \cdot \left(1 - \frac{1}{n}\right) \hspace{3cm} (4.3\text{-}19)$$

applies, whereby d characterises the thickness of the plate and n the refractive index of the plate.

Refocusing

In order to measure measuring marks which are arranged in varying distances from the telescope, the telescope must be refocused. This normally takes through a change in distance between the eyepiece and the objective. It is important to note that the two optical axes, and thus the resulting telescopic target axis, are not changed. This can, however, only rarely be ensured with sufficient accuracy, as the appropriate guides and optics can only be produced with great difficulty.

As a remedy, a focusing lens can be inserted between the eyepiece and the objective (**Figure 4.3-49**). As long as the refractive power of this focusing lens is small in comparison with the refractive power of the objective, guidance errors have only a very small influence.

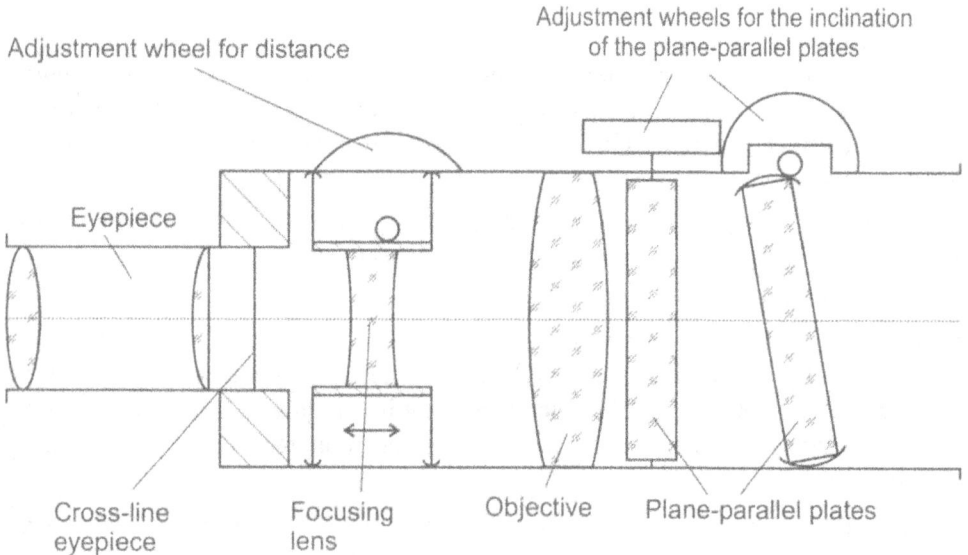

Figure 4.3-49: Alignment telescope with focusing

Measurement Uncertainty

The uncertainty, with which the lateral offset of the measuring mark from the target axis of the telescope can be indicated, depends on the resolution of the eye, the magnification of the telescope and the distance of the measuring mark from the telescope. According to [Tiz 91b] the measurement uncertainty can be determined according to the following formula:

$$\Delta y = \pm\left(a + \frac{L}{b}\right)\mu m \qquad\qquad (4.3\text{-}20)$$

A telescope with 30 times magnification is specified as an example, where $a = 5$ and $b = 3 \cdot 10^5$ are determined empirically. The capture uncertainty of the eye and the guidance errors of the focusing lens are representatively contained in b.

4.3.4.2 Autocollimating Telescope

The autocollimating telescope (ACT) is used for directional testing. *Angle deviations* are measured against the optical axis of the telescope.

a) Structure of an autocollimating telescope

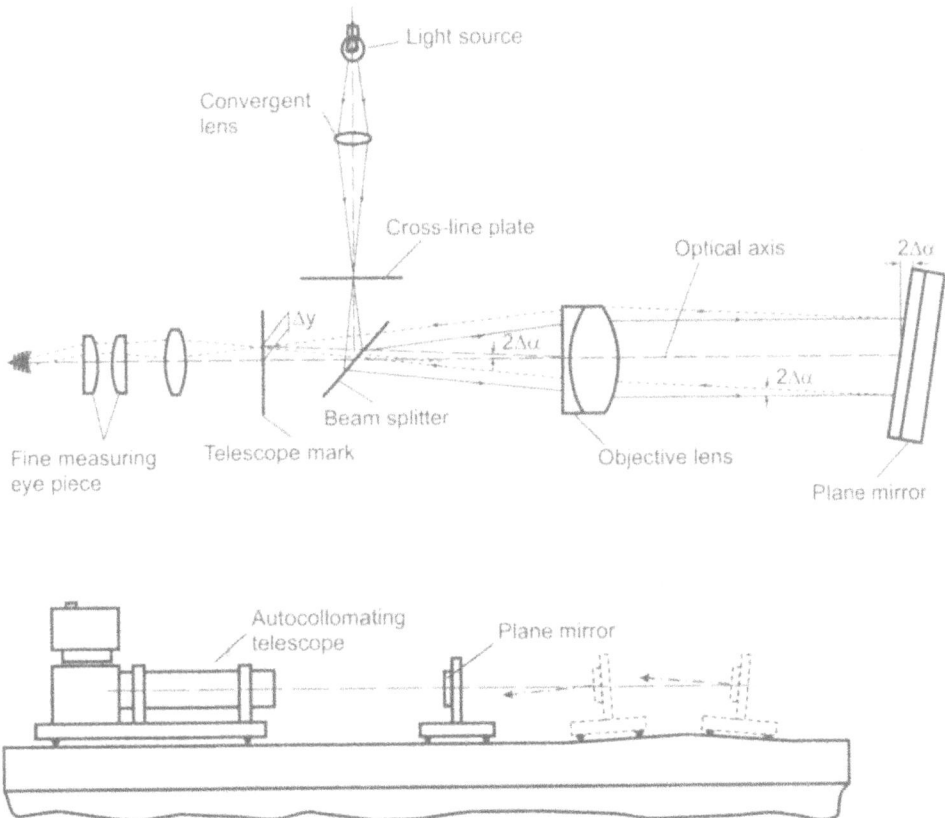

Figure 4.3-50: Principle of the autocollimating telescope

The ACT is particularly well suited for:

• Parallel alignment of surfaces, as well as

- Parallelism testing.

An illuminated mark is projected from the ACT into infinity. In the ray path, a highly precisely manufactured flat mirror is placed perpendicularly to the surface being tested so that the image of the illuminated mark is reflected back to the ACT. If the surface is not parallel with the path of the rays, the projection beam is not reflected back to itself, but slightly away from its initial position in the ACT. This deviation is a measure of the tilt of the tested surface against the optical axis of the ACT. The angle of tilt is doubled through the reflection (**Figure 4.3-50**).

Therefore the ACT has double measurement sensitivity for angular deviations. The maximum distances of the flat mirror from the ACT is limited, amongst other things, by the angle of tilt.

Mirror Tilt
With a large distance between the mirror and the ACT as well as a too large tilt of the mirror, the projection beam can be reflected past the ACT. This means that it is no longer possible to perform measurements.

Measurement Uncertainty
The uncertainty with which the angular deviation $\Delta\alpha$ can be determined from the optical axis of the ACT depends essentially on the objective focal length f_{OB} and the uncertainty with which the offset Δy of the displayed projected mark in the eyepiece level can be determined. The following relationship applies:

$$\tan(2\Delta\alpha) = \frac{\Delta y}{f_{OB}} \Rightarrow \Delta\alpha_{rad} \approx \frac{\Delta y}{2f_{OB}} \tag{4.3-21}$$

4.3.4.3 Measuring Microscope and Profile Projector

Measuring microscopes and profile projectors are among the most versatile optical inspection devices for determining geometrical dimensions of workpieces.

With both procedures, the fast and economical determination of two-dimensional coordinates of preferably 2D and 2½D objects are possible on one measuring plane. Such measured objects are, for example, stamped and flexible parts, plastic parts, cam plates, cams, outer threads, gear wheels, gauges and seals. The illuminated measured object is magnified with display optics and observed with a measuring microscope, through a measuring eyepiece, as shown in **Figure 4.3-51,** or with profile projectors where it is made visualised on a projection screen (**Figure 4.3-52**). In order to avoid measurement deviations due to the change in the display scale through insufficient focusing of the object on the measurement plane, optics are used with telecentric paths of rays [Srö 90].

In order to set the optimal suitable illumination for a respective measuring function, transmitted light as well as incident light - bright field - (illumination vertical to measurement plane) and incident light - dark field – (illumination under an angle to the measurement plane normal) are available to the user. It is preferable to use transmitted light illumination, where the profile of the measured object is displayed using the shadow image procedure, as smaller measurement uncertainties occur. Topside contours, such as grooves and countersunk holes can, in contrast, only be inspected in incident light and therefore less accurately.

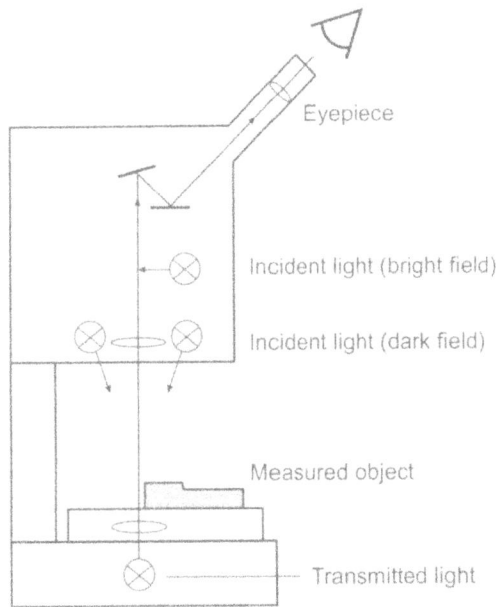

Eyepiece

Incident light (bright field)

Incident light (dark field)

Measured object

Transmitted light

Figure 4.3-51: In principle structure of a measuring microscope

The physical principle of the measuring microscope corresponds with that of a conventional microscope. With a microscope, the objective produces a magnified real intermediate image of the object. This intermediate image is viewed with an eyepiece, which has the function of a magnifying glass, and further magnifies the intermediate image (**Figure 4.3-53**).

The total magnification of the microscope results from the product of the objective and eyepiece magnification. The resolving power of a microscope, i.e. the distance Δx of two points which can still be observed separately, depends on the wavelength λ of the light and the numerical aperture NA of the objective according to the following relationship [Her 89]:

$$\Delta x \geq 0,61 \frac{\lambda}{NA} \tag{4.3-22}$$

From the above equation, it follows that the resolving power increases with shortening wavelengths and increasing numerical aperture. For this reason, lasers

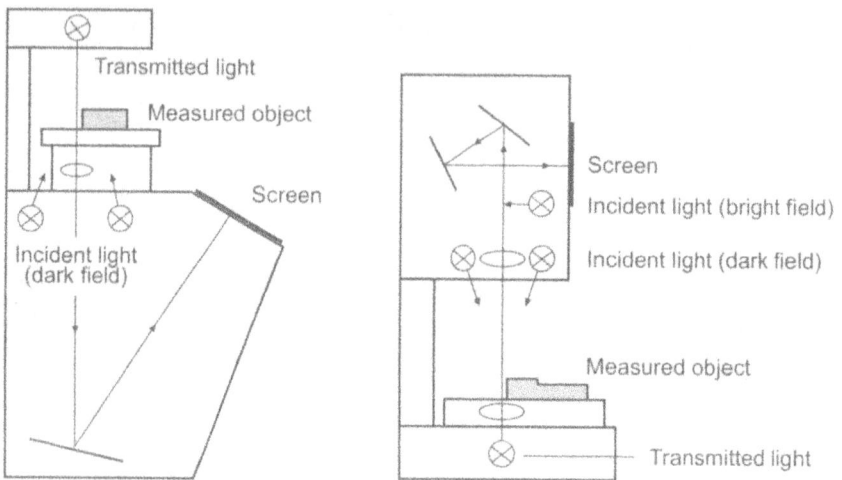

Figure 4.3-52: In principle structure of profile projectors

which emit beams reaching into the ultraviolet range are used as sources of light in modern confocal microscopes.

The confocal principle also permits the determination of the missing third coordinate in spatial structures. Here a point form source of light is displayed on the object through the microscope lens (**Figure 4.3-54**). If the object lies exactly in the focal point, the light is reflected from the object along the same path through the lens and a beam splitter onto a detector. Conversely, if the object is outside the focal point, a pinhole screen holds the reflected light back from the detector. The displayed light point is moved very quickly through a mirror system, point by point, line by line, over the object, thus scanning a plane of measurement [Lic 94]. Subsequently the object is moved slightly in height and scanned again. As a result of laying these images on top of each other, the three-dimensional structure of the object results.

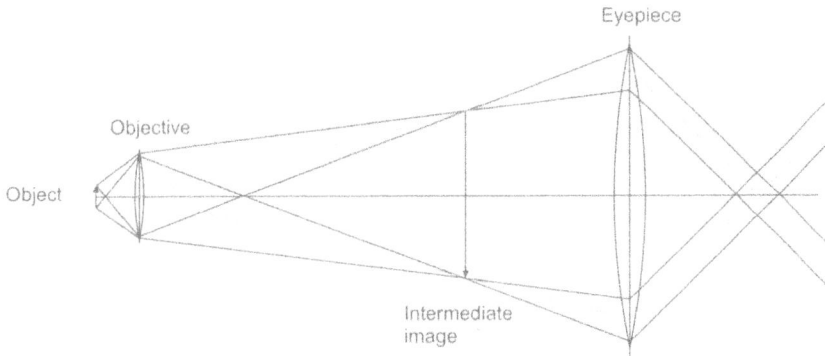

Figure 4.3-53: In principle path of rays with a microscope

Modern profile projectors are used in two different designs, whereby the image of the measured object is deflected in each case by highly precise mirrors and displayed in magnified form on a projection screen (**Figure 4.3-52**). The measuring range of profile projectors is generally larger than that of measuring microscopes, so that even larger objects can be directly measured without using linear modules. Thus, for example, a gauge inspection can easily be carried out manually by a user applying a template to the projected image.

For measuring object geometry with measuring microscopes, or profile projectors, the measured object is clamped on two movable tables which are perpendicular to one another and 'scanned' optically. Here the inspector brings a mark, e.g. a cross hair, or reticule into a defined overlap with the edge of the magnified display of the measured object to be probed through the appropriate positioning of the moveable tables (coincidence position). With micrometers or incremental positional displays which are coupled with the tables, the two-dimensional point coordinates are measured and these then, for example, linked to a distance between two points (Section 4.4). The devices used in production metrology are thereby equipped with different line plates adapted to the respective measuring function (**Figure 4.3-55**), whereby the magnification can be variably adjustable between 5 and 400 times through the appropriate objectives.

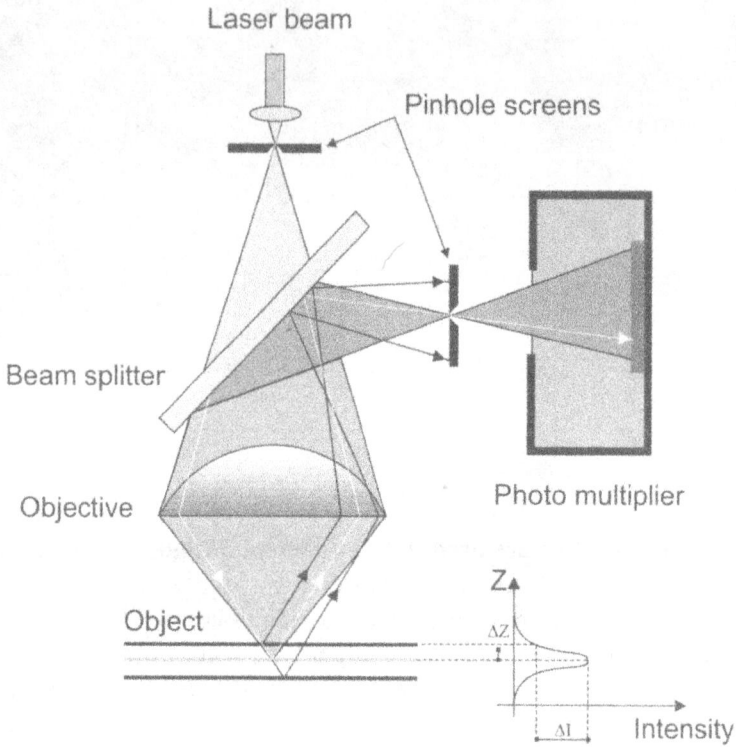

Figure 4.3-54: In principle path of rays with a confocal microscope

As only subjective measurement value acquisition is possible with visual probing, modern measuring microscopes and profile projectors are equipped with optoelectronic sensors. These optical edge sensors or 'annulus circular ring detectors' are photo receptors and thus correspond, in principle, with a light barrier which produces a switching impulse in the coincidence position for the selection of path scales. This enables an automated, objective and dynamic measurement value acquisition, where measurement uncertainties the µm range are possible.

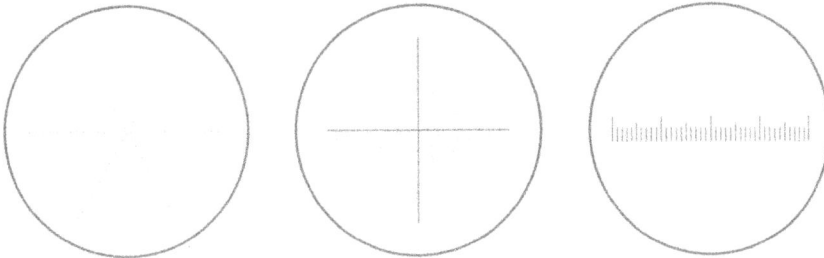

Figure 4.3-55: Various function-specific reticule plates

The advancements of the past years in the area of measuring microscopes and profile projectors have particularly led to extremely efficient optical three-dimensional coordinate measuring machines, through the integration of image processing (**Figure 4.3-56**) [Chr 91].

The measured object is magnified and displayed on a CCD camera. With appropriate image processing algorithms, the edge points from these grey tone images are automatically measured with a measurement uncertainty down to the sub micrometer range. Several edge points can be measured either, as with tactile coordinate measuring machines, through sensor technology procedures and measurement of one point per image (measurement of the image) or with smaller objects, directly on the image, thereby speeding up measurement. Disturbances through reflexes and dust can be eliminated with the aid of appropriate image improvements (filter operations), so that these devices are also used directly in the production area.

By integrating autofocus sensors, measurements can also be carried out in the third coordinate direction Z. One focuses on the object edge or object surface, by determining the contrast or gradient of an image window during a search function in Z direction. At the Z position of the maximum contrast/gradient, the focus point is reached and the object is a known distance in front of the sensor.

The efficiency of optical coordinate measuring machines corresponds to tactile devices in terms of flexibility, automated operational sequence of measuring programs, programmability and measurement data documentation. The device-specific parameters, which are determined according to VDI 2617, sheet 6, are also comparable with those of tactile devices, whereby, however, faster measurement data acquisition is achieved and particularly small, flat or ductile objects can be measured with these devices.

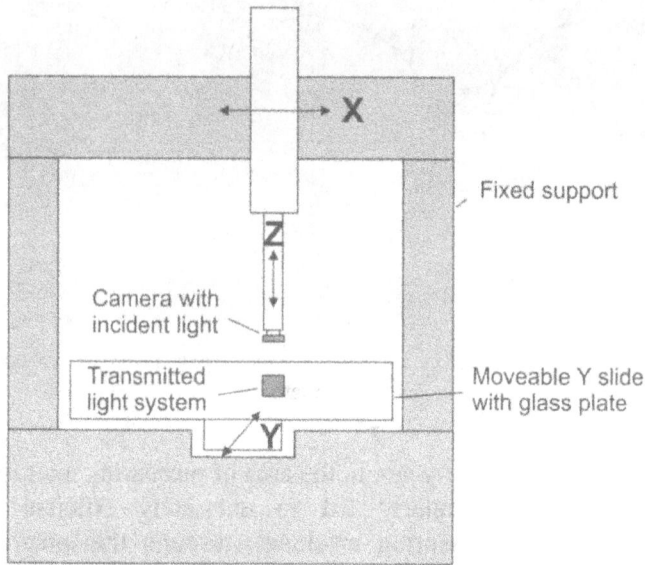

Figure 4.3-56: Principle diagram of an optical three-dimensional coordinate measuring machine

The probing deviations with optical coordinate measuring machines are substantially attributable to error influences in the optical representation of the object [Sha 96]. These arise, for example on curved surfaces (horizontal cylinders), due to the form of the edge and the height of the measured object, as well as with incident light measurements, particularly through micro surface texture. These can, however, be minimised by setting the optimal illumination in each case as well as through measuring function-specific calibration.

4.4   Coordinate Measuring Technology

4.4.1 Fundamentals of Coordinate Measuring Technology

In the course of rising quality demands on industrial products, it is necessary to monitor the product quality as well as the production process ever more precisely, but foremost faster and more economically. With rising product complexity, the measuring inspection becomes increasingly meaningful. Devices and measuring procedures using coordinate measuring technology have emerged as extremely flexible and efficient tools for the geometrical inspection of workpieces.

4.4.1.1 Principle of CMT

A characteristic of the procedure for the coordinate measuring technique is the idealised definition of individual workpiece geometries (e.g. bore hole, flange surface) by mathematically describing individual surface points in marked areas, which are mostly important for the proper operation of the workpiece, in a common coordinate system (**Figure 4.4-1**) [Neu 93], [Pf 92]. The recording of these spatial points can be realised using different principles which are detailed further in Section 4.4.2.2.

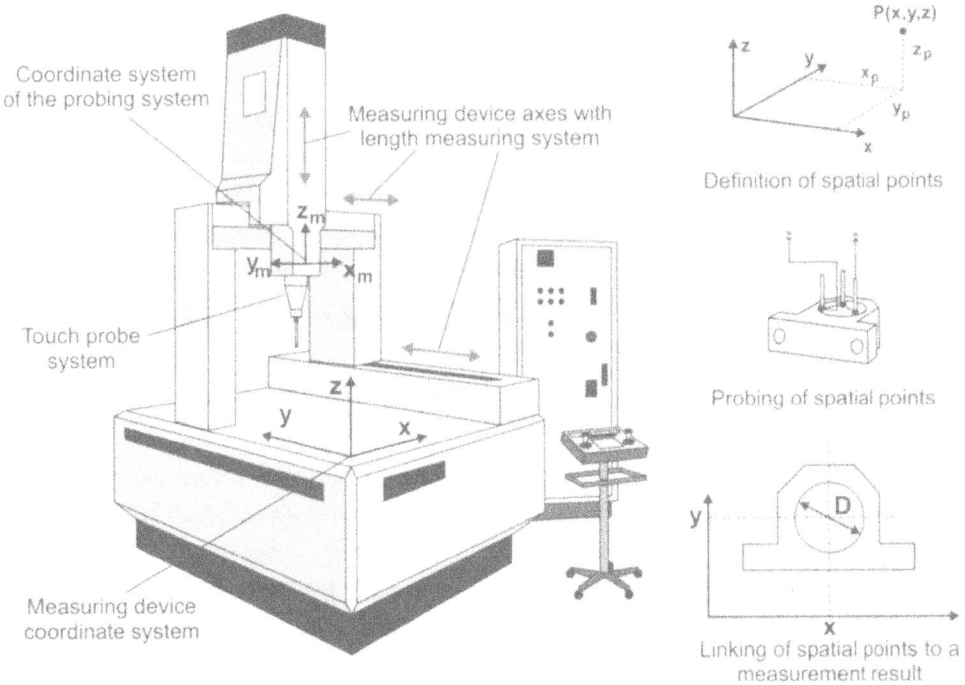

Figure 4.4-1: Principle of coordinate measuring technology

Basically, all measuring devices which use the above mentioned principle of coordinate measuring technology to acquire the actual geometry of real workpieces and determine their deviations in comparison to the specified geometry can be defined as coordinate measuring machines (CMM). In the following, however, the usual general linguistic sense will be used, according to which a coordinate measuring machine is understood to be a measuring device with limited measuring volume which measures in a three-dimensional space.

In combination with the appropriate measuring and evaluation software, CMM's offer measuring possibilities for geometrical features on prismatic, rotationally symmetrical and rotationally asymmetrical workpieces.

In addition to the traditional CMM, as an important representative of a certain type of coordinate metrological devices, two further coordinate measuring systems will be mentioned here, which are mainly used when the measurement volume of usual CMM's is not sufficient for the size of the measured object, or when a large number of measuring points needs to be acquired within a short period of time:

- Triangulation measuring systems (Section 4.3.3.1)

- Photogrammetric measuring systems (Section 4.3.2.4)

4.4.1.2 Terms for the Description of Workpieces

The most frequent function of industrial metrology is to determine deviations of the real form from the ideal form by measuring the workpiece, i.e. determining dimensional, positional and form deviations. Here the form of a workpiece can be described by its individual form elements and their spatial position with respect to one another. The following terms are used in coordinate metrology:

- The *specified geometry* defines an ideal form, as it is theoretically preset by design engineering and produced in manufacturing.

- In contrast, the *actual geometry* describes an idealised form, as it has been metrologically acquired.

- The *deviation,* which represents the difference between actual geometry and specified geometry, is the third element of the status description of workpiece geometry.

As the real workpiece surface is only represented by individual points during measurement, the actual geometry can only approximate the real feature of the workpiece form. In order to clarify that the measurement of the workpiece form only concerns a representative illustration of the real form, which describes the form with mathematical description mechanisms, the term workpiece geometry is used (**Figure 4.4-2**).

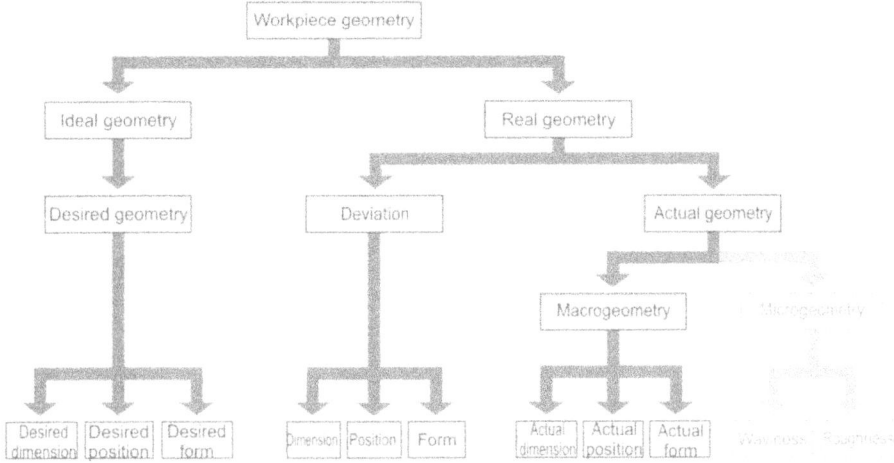

Figure 4.4-2:   Ideal and real workpiece geometry

4.4.1.3 Geometry Elements and their Parameters

The form elements of workpiece geometry are defined by features and attributes. With the geometry measurement of a bore hole it can be seen that an ideal geometrical element is assigned to the technically produced bore hole (**Figure 4.4-3**). In the case of a bore hole, this ideal geometrical element is a cylinder. The mathematical description of a cylinder provides the geometrical parameters with which the features of dimension, position and shape of this form element are defined.

In order to be able to judge the deviations of a form element in a nominal/actual value comparison, both the specified and the actual geometry are defined by a geometrically ideal equivalent element. An equivalent element is described by regression equations, the parameters of which are calculated for discrete measurement values (spatial points) and compared with the parameters of the specified geometry. The required compensation calculation is executed numerically. The compensation cylinder calculated from the measurement points is thus generally accepted to be a compensation element, the parameters of which determine the actual geometry of the bore hole. Finally, of interest for quality testing, is the comparison of the individual parameters or parameter values of a form element with the predefined specified values (specified dimension, position and form) and their tolerances.

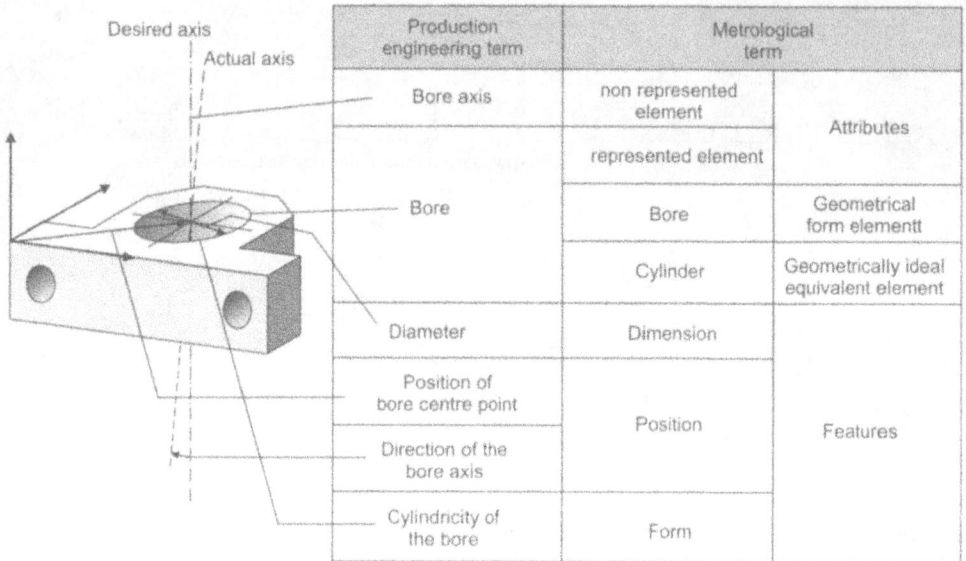

Figure 4.4-3: Geometry element features

The example of a bore hole illustrates that with the calculation of the compensation cylinder, the dimensional deviation of the diameter and the deviations of the individual measuring points of the compensation cylinder describe the form deviations of the real bore hole from the specified geometry (**Figure 4.4-4**).

Measurement task- Bore diameter

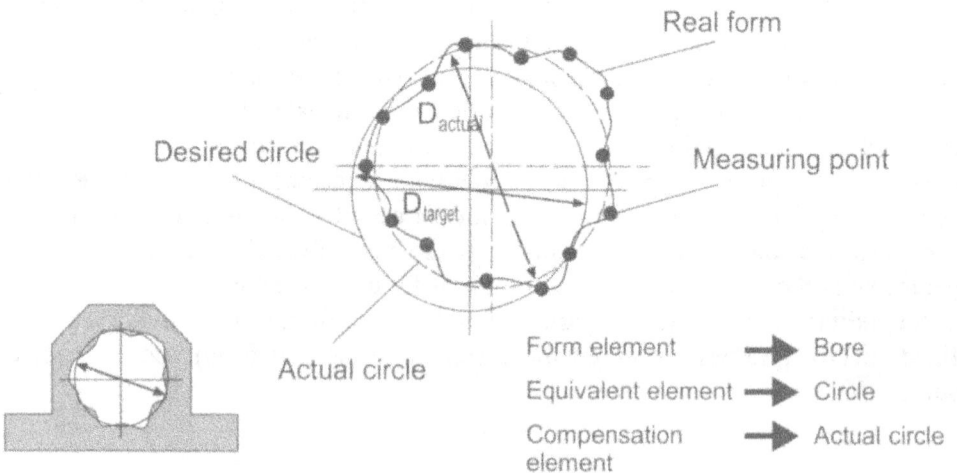

Figure 4.4-4: Measuring function of bore hole diameters

However, this does not initially express anything yet about the position of the cylinder, respectively, the bore hole in the workpiece. In order to acquire the position also metrologically now, a reference system must be defined to which all form elements refer. In coordinate metrology, the 3-plane reference system is typically used (**Figure 4.4-5**). It results from the relationship with the measuring device coordinate system which is mostly realised in the measuring device as a Cartesian system and conforms with the coordinate system which is used during numerical evaluation in the vector analysis. This reference coordinate system enables the exact definition of the cylinder position in space and concomitantly also with espect to other geometries measured in the reference coordinate system. Features can also be derived from linkages from several geometry elements, e.g. parallelism of two axes in a gear box.

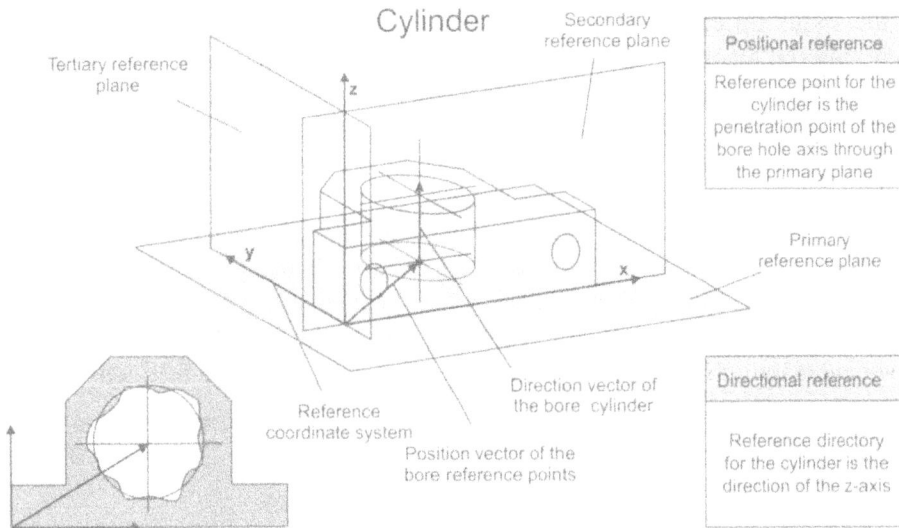

Figure 4.4-5: Reference element – reference coordinate system

Coordinate metrology is typically able to measure geometrical deviations which can determine several form proportions of irregularities, in addition to macro geometrical characteristics such as dimension and position. Typically, irregularities of the first and second order, i.e. according to the definition of DIN 4760 Form Deviations and Ripples, can be acquired with CMM's. Which resolving power the coordinate measurement has with respect to the deviation regulation, depends very strongly on the probing principle and the probing system.

4.4.1.4 Determining Geometrical Basic Elements

In order to determine the geometrical quality criteria of a workpiece, spatial points are acquired in a defined coordinate system and transformed into parameters of geometrical basic elements using stylus tip radius correction.

Acquisition of a Spatial Point

The acquisition of individual spatial points can be performed using different measuring principles (section 4.4.2.2). In the following, all procedural descriptions refer to tactile, mechanical probing systems. The measurement of the workpiece geometry is performed by touching with a probe, which is in principle available in various forms – as a ball, disc, cylinder or cone. The following will always assume a ball-shaped probe, as it is the most universal probe element which is most often used on CMM's. The probe geometry (i.e. the stylus tip diameter and the position of the stylus tip centre point) is determined prior to measurement by calibration with material standards (usually a high-precision ceramic ball) with reference to the coordinate system. With the aid of linear modules, a point is probed by the relative movement between the probing system and the workpiece, and the stylus tip centre point is transferred to the evaluation software as a measured coordinate in the measuring device coordinate system. By the measurement of several points and their assignement to a geometrical element, the measurement data, i.e. the parameters of the computed compensation geometry, is made available in the workpiece coordinate system (WCS). The actual geometry can subsequently be calculated from the measuring data. In order to obtain the actual geometrical element when evaluating the measured points, three basic tasks must be fulfilled:

- coordinate transformation from measuring device coordinate system into the WCS
- stylus tip radius correction (probing point calculation)
- calculation of the equivalent element (calculation of the actual geometry)

The actual geometry is not only used for determining the deviations of the real workpiece, but also, e.g. to form a workpiece-specific coordinate system.

Coordinate Systems and Transformations

Workpieces to be inspected are described by design engineering in a uniform system of units. The production drawing determines the reference system, in which the position and dimensions of the individual form elements are defined. When measuring, it must now be ensured that the actual geometry and deviations of the features are also output in this reference system. For this purpose, when measuring, at first a coordinate system is metrologically defined which corresponds to the required reference system of the structural specification (**Figure 4.4-6**) .

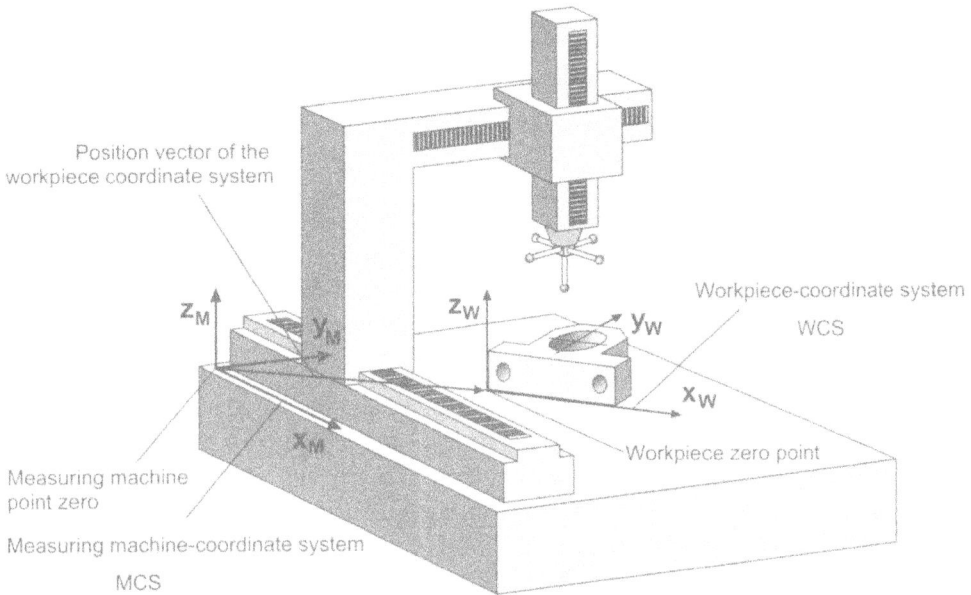

Figure 4.4-6: Measuring device and workpiece coordinate system

Cartesian coordinate systems are not the only coordinate systems which are used **(Figure 4.2-6)**, but also spherical and cylindrical coordinate systems. Cylindrical coordinate systems are primarily used where the rotational axis represents the main axis of the CMM, e.g. with special gear measuring devices, shaft inspection devices and form measuring devices. The measuring coordinate systems are, however, mostly designed as Cartesian coordinate systems. Thus, the user also appropriately defines a coordinate system of the same type as the workpiece coordinate system. The following deals solely with Cartesian coordinate systems. In principle, the procedure in coordinate measuring technology is, however, fully independent of the type of coordinate system in which the spatial position of a point is measured and represented. Here it will only be pointed out that other coordinate systems exist. All coordinate systems can be converted to the Cartesian coordinate system through coordinate transformation.

In conventional metrology, an inspected object must be aligned in a mechanically complex manner, as the measuring axis of the measuring instrument must be parallel with the reference direction of the dimension to be measured. In practice, with complex geometries, this often results in the use of expensive devices. A great advantage and a significant saving of expenses using coordinate measuring technology is that mechanical alignment is replaced by computational alignment. By measuring the reference surfaces of the workpiece in the measuring device coordinate system, the workpiece coordinate system is determined prior to the actual execution of the measuring function. This involves determining the origin and alignment of the workpiece coordinate system in relation to the measuring device coordinate system.

Stylus Tip Radius Correction

After probing the workpiece surface, the stylus tip centre is available as a measuring value. The actual point of interest with probing is, however, not the stylus tip centre, but the contact point when probing, as it is a point of the real workpiece geometry. The procedure to calculate the probing point based on the stylus tip centre is called stylus tip radius correction.

When probing the workpiece at one point, there are various defined terms which are required to describe the different statuses of the workpiece (**Figure 4.4-7**). The point which describes the designed geometry is called the specified point. This is approached from outside the workpiece, in the ideal case, in line with its direction normal – the specified normal direction. The stylus tip touches the workpiece surface when probing. This point is called the contact point. The point, which is provided by the measuring device in Cartesian coordinates, is called a measuring point and is, under ideal conditions, i.e. without deformations of the stylus tip, identical with the stylus tip centre. A probing point is calculated from the measuring point with the aid of stylus tip radius correction. It need not be physically present on the surface and is also, in most cases, not identical with the contact point. The calculated probing point thus represents an approximation of the real contour and defines a point of its metrological representation, or the actual geometry. This point is therefore also called the actual point.

The selection of the sequence of stylus tip radius correction and equivalent element calculation has a crucial influence on the computational operation. There is a differentiation between two versions:

- compensation at the point and
- compensation at the equivalent element.

When compensating the stylus tip radius at the point, the appropriate probing point is calculated for every measuring point with the aid of the stylus tip radius and a correction direction. The calculated probing points represent the actual geometry of the workpiece. The accuracy of the correction thereby crucially depends on the correction direction.

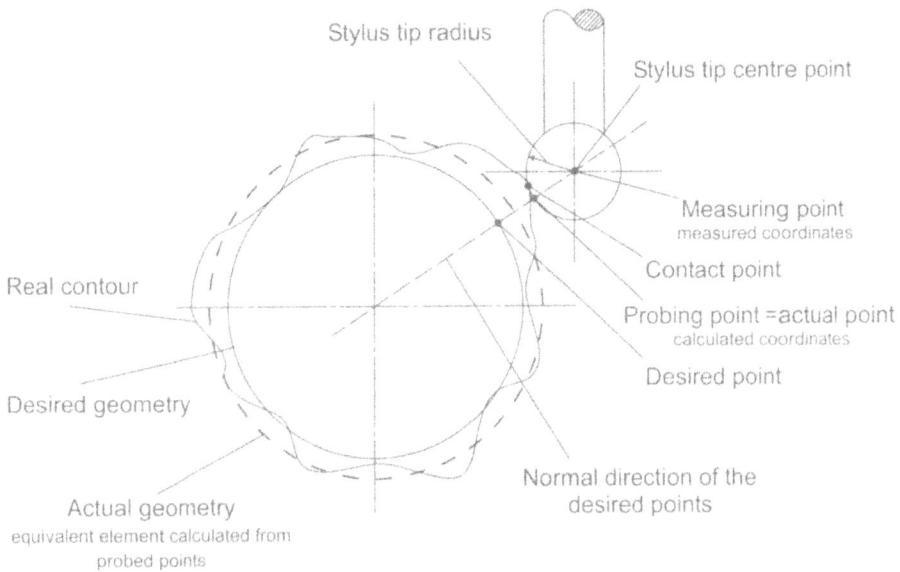

Stylus tip radius

Stylus tip centre point

Measuring point
measured coordinates

Contact point

Real contour

Probing point =actual point
calculated coordinates

Desired point

Desired geometry

Normal direction of the
desired points

Actual geometry
equivalent element calculated from
probed points

Figure 4.4-7: Definition of terms for point measurement

The procedure of compensating at the equivalent element is the most important, as it is used in most measuring functions, namely the measurement of standard geometric elements. First an ideal geometric equivalent element is calculated using the stylus tip centre and then the stylus tip radius correction is performed with the equivalent element.

Geometric Basic Elements and Computational Procedures
Workpiece surfaces can be described with geometric form elements, which can often be relatively simply defined through the elements of plane and spatial trigonometry. The mathematical handling of these elements is the prerequisite for solving the measuring tasks, such as determining the distance of two bore holes or determining the diameter of a shaft. There is a differentiation between the following basic geometric elements in coordinate metrology:

- 1-Dimensional :   point
- 2-Dimensional :   straight line, circle
- 3-Dimensional :   plane, sphere, cylinder, cone

The four basic elements - plane, sphere, cylinder and cone – are three-dimensional distinct elements which are represented by their surface. Therefore, the calculation of their actual normal and an evaluation is always possible without further auxiliary elements. A characteristic of the elements point, line and circle is that they can not be uniquely determined as an element through probing alone. For example, straight lines are only represented as edges, which can not be uniquely probed. The circle is also only embodied as an edge. However, it is often used on cylindrical surfaces to, for example, obtain auxiliary elements for further calculations. In place of this circle, it is preferable to use a cylinder for the form and positional inspection of bore holes and shafts.

Vector algebra methods are used to represent the basic geometric elements [Neu 93], [Pf 92].

4.4.1.5 Measurement and Evaluation Strategy

Mathematical and metrological reasons demand a specific minimum number of points for every geometric element:

- The mathematical minimum number of points results from the number of degrees of freedom which an element has. There are also secondary conditions regarding the arrangement of measuring points. Therefore, e.g., the four points of a sphere must not be situated on one level and points can not be identical.

- The metrological minimum number of points results from the fact that the influence of small form deviations on the result should remain small. It must therefore be larger than the mathematical minimum number of points. Depending on the form deviation of the real geometry and the requirements resulting from the order of magnitude of irregularities which still need to be resolved, a multiple of the mathematical minimum number of points mostly needs to be measured as a real used number of points.

If more than the mathematical minimum number of points is available for data evaluation, the geometric equivalent element is overdetermined by measuring points. The selection of the regression procedure has a substantial influence on the calculated compensation element. The regression procedure which is primarily used in coordinate metrology is the Gaussian condition, with which nearly all algorithms for compensation element calculations operate. In addition to the selection of a compensation procedure, the selection of the compensation condition is also a deciding factor in determining whether the measurement result reflects the functional features of the workpiece. Depending on the form of the geometric element and its specific function in its structural unit, it also makes sense to select other compensation conditions than the Gaussian. The four usual compensation conditions are:

- Gaussian condition or also L_2 standard

- Tschebyscheff condition or also L_∞ standard

- Pferch condition

- Envelope condition

Further reading detailing the calculation procedures and compensation algorithms can be found in [War 84], [Hmd 89], [Wol 84], [Krm 86] and [Wol 75].

Evaluation of Measuring and Calculation Results

The application of statistics is an aid in evaluating measuring results. As the total surface of a measured object is not acquired in coordinate metrology, but only a finite number of measuring points as a random sample, this opens up the opportunity to regard the deviations d_i of the measuring points more closely with reference to the compensation element. As many deviations are of a systematic nature, e.g. an n-sided lobing on a shaft, necessary statistical conditions are often violated, so that when interpreting a measuring result, apart from the standard deviation, the smallest and largest deviations should also always be taken into account.

4.4.1.6 Measuringt Deviations and Measuring Uncertainty of Coordinate Measuring Machines

The quality of the measuring results is crucially influenced by environmental conditions (Section 2.3), which influence the coordinate measuring machine, or the measured workpiece, as well as by device behaviour and components themselves. Device-specific causes of the measuring deviations of a coordinate measuring machine are e.g.:

- imperfect measuring systems

   -material measure

   -sensors

   -interpolation mechanism

- friction

- probing behaviour

- static and dynamic deformations

- device software

For users of a coordinate measuring machine it is of particular interest to know which measuring uncertainty (Section 2.3) can be expected when performing a measuring function. Generally, calibrated material standards are used for comparative measurements. The value of the measuring deviation is dependent on the measuring function and can therefore only be interpreted as a length-dependent measuring deviation in coordinate metrology [VDI/VDE 2617 1-6], [DIN EN ISO 10360-2]. DIN EN ISO 10360-2 concerns an international guideline for the monitoring of the measuring uncertainty of coordinate measuring devices for the spatial length measuring function. Here gauge blocks or stepped gauge blocks are used as calibrated length standards. The guidelines are used to enable a periodic monitoring of the coordinate measuring machines or a final inspection of the devices (Chapter 6).

The measuring deviation of coordinate measuring machines is substantially influenced by the components of the deviations of the linear guideways [VDI/VDE 2617 1-6]:

- positional deviations
- trueness deviations
- straightness deviations
- rotational deviations
- probing deviations

These deviation components can be acquired with the help of artefacts or laser interferometers (Section 4.3.3.6) and partially corrected computationally [Neu 88]. The description and correction of systematic deviations with spatial length measurements is an initial starting point for decreasing the uncertainty of measuring results.

In addition to computational correction of systematic measuring deviations, further uncertainty influences can be estimated computationally. This principle, also known as virtual CMM, pursues this initial starting point. The concept of the virtual CMM consists of three steps [Tra 96]:

- metrological determination of systematic and coincidental deviation components of the CMM
- estimation of non-metrologically determinable uncertainty contributions
- simulation of the measuring function on the virtual CMM to estimate the measuring function-specific uncertainty of the results

The virtual CMM has been integrated into existing software by some manufacturers, so that feature-specific measuring uncertainty estimation can be performed and documented together with the measuring results recorded on the coordinate measuring machine.

The length measuring uncertainty determined with the above mentioned procedures can, however, only be transferred to the actual measuring uncertainty arising during a particular measurement in a limited manner. This becomes particularly clear when the measuring function differs substantially from a length measurement. In order to estimate measuring uncertainty in this case, a calibrated artefact, a so-called comparator, is required on which the measuring function can be carried out in the same, or at least in a similar, form. The measuring deviations are determined from the comparison of the calibrated values with the measured values.

A typical example for the comparator procedure is also found with gear measurements on coordinate measuring machines. Here, a special artefact, the bevel gear standard, has been developed and tested for estimating the uncertainty of bevel gear measurements [Pf 96].

4.4.2 System Components and Designs of Coordinate Measuring Machines

In the mid-1960's coordinate measuring machines were developed from machine tools equipped with length measuring systems. Thus, the acquisition of almost any spatial geometry became possible within a short period of time and with low measuring uncertainty. The measuring routine initially was performed by manual guidance of the measuring device and through reading off the values from the measuring systems by the operator. Apart from that, with the first devices, the workpiece had to be manually aligned with the coordinate system of the device. The further development of coordinate measuring machines mainly concerned the automation of measuring value acquisition, particularly measuring data processing. By the application of measuring device computers, computerised numerical control of the CMM's was realised, as well as substantially accelerating measuring value acquisition and evaluation.

The problem-oriented application software which also has a crucial influence on the solution accuracy of the set measuring functions influences the utility value of the CMM. In the 1980's, special attention was placed on the realisation of measuring function-specific software. Probing systems were optimised to improve measuring device hardware and optical sensors were adapted in addition to mechanical probing systems. Development in the 1990's is particularly marked by efforts to integrate CMM's into the production flow and by computer-supported correction of error influences during measurement (Section 4.4.1.6).

The following will first describe the structure and different designs of coordinate measuring machines.

4.4.2.1 Designs

The structure of the coordinate measuring machine essentially consists of mechanical machine assemblies, drives, length measuring and probing systems, the

control and operating console and the measuring machine computer with periph-
eral devices for the output of measuring results as well as problem-oriented measuring software. Depending on the model of coordinate measuring machine, there
can be further supplementary devices such as mobile or base plate-integrated rotary tables, probing system change mechanisms, temperature sensors and workpiece loading or clamping mechanisms, in addition to the "basic equipment" mentioned.

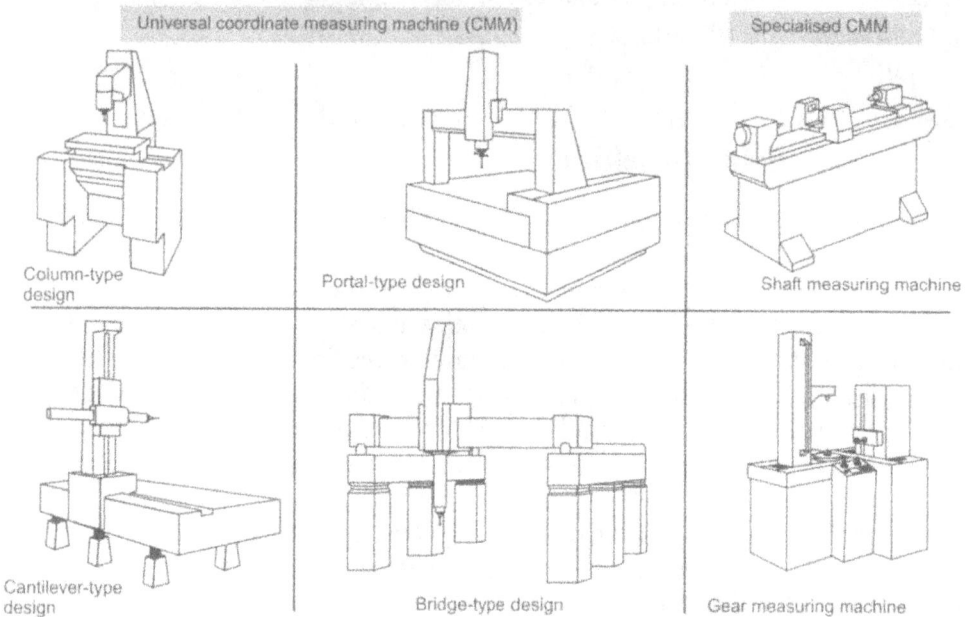

Universal coordinate measuring machine (CMM)   Specialised CMM

Column-type
design   Portal-type design   Shaft measuring machine

Cantilever-type
design   Bridge-type design   Gear measuring machine

Figure 4.4-8:   Types of coordinate measuring machines

The different designs of universal coordinate measuring machines can be divided
into four basic designs: (**Figure 4.4-8**):

- column-type design
- cantilever-type design
- portal-type design
- bridge-type design

In addition to the four basic designs, a new design has been developed in recent
times, which is based on the articulated arm principle. Its kinetic design is similar
to that of a robot, with two steering wheels and vertical joint axes. A coordinate
measuring machine constructed according to this principle consists of a height-
adjustable joint system with two highly exact angle measurement systems and a

3D measuring probe. Small dimensions and simple manual operation distinguish this measuring device. It is particularly used near to production for the inspection of dimensional, form and positional deviations on workpieces with small and medium deviations [Lot 96].

Upright Design
Devices of this design typically make available a measuring volume of up to 0.25 m^3. They are characterised by good accessibility and offer very high rigidity due to their usually short axes. Through the rigidity and the arrangement of the device axes, where the Abbé principle is adhered to relatively well, a very low measurement uncertainty to less than 1 μm can be realised with these devices. The fields of application range from gauge inspections over small prismatic and shaft-formed workpieces right up to gearing.

Horizontal Arm Design
The horizontal arm design is characterised by maximum accessibility and the fact that it is produced in the most varied size ranges, from the tabletop device right up to measurement volumes which are room-sized. Due to the short guidance length in relation to the cantilever length, the static and dynamic rigidity of this type of device is the lowest. The minimum attainable measurement uncertainties are accordingly distinctively higher than with the other designs. Fields of application are, e.g. the measurement of semi-finished products and workpieces from sheet metal processing as well as geometry inspection of entire assemblies in vehicle, aircraft and plant construction.

Gantry Design
The gantry construction method is meanwhile the most frequently represented design. It covers a majority of the measurement volumes required in production metrology. Typical representatives provide a measurement volume of 1 to 2 m^3. They are characterised by great rigidity. There is a differentiation between two variations of the gantry construction method:

- with one, the gantry is a mobile assembly and the workpiece is fixed on the device table,

- with the second, the device table is realised as a mobile assembly with the workpiece fixed on it, whereby the gantry is stationarily connected with the basic structure of the device.

CMM's with a gantry design are typically used for measuring workpieces and small to medium assemblies in machine, device and vehicle construction.

Moving Bridge Design

The moving bridge design is conceived in order to measure large objects. CMM's with a measurement volume of up to 16 x 6 x 4 m^3 have already been realised. This design is characterised by the fact that the mobile assemblies are not set up on a common device base table, but the base is formed by a foundation. The areas of application for this type of device are large and heavy mechanical engineering and also vehicle body measurement or the measurement of large tools for metal forming as well as aircraft construction and turbo machine components.

Despite the varying designs of CMM's, all types can be attributed to a common device engineering concept. With all designs, the device base carries the fixed and mobile mechanical modules. The device base must be sufficiently rigid to minimise load-sensitive deformations and have vibration-isolating bearings, if possible, on three points. Hard rock, steel or casting are used a preferable material for the device assemblies. The device table often consists of hard stone, as its natural aging has been completed. Hard stone displays low susceptibility to corrosion and at the same time is cheaper and lighter than steel. It also has a lower temperature expansion coefficient than steel, so that variations in temperature cause smaller length variations. Through the lower heat conductivity, the adaptation of the assembly to temperature variations, however, is substantially slower, so that internal tension in the component can lead to deformations. In order to better eliminate the influence of short term temperature gradients on the measuring device, the hard stone is today often partly replaced by aluminium for mobile modules (e.g. gantry and guide). With temperature variations, the aluminium component changes its outside measurements much more strongly than that of stone, however, due to its better heat conductivity, the inside temperature gradients and thus the deformations, are smaller. Additional materials which are used in coordinate metrology are ceramic (lower temperature expansion coefficient) and carbon fibre composite materials (low weight with high stability at the same time, low mass inertia of the components).

4.4.2.2 System Components

The attainable measurement uncertainty of a coordinate measuring machine is influenced by its system components. Apart from the mechanical structure and precision of the mechanical components, consisting of guides, bearings and drives, the length measurement systems of the mobile axes and the probing system are of crucial importance.

Guides – Bearings – Drives

The definitive components for the mechanical accuracy of the measuring device are the straightness of the guide rails and the perpendicularity of the guides to one another, which embody the directions of the measuring device or reference coordinate systems.

Hard stone and steel are used as guides. Air bearings are used in connection with hard stone and anti-friction bearings with steel. Air bearings have the advantage of not displaying a stick slip effect, taking up large loads statically and dynamically having self-cleaning characteristics. Anti-friction bearings can also take up large loads dynamically with good accuracy and display lower friction as well as low wear.

Today, the assemblies are generally moved with electrically propelled servo motors, depending on the automation level of the CMM. The control of the motors can either take place through the operator via the control desk of the CMM or through control programs which are processed in the measuring machine computer. The transfer of force from the drive to the mobile assembly takes place in varying ways with the aid of recirculating ball screws, friction wheels, and bands or, also more rarely, with gears and chains. The selection of the transfer method depends on the costs, as well as the various demands on characteristics such as rigidness, dynamic behaviour, slip, reversal tolerance, wear and friction.

Length Measurement Systems
The device axes are assigned to length or positional measurement systems. They consist of the measure embodiment and sensors to read the position or path modification. The following coupling mechanisms are used between measure embodiments and reading systems in coordinate measuring machines:

- rack - spur wheel
- trapeze - elevating spindle
- recirculating ball screw
- inductosyn procedure (inductive procedure)
- glass and metal line scale (optoelectronic procedure)
- laser interferometer

Incremental measurement systems are mainly used in CMM's, as knowledge about the absolute origin of the reference measurement coordinate system is usually not required for measurement. Length measurement systems are a crucial component for the accuracy of a coordinate measurement machine and are therefore laid out an order of magnitude higher in their resolution, than the length measurement uncertainty of the CMM. .

In order to minimise measurement deviations, a meaningful arrangement of length measurement systems must be realised in the CMM. According to the Abbé comparator principle (Section 2.3.2.4), the measure embodiments must be aligned with the object lengths to be measured or at least have as small a distance as possible. For structural reasons, guides and measurement systems can not naturally align, but only be arranged in parallel and closely adjoining with technological accuracy. Due to the violation of the comparator principle and the imperfection of the guides and axial directions, measurement deviations are induced which are divided into 21 components: 9 translational, 9 rotational and 3 right angle deviations. The 21 component deviations can, to large degree, be compensated computationally [Neu 88].

Probing Systems
The probing system of the CMM establishes the reference between the measuring point on the object and the device coordinate system. The principle of measuring point acquisition consists of probing the surface with a probing element and picking out the length measurement system of the CMM in this position at the same time. Probing systems are used in CMM's which are based both on the optical as well as the mechanical (tactile) probing principle (**Figure 4.4-9**).

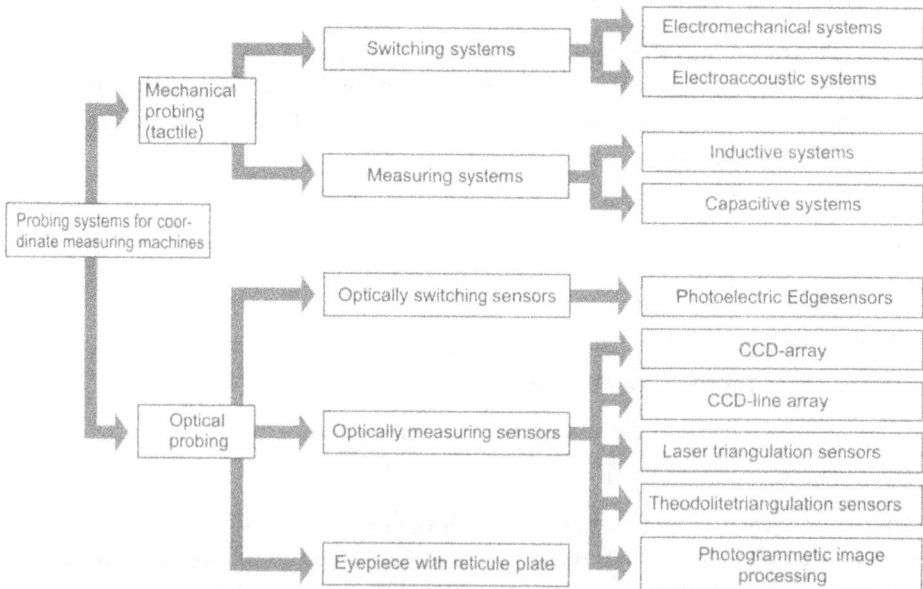

Figure 4.4-9: Allocation of probing systems for coordinate measuring machines

Most systems work with mechanical probing, whereby there is a differentiation between switching and measuring probing systems. The measurement of individual points takes place by probing with a probe which, in principle, is available in various forms – a ball, disk, cylinder or cone probe. With mechanical probing systems, probe tips with balls (ball materials – ruby, glass, steel) are also used as probing elements, which are as rigid and wear resistant as possible and their diameters and spatial positions can be very precisely calibrated. For the past several years, optical triangulations sensors and CCD cameras with image processing have been used (Section 4.3.2 and 4.3.3). The objective of integrating optical systems is mainly the reduction in measuring time. The following will only deal with mechanical probing systems [Pre 97].

Kinetic (Switching) Probing systems

Due to the reaction time of the control during probing and for collision protection reasons, a deflection of the probe is required during probing. With kinetic systems, this is mostly realised through a "bend" (Figure 4.4-10). Kinetic probing systems basically operate dynamically, e.g. at least one device axis is moved during probing. The probing recognition takes place through an impulse which triggers the selection of the length measurement system and the storing of the coordinates in the measuring machine computers.

Measuring Probing Systems

With measuring 3D probing systems, the probe is connected with the device axis through three spring parallelograms which are arranged at right angles to each other. For each parallelogram, there is usually an inductive positional sensor which converts the shift from the resting position into a distance-proportional electrical signal **(Figure 4.4-10)**. The three parallelograms are arranged orthogonally to one another and thereby enable the precise determination of the spatial deflection of the probing systems in the form of Cartesian coordinates. The described functionality of the mechanical spring parallelograms can also be realised through electronic springs. The probing recognition with measuring systems takes place through deflection of the probe, whereby the determination of the measuring point coordinates takes place by offsetting the coordinates of the selected length measurement system of the CMM from the coordinates of the deflected probing system.

Devices with measuring probing systems enable a multitude of different probing procedures.

Single point probing

- With static measurement, the movement is stopped after probing recognition and after the natural oscillations of the measuring device fade away, the measuring systems are selected. This probing procedure delivers measurement data with the least measurement uncertainty.

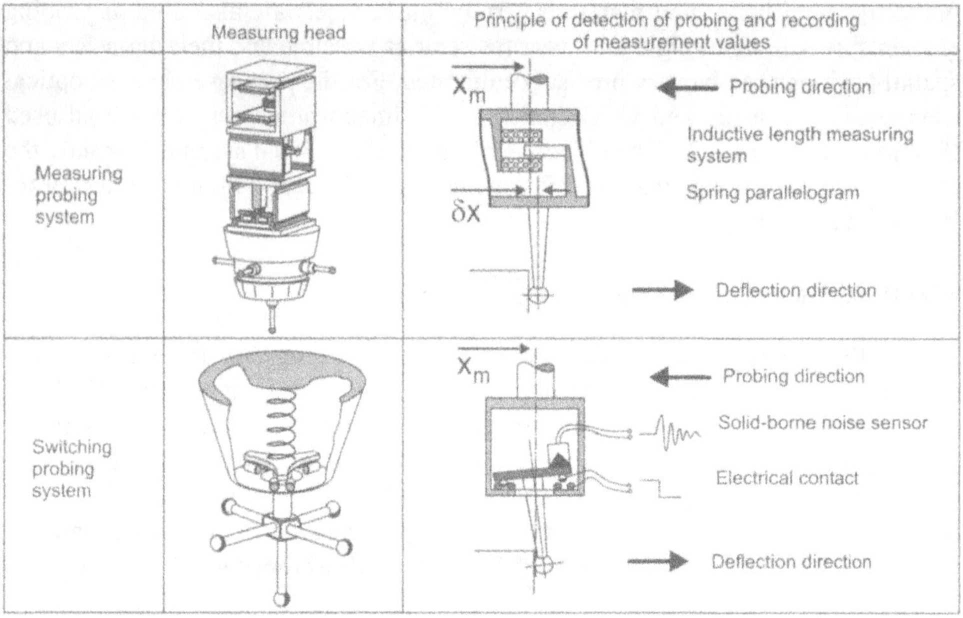

Measuring head	Principle of detection of probing and recording of measurement values
Measuring probing system	
Switching probing system	

Figure 4.4-10: Mechanical probing systems

- With dynamic probing, the measuring point is taken up during the deflection movement. After probing recognition, the measurement system in the probe and the length measurement system are continuously selected and those coordinates stored as measuring points which are either attained at a given probing force or are supplied by the measuring device axes at a probe deflection of "zero". Dynamic probing has the advantage of higher measuring rates than static probing, but the attainable measurement uncertainty is somewhat higher.

Scanning Procedure

- In contrast to single point probing, with scanning, the surface of the workpiece is continuously driven along. With scanning, the probing system is guided with motors in such a way that it is always deflected, so that probing takes place. The path which the probing system describes is defined by the target geometry and, in the case of large deviations, is adjusted in the target normal direction until a sufficient probe deflection is given. The measuring points are taken up either path-dependently or time-dependently during the dynamic measuring procedure. With scanning, geometrical elements can be taken up quickly and with a very high number of points (several thousand points per element are possible). The measurement values are, however, subject to dynamic influences which result from, among other things, machine dynamics and friction effects. For the optimal application of scanning technology, the precise knowledge of relationships between measurement uncertainty, scanning parameters and other factors specific to measuring functions are of great importance. A substantial advantage of this technology is that, in addition to dimensional and positional inspection, form deviations can also be acquired on the same measuring device.

4.4.2.3 Operating Modes and Levels of Automation

Three points of emphasis can be derived from the multitude of individual functions of a CMM, which enable the movement along the axes and probing of the work-piece, right up to data evaluation and programming of the operational measuring routine:

- axial drive and object probing
- programming and control of the measuring routine
- data acquisition and evaluation

These points of emphasis can essentially be assigned to three different automation levels of coordinate measuring machines which are available on the market today. The versions can be separated into:

- hand-guided coordinate measuring machines with positional indication of individual axes,
- hand-guided coordinate measuring machines with computer supported measurement data evaluation and logging, as well as
- CNC coordinate measuring machines

Due to a high level of flexibility which is generally aimed at and a concurrent high degree of automation, even with measuring and testing facilities, CNC controlled CMM's are being ever more frequently used for fully automatic operation.

Software

Software is divided into components for evaluating and logging measurement data, for measurement sequence planning and control and also for communication with external data processing systems. In order to produce a measuring program which runs automatically, for example, a menu surface is opened for the user in which a pre-defined operational sequence is selected and the specialised execution only specifies the input of parameters.

For programming coordinate measuring machines, e.g. converting the set measurement function into controlling and evaluation functions, manufacturers now offer numerous software packages. Due manufacturer-specific software systems, there is a disadvantage of non-uniform handling. Measurement programs for the same measurement function are thus not always transferable from one coordinate measuring machine to another.

The use of coordinate measuring machines in batch testing requires a rational method in order to carry out frequently repeated measurement functions quickly and economically. This does not require repeated manual measurement sequences on the CMM, but measuring workpieces of the same geometry in the CNC sequence through so-called parts programs. These control programs substantially reduce the operating complexity of geometry testing on coordinate measuring machines. Despite the partly greatly differing software concepts, the available software systems make comparable functions available. Basically, there are three different programming methods for generating the information required for carrying out a measurement on the CMM [Pf 92]:

- Manual Programming

This is understood to be the manual creation of control data with the aid of an appropriate programming language. This procedure is often used for subsequent modification of CNC measurement programs and can be producer-specifically applied directly to the control computer of the CMM.

Learning Programming or Teach-in Procedure

-

This procedure, already well-known from the programming of automatically running motion cycles and function commands in robot programming, enables the creation of a measurement program through the automatic storage of all motion sequences as well as the probed form elements required to carry out measurement during a manually controlled measurement

- Offline Programming Procedure

Offline program creation is applied away from machines at an appropriate programming workstation without a measuring device. The systems used here are structured on CAD functionalities and workpiece information there available.

4.4.3 Application of Coordinate Metrology

By integrating coordinate measuring machines into the production environment and the optimising the interface of coordinate metrology with preliminary areas of product development and work planning, the economy and possible fields of application of coordinate metrology can be expanded.

4.4.3.1 Fields of Application

Coordinate measuring machines were initially conceived primarily for use in the measuring room. The field of application of coordinate measuring machines has been substantially expanded in the course of more modern, in particular, auto-mated and computer controlled production equipment. Due to the flexibility and universality of the coordinate measuring principle, it is possible to realise different execution forms of the CMM through appropriate adaptation, expansion or further development of hardware and software components, which can meet up to tasks in practically all areas (**Figure 4.4-11**).

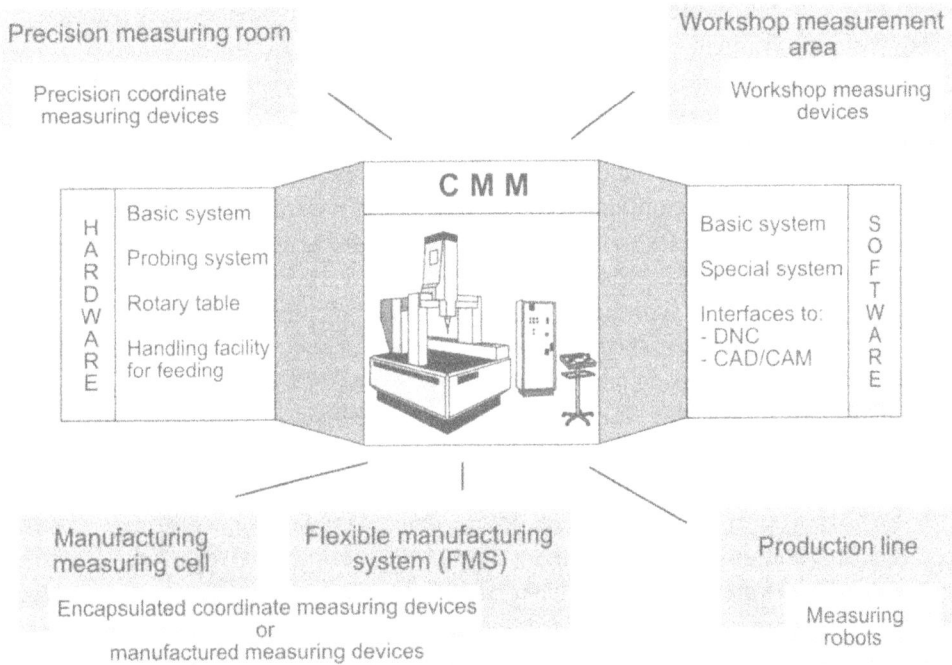

Figure 4.4-11: Fields of application for coordinate measuring machines

Precision coordinate measuring machines are applied in the precision measuring room. They are structurally configured for the lowest possible measurement uncer-tainty, which also requires air conditioned and vibration-free surrounding condi-tions. The measuring possibilities are very universal. Highly precise, tactile meas-uring probing systems are mainly used. Due to the diversity of probe exchange mechanisms, turntables, etc. according to varying devices, the degree of automa-tion is generally relatively low. However, through computer-guided CNC control, the flexibility and universality is very high.

For a normal shop floor, faster and more economical coordinate measuring machines are required. The mechanical structure of devices is durable and as insensitive as possible to usual surrounding conditions on the shop floor.

In highly automated production facilities such as production cells and production systems, the CMM lends itself to being integrated directly into the system. With such systems, the function of the CMM is coupled with the flow of material and information, thus enabling the automatic driveing of the CMM with palettes as well as enabling measurement to be controlled from the production guidance level [Pc 97]. Coordinate measuring machines are used from the area of shop floor measuring devices, which can be further protected from surrounding influences through air conditioned casings.

On the software side, finished measuring programs are offered which carry out the complex measuring routine and measurement data evaluation as automatically as possible. Measuring instrument manufacturers offer measuring programs for standardised machine elements such as gears or camshafts and also for tools which are often used, such as milling cutters and trimmers. With increasing complexity of workpiece geometry, as with bevel gear teeth or turbo rotors from the area of fluid flow machinery, it is necessary to use a fourth device axis. When a measuring machine coordinate system, which is usually configured as a Cartesian system, is extended through a turntable, the measuring possibilities are expanded, particularly for workpiece forms which are difficult to access. As gears are very frequently used in mechanical engineering, there are special gearing measuring devices available for this type of workpiece. The device concept regarding workpiece acquisition and the rotational axis as the main axis of the measuring device are adapted to the type of workpiece. Similarly specialised device concepts also exist as shaft testing devices and camshaft testing devices.

4.4.3.2 Economy

As coordinate metrology and the coordinate measuring machine do not represent the only system for solving a measuring task, several prerequisites must be fulfilled in order to realise an economical application in comparison with other measuring instruments and measuring methods. Coordinate metrology will always need to be used when the workpiece geometry is distinctly three-dimensional and particularly when features to be measured, e.g. drill hole axes are not physically present. The use of coordinate measuring machines is meaningful when the number of features per workpiece increase, e.g. a very low measurement uncertainty is required for the measurement result. An economical application of coordinate metrology can be obtained if the measurement results are not used solely for quality inspection, but serve the continuous adaptation of production and construction processes in a control loop.

The advantage of using coordinate measuring machines, in comparison with conventional measuring and testing equipment, lies in the fact that a multitude of different measuring functions can be partly or fully automatically processed in one, or a maximum of two, clampings. In conventional production testing there are several alternative measuring devices for almost every measuring function. This requires extensive experience during test planning and selection of the measuring instruments. The use of conventional testing instruments frequently cause high downtimes, as the measuring routine can not usually be fully automated.

These disadvantages of conventional measuring and testing technology are faced by the disadvantage of high acquisition or operating costs in the area of universal coordinate metrology. An economical application of coordinate measuring machines can only be obtained if substantial measurement time savings can be attained in comparison with conventional measuring procedures, a good utilisation of the device is achieved and also with frequently changing measuring functions.

4.4.3.3 Integration of Coordinate Metrology

With regard to the efficient application as well as the integration of coordinate metrology in the product development process, several communication interfaces must be established between the CMM and inspection preparation, or evaluation, processes, as well as the user. At present, there only exist partial solutions which enable data exchange between individual components of the product development process. An integral system solution is being aimed at for the future.

Construction CMM

A distinct contribution to reducing programming complexity results from an information-technological coupling of computer aided drawing (CAD) and offline programming for production (NC programming), as well as handling (robotics) and the generation of CNC measurement programs. The objective is integrated information processing through reusing data already used once in the production development process.

There are different producer-specific interfaces as well as standardised and partly standardised interfaces available for exchanging data between systems. Available interfaces which are often used and standardised for CAD data are

- IGES,

- VDAFS   [DIN 66301]

Today's state of the art enables data transfer for prismatic and rotation symmetrical workpieces, as well as for workpieces with free formed surfaces.

With the generation of CNC measuring programs away from the measuring device, the generated control data is transferred through an interface to the measuring machine computer where it is converted to a measuring routine. The interpretation and execution of the control data takes place on the measuring machine through a producer-specific post processor. In addition to the producer-specific solutions, there so far exists one interface which is standardised through an American norm,

- DMIS,

with which data for controlling coordinate measuring machines can be exchanged between programming systems away from the measuring machine and coordinate measuring machines [DMIS 3.0d]. The DMIS vocabulary consists of elements for defining measuring functions and for describing geometrical relationships, e.g. form elements, tolerances and coordinate systems.

Test Planning – CMM

In contrast to other measuring and inspection devices, with test planning using coordinate measuring machines it is absolutely necessary to set a measuring strategy. Currently, this is usually carried out by the expert who has also created the measurement program, so that there is no uniformity and subsequently, no comparable measuring strategies are used. Depending on the measuring function, the measuring strategy contains information about the probing method, probing point distribution on the workpiece, evaluation procedures and representation of the measurement results. However, this is precisely the information which can only be provided by specialists with the appropriate know-how.

An interface between design engineering (CAD), coordinate measuring machine and test planning can be used for the automated definition of measuring strategies for individual measurement functions of a workpiece, assigning features and, building on that, creating a measuring routine.

User Interface

Due to a high level of operating and programming complexity, coordinate measuring machines require a highly qualified specialist, who is familiar with the machine and the available software. The operation and program creation for coordinate measuring machines from different manufacturers have no uniformity and can currently only be operated by specifically trained specialists.

The representation of measurement results usually takes place in the form of metrology records, thus not enabling sufficient visualisation. The future objective here must be a user friendly operating control as well as a clear and feature-orientated representation of the measurement results.

Measurement Data Supply

The exchange of data between systems from different fields of application is only problem-free if functions are standardised as well as data structures. Today, this is generally not yet implemented. There are specialised solutions, e.g. for gearing, where a data feedback of the actual geometry to the design engineering and manufacturing process is implemented. The goal of integrating coordinate measuring machines into the quality control loops of a modern production process is not yet implemented for every geometry in coordinate metrology.

While the flow of data from design engineering to coordinate metrology is already automated to a large extent, gaps still exist today with measurement data feedback. There is generally no automatic storage of measurement results in an appropriate data storage system; however it is of interest in the context of product liability and subsequent processing with other systems (e.g. statistical process control SPC, Section 5.2). The results of coordinate metrology are still often bring in manual intermediate steps, such as controlling or regulating data in the production process. There is an attempt to close this gap today with further software developments and standardisation efforts.

4.5   Form and Surface Testing Technology

4.5.1 Form Testing Technology

The function of form testing technology is the metrological acquisition of form deviations on a workpiece and making a predication about the quality of manufactured components by comparing the determined form parameters with the tolerated dimensions.

4.5.1.1 Introduction

According to VDI 2601, deviations from geometrically ideal form occur with manufactured workpieces for the following reasons (**Figure 4.5-1**):

- machine-dependent causes

- workpiece-dependent causes

- environmentally-dependent causes

Additionally, changes in the contact and friction conditions during processing, as well as operating errors can cause form and positional deviations in manufactured workpieces. Several different causes usually affect the manufacturing result at the same time. It is generally only partly possible to separate them [War 84].

VDI/VDE 2601: "Every workpiece, no matter how precisely it has been manufactured, displays deviations from the geometrically ideal form"		
Machine-dependent causes	Workpiece-dependent causes	Environmentally-dependent causes
• static and dynamic deviations of form due to guide reails and bearings of mobile machine components • positioning deviations of these mobile components • elastic deformations of the machine, the guide rails or the tool • tool wear • bearing play • vibrations between the tool and the machine	• material inhomogeneities • deformation of the workpiece during processing • differing local temperature distribution during the production process • subsequent shrinkage after processing • releasing of inner stress after processing • deformation due to hardening	• local temperature fluctuations • temporal temperature fluctuations • vibrations transfered to the machine from the surroundings via the foundation

Figure 4.5-1:   Causes for form deviations according to VDI 2601

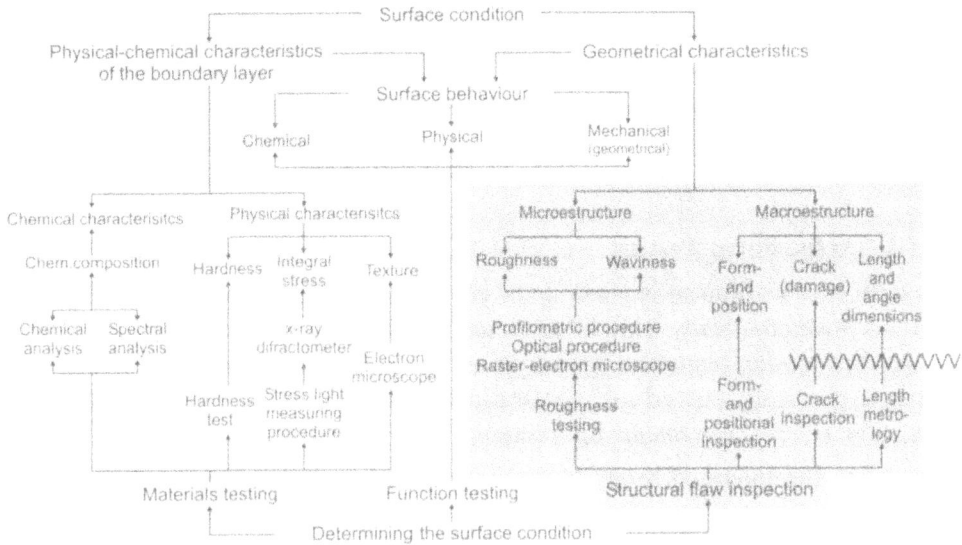

Figure 4.5-2: Surface features and their testing according to VDI 2601

In addition to the other surface feature functions mentioned in VDI/VDE 2601, which are represented in **Figure 4.5-2,** the macrostructural deviations will be examined in this section and the micro structural deviations in Section 4.5.2.

Structural deviation (superelevated representation as a profile cutaway diagram)	Examples fot the type of deviation	Examples fot the original cause
1st order: form deviations	straightness, flatness, roundness deviations among others	Errors in tool machine guidance, bending of the machine or the workpiece, incorrect clamping of the workpiece, deformation due to hardening, wear
2nd order: waviness	waves (see DIN 4761)	Off-centre clamping form or path deviations of rotatory grinder, machine or tool vibrations
3rd order: roughness	grooves (see DIN 4761)	Form of the tool cutting edge, feeding or submission of the tool
4th order: roughness	ridges scales cones (see DIN 4761)	Procedure of chipping formation (ripping chippings, shear chipping, built-up edge), material deformation during shot-blasting, bud formation during galvanising treatment
5th order: roughness no longer easily representable	textural structure	Crystallisation procedures, change in surface due to chemical influences, corrosion procedure
6th order: no longer easily representable	lattice structure of the material	

The structural deviations from 1st to 4th order are generally superposed to the real surface

Example

Figure 4.5-3: Order system for structural deviations [DIN 4760]

In the order system for form deviations according to DIN 4760, form deviations are indicated as structural deviations of the 1st order and thus macrostructural deviations (**Figure 4.5-3**). Conversely, structural deviations of the 2nd to 5th order rank as microstructural deviations. Accordingly, different production measurement techniques are used for the acquisition of these deviations.

4.5.1.2 Tactile Form Testing

The form of a workpiece is made up of geometrical form elements. This can concern the mathematically simple to describe standard form elements of point, straight line, circle, plane and sphere, as well as complicated geometrical elements such as aspherically curved surfaces or cams. Ideal geometrical forms are indicated as *nominal reference elements* for tolerating form elements.

In contrast, the tolerated form elements are called *actual reference elements*. By indicating form tolerances in a drawing, a *tolerance zone* is determined within which all points of an actual reference element must lie. Depending on the type of form tolerance and its specification in the drawing, the tolerance zone can be planar or spatial. The tolerance zone is limited by geometrically ideal surface or line elements, as described in detail in Section 2.5.2.2. *Form deviations* are those structural deviations which can be recognised in their entire expansion when viewing the entire surface or one of it partial surfaces.

The relationship of the form deviation distances to the depth is, as a rule, greater by 1000 : 1. The form deviation is the value of the largest deviation of a form element from its ideal form. By specifying the *form tolerance* , the greatest permissible form deviation is determined in a shop floor drawing. In ISO 1101 form tolerances are defined as (**Figure 4.5-4**):

- *straightness*
- *flatness*
- *roundness*
- *cylinder form*
- *line form and*
- *surface form*

Form deviations are those structural deviations which can be recognised in their entire expansion when viewing the entire surface or one of its partial surfaces

Description	Straightness	Evenness	Roundness	Cylinder form	Line form	Surface form
Symbol						
Tolerance zone						
Drawing specification						

Figure 4.5-4: Form deviations and form tolerances

In line with the variety of workpieces which can be inspected with regard to the form of the actual reference elements, there are a great number of different testing principles and procedures. A detailed representation can be found in [War 84]. The procedures generally consist of a comparison with a test normal, which can either be an embodiment (e.g. ruler, measuring plate) or not (e.g. laser beam). The tolerated geometry elements (straight line, plane, roundness) are determined either with distance or angle measurements to the reference element (Section 2.5.2.2).

The testing devices are to be selected in such a way regarding their measurement uncertainty, that the measurement uncertainty of the testing method is in a justifiable relation to the given tolerance. The measurement uncertainty of the device is, as a rule, not more than 20% of the respective tolerance (Section 2.3.3).

For the alignment of workpieces, the minimum condition generally applies: when measuring form deviations, the workpiece should be aligned in such a way that the greatest distance between the reference line, or surface, and any point of the measured surface is minimal.

As examples, tactile roundness testing, interferometric form testing on smooth surfaces and interferometry on rough surfaces will be discussed in this section and in the following section.

4.5.1.3 Tactile Roundness Testing

In addition to straightness and flatness testing, roundness testing represents the most often used method in production testing and will be discussed here in a representative function. **Figure 4.5-5** shows the execution form and principle of the form testing device (form tester MFU7, Mahr GmbH, Göttingen) for roundness testing. The workpiece must either rotate around its own axis or the probe element must rotate around the standing workpiece. In any case, it is necessary to produce a rotational axis which is vertical to the desired plane of measurement and runs through the centre point of the workpiece profile. Positional changes of the measuring surface and the rotational axis overlap the measurement as a disturbance and thereby result in a measurement error. The inspection is carried out on the required number of measurement cross sections. The greatest of the roundness deviations determined on the individual cross sections is compared with the indicated tolerance value.

If a *roundness inspection is carried out by measuring radius deviations*, the radius changes determined on the dimension of a rotation symmetrical geometry element are evaluated according to one of the methods specified in **Figure 4.5-6**. In each of the cases represented, the roundness deviation f_k results in:

$$f_k = R_a - R_i \qquad\qquad\qquad (4.5\text{-}1)$$

where R_a is the largest and R_i the smallest determined radius.

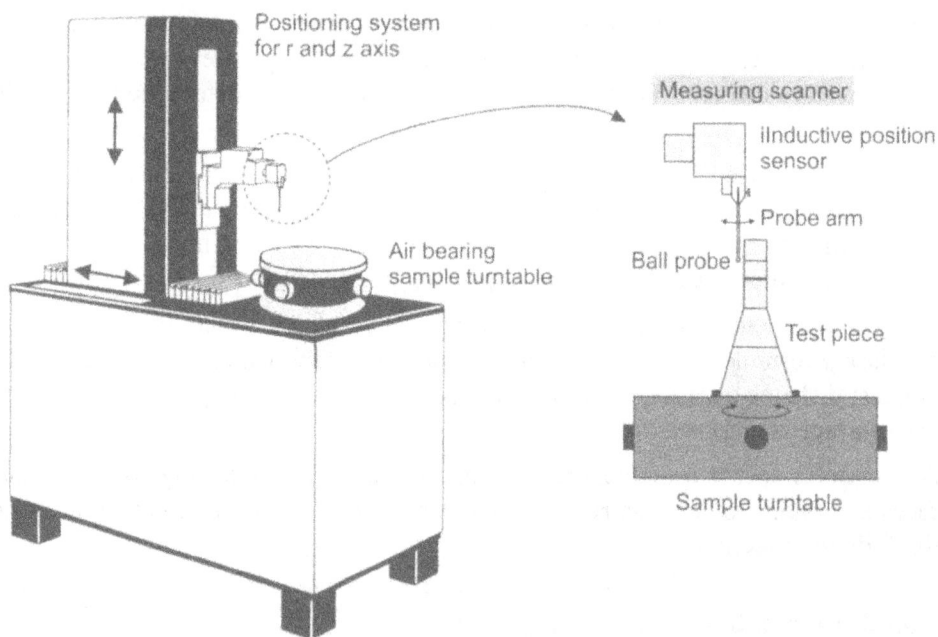

Figure 4.5-5: Tactile form testing device

With the *Minimum Circumscribed Circle* procedure (MMC), an enveloping circle is calculated which represents the smallest circle encompassing the roundness profile. The *Maximum Inscribed Circle* (MIC) represents a packed circle which can be registered as the largest circle on the roundness profile. According to the *Minimum Zone Circles* method (MZC) two concentric circles are calculated which encompass the roundness profile with minimum radius distance. The *Least Square Circle* (LSC) runs through the measured roundness profile and the minimised squared sum of the profile deviations.

For *single point probing*, represented in **Figure 4.5-6 a),** the test piece is arranged coaxially to the measuring instrument on a round table. It is advantageous for the alignment if the round table has a centring or levelling mechanism. The radial deviations with reference to the rotational axis are taken up during a full rotation and used to determine the roundness deviation.

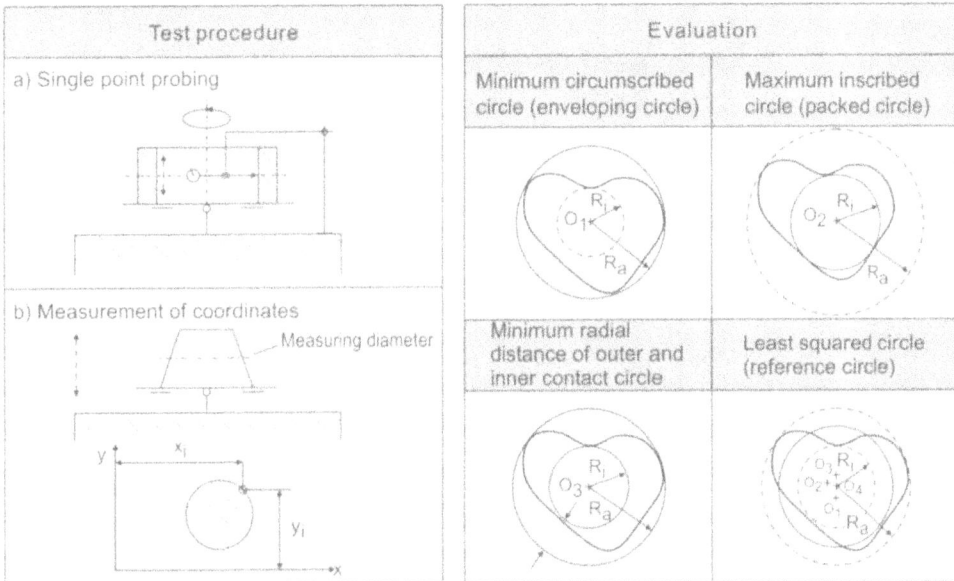

Figure 4.5-6:   Roundness testing by measuring radius deviations

Determining radius deviations for roundness testing is also possible by *measuring coordinates*, as shown in **Figure 4.5-6 b).** For this, the test object is aligned with its rotational axis parallel to a coordinate axis. For the required number of measurement points of a minimum of 4 on a measurement cross section, both of the coordinates x_i and y_i are determined. Subsequently, the roundness deviation of f_k can be calculated according to one of the presented evaluation procedures. If a roundness test is carried out through *two-point probing*, the tested object must be fixed according to the alignment. The maximum display difference is determined by rotating the tested object. The roundness deviation f_{ak} is then equal to half of the maximum display difference. The procedure is used to determine form deviations which are due to even numbered polygons (**Figure 4.5-7 a)**). Form deviations which are due to odd numbered polygons (orbiform curves), can only be determined with *three-point probing* (**Figure 4.5-7 b)**. The factor k occurring in the formula for roundness deviation depends on the opening angle α of the v-shaped prism as well as on the number n of the elevations of the profile line (**Figure 4.5-7 c)**. In practice, the most common prism angles are $\alpha = 90°$ and $\alpha = 108°$. Form deviations which are due to even numbered polygons can not be acquired with three-point measurement.

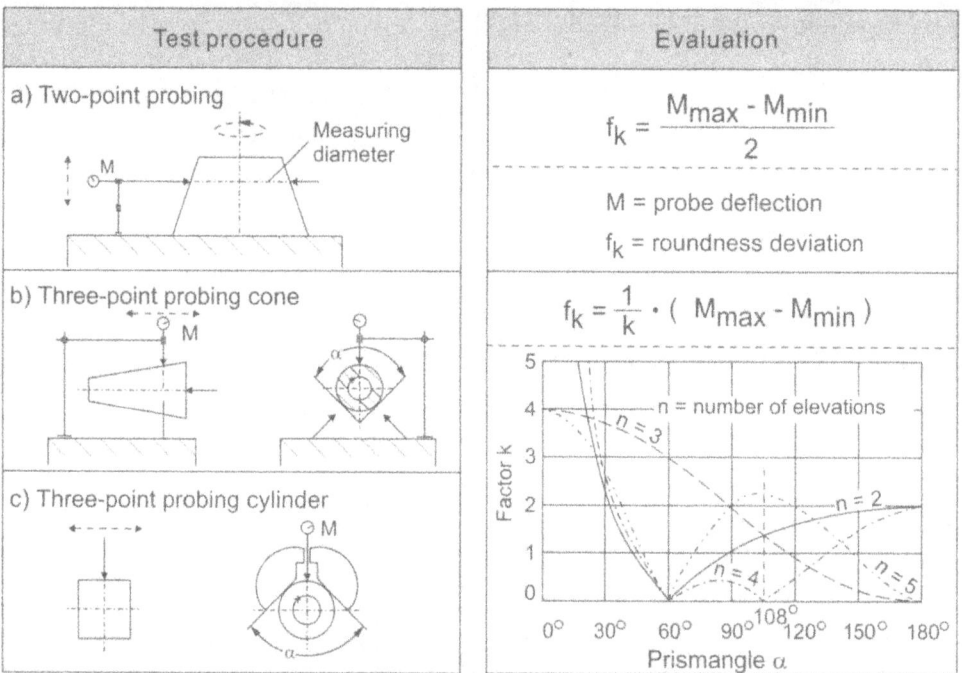

Figure 4.5-7: Roundness testing through two-point and three-point probing

Figure 4.5-8 left shows the metrology record of a measurement on a camshaft. The unfiltered representation makes it clear that several structural deviations overlap in the measurement result, so that clear predications are not easily possible. Particularly high frequency oscillations must first be eliminated by low-pass filtering. The substantially more informative result after filtering is represented in **Figure 4.5-8** right, with a modified scale.

Left panel:

Roundness		
Centre point	x	0.0524
	y	0.0490
	z	8.7071
Spatial position	nx	0.0000
	ny	0.0000
	nz	1.0000
Diameter		23.9041
Min. deviation		-0.0114
Max. deviation		0.0150
Straggling		0.0035

0.0150

t = 0.0050	Num: 2	Bez	N= 1024	Magn. = 900.00
Ist = 0.0265	FORM AND POSITIONAL INSPECTION DIN-ISO 1101			
Inspector: Bar	Part no.	WZL/DATA	Mark:	

Right panel:

Roundness		
Centre point	x	0.0524
	y	0.0490
	z	8.7071
Spatial position	nx	0.0000
	ny	0.0000
	nz	1.0000
Diameter		23.9040
Min. deviation		-0.0022
Max. deviation		0.0019
Straggling		0.0010

0.0019

t = 0.0050	Num: 2	Bez	N= 1024	Magn.= 6000.00
Ist = 0.0041	FORM AND POSITIONAL INSPECTION DIN-ISO 1101			
Inspector: Bar	Part no.	WZL/FILTER	Mark:	

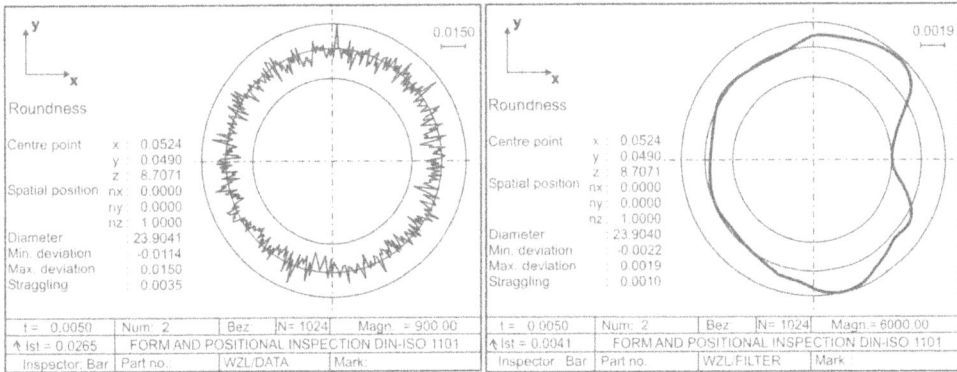

Figure 4.5-8: Roundness measurement before (left) and after (right) low-pass filtering

4.5.1.4 Form Testing Interferometry on Optically Smooth Surfaces

In addition to the probing methodology of form testing, automated form testing interferometers have established themselves firmly in production metrology, with the development of computer technologies [Tiz 91a][Mal 92][Tu 95]. Everywhere where, on the one hand, measurement uncertainties in the sub micrometer range are required (glass, optics, semiconductor industries) and on the other hand, the roughness of the inspected surfaces lies below the laser light wavelength of 633 nm typically used in interferometry, form testing interferometers represent the state of the art, so that they will also be considered within the context of this book. The concept of form testing interferometry was already explained in Section 4.3.3.6. Appropriate realisations for the measurement of flat, spherical and aspherical surfaces will be introduced here.

Form Testing Interferometry on Surfaces with a Basic Flat Form

As already explained in 4.3.3.6, high quality reference plates are used in this case, with form deviations of less than approx. 15 nm ($\lambda/50$). As it generally concerns relative measurements, which determine the deviation of the test piece relative to this reference surface, the attainable measurement uncertainty is accordingly limited. In practice, measurement uncertainties of $\lambda/10$ can be achieved [Tu 95]. The structure of a commercially available form testing interferometer following Interferometer V-100 produced by the company Möller-Wedel is shown in **Figure 4.5-9**.

Beyond the measurement principle already described in Section 4.3.3.6, the module for adjusting the reference surface and the test piece between the adjustment and measurement modes can be particularly identified here. By a motorised swiret-type glass plate can be switched between adjustment and measuring mode. In the adjustment mode, the light reflexes of an LED are detected with a camera on the collimator and reference, or test piece, and brought into compatibility. The fine adjustment then takes place by minimising the interference line density in the measuring mode.

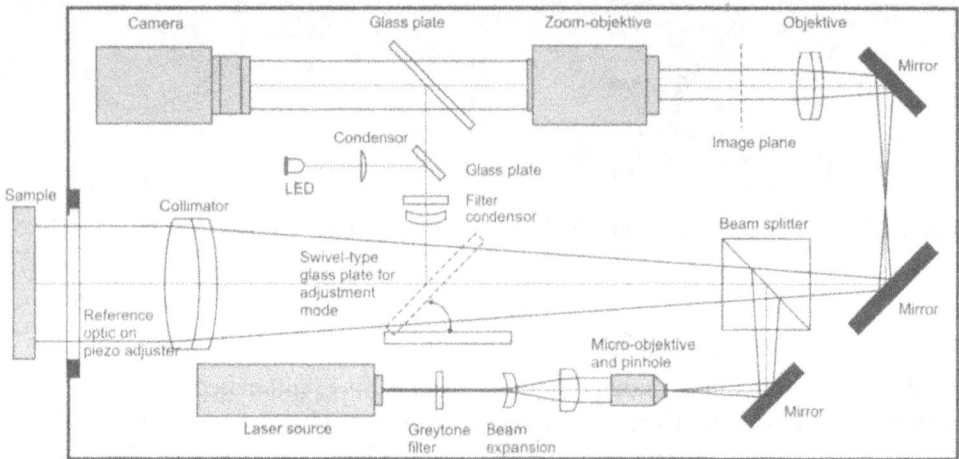

Figure 4.5-9: Structure of a form testing interferometer for inspecting flat surfaces

For detecting interferograms, commercially available CCD sensors are used with 8 bit grey tone resolution. Accordingly, phase resolutions of better than $\lambda/100$ can be realised, taking into account a non-optimal contrast. The lateral resolution is determined by the number of pixels on the CCD chip used (typically 768x576) and the magnification ratio of the Kepler arrangement when taking into account a zoom objective which could possibly be used.

With digitisation, the sampling theorem is valid, whereby a light-dark period of the stripe pattern must be detected by at least two probing points. These correspond to a maximum theoretical local gradient of the test piece surface of:

$$\Delta z_{max} = \frac{1}{2}\frac{\lambda}{2} \tag{4.5-3}$$

If this limit is exceeded, e.g. when measuring aspheres, an under-probing of the signal occurs; so-called aliasing. Additionally, the modulation transfer function (MTF) is disturbed, which leads to a deterioration of the contrast. Therefore the measurement range of commercially available interferometers is limited to approx. 25 µm deviation of the test piece from the respective reference.

Date: 03/13/97
Time: 09:48:49
Wavelength: 632.8 nm
Wedge: 0.50 wv/fr
Size: 512 X 480
Pupil: 100.0 %
Aperture: 50mm

Surface Stats:
RMS: 54.908 nm
PV: 352.549 nm

Terms Removed:
Tilt
Filtering: None
Restore: No
Ref Sub: No

3D Plot

2D Analysis

X / 2 Pt / Radius Scan
29.32 nm
-20.00
-60.00
-100.00
-140.00
-174.95

Y / Circumference Scan
122.38 nm
50.00
0.00
-50.00
-112.64

Fraunhofer Institut
Produktionstechnologie

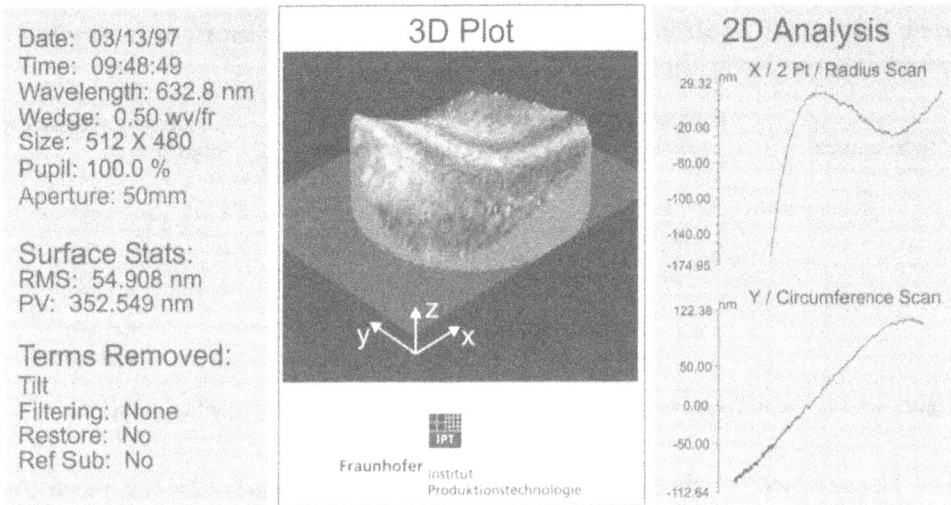

Figure 4.5-10: Measuring a glass sample with a laser interferometer (WYKO6000, WYKO Corp., Tucson, USA)

As an example of a metrology record, the result of measuring a polished BK7 glass sample can be seen in **Figure 4.5-10**. While in a 3D representation, the maximum form deviation (peak-to-valley, P-V) and the visual representation of the form are of particular interest, with 2D representation, the quantitative profile analyses along cut straight lines are interesting.

Form Testing Interferometry on Surfaces with a Basic Spherical Form

Frequently inspected specimens, e.g. glass lenses, display more of a basic spherical form than a basic flat form. As already mentioned, when using flat references, this quickly leads to aliasing and an evaluation of the measurement becomes impossible. In this case, spherical reference surfaces that are optimally adapted to the test pieces can be used, which convert the flat wave front into a spherical wave front of high quality (better than $\lambda/10$) (**Figure 4.5-11**). With these precision optics constructed from several convex-concave lenses, the last surface is usually used, which is not anti-reflection coated for this reason [Tu 95].

In commercial systems, (compare **Figure 4.5-9**) the exchange of the flat reference optics with these spherical reference optics, which are also referred to as measuring objectives or *Fizeau objectives* can take place very quickly and easily, as the optics are held in appropriate quick release mechanisms and the maladjustment can be very quickly corrected in the adjustment mode. In this way, test pieces with form deviations of up to 25 µm can be measured relative to the spherical reference used. This option is particularly interesting for lenses and mirrors, as mostly basic spherical forms are applied in addition to flats.

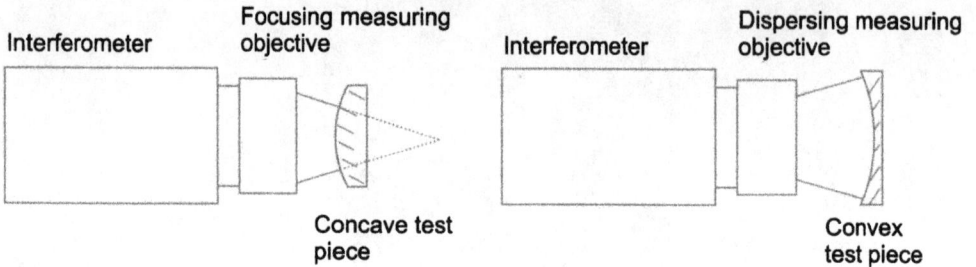

Figure 4.5-11: Use of Fizeau measuring objectives for the measurement of spherical test pieces

The Fizeau objective should, however, be adapted to the particular test piece. As **Figure 4.5-12** shows, the usable aperture, or lateral resolution otherwise remains limited. Generally, the aperture ratio of the measuring objective, e.g. diameter divided by the focal distance, correspond to half of the aperture ratio of the test piece.

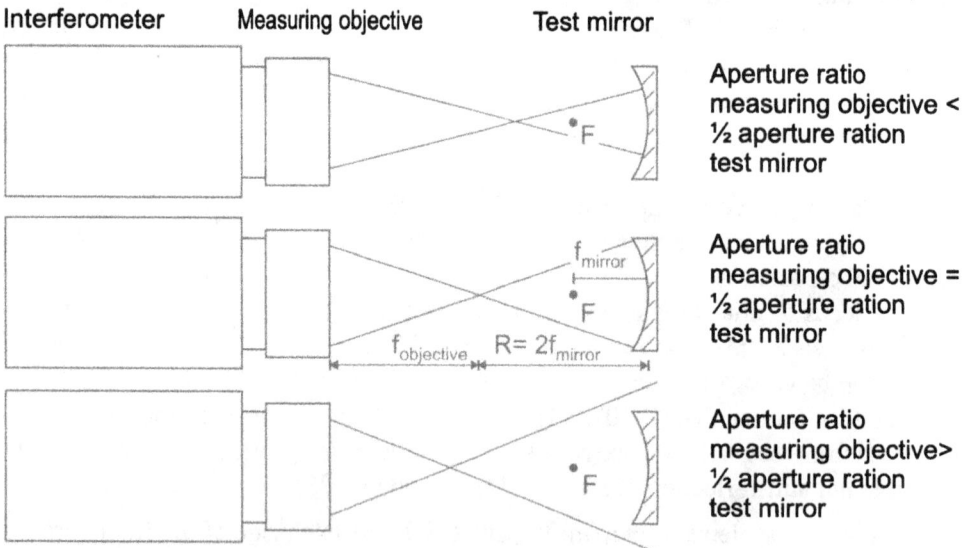

Figure 4.5-12: Adaptation of Fizeau measuring objectives to convex test pieces

In addition to spherical and flat surfaces, aspherical surfaces are continuing to gain importance. Their interferometric inspection is possible in special cases [Tu 95][Pf 97] and also takes place in industry.

Interferometry on Aspheres with Synthetic Holograms

Just as planes and spherical surfaces are still the most common basic forms of optical components, particularly cylindrical, toric and rotation symmetrical aspherical surfaces are of interest in industrial applications such as scanner optics, cylindrical collimation lenses or beam formation optics in laser technology. For the case of cylindrical and toric lenses, the use of interferometers with synthetic holograms as diffractive reference optics already offer a commercial possibility for interferometric testing. The design of such a measuring system (interferometer Zyl-Mess45, Berliner Institut für Optik GmbH, BIFO, Berlin) is represented in **Figure 4.5-13**. While in Section 4.5.1.3, Fizeau objectives were still used to produce a refractive reference wave adapted to the test piece, in this case this takes place through optical diffraction by using asynthetic hologram [Shw 76][Tiz 88]. The core of the represented system is a small form testing interferometer with integrated phase shifting and a CCD camera (µphase compact interferometer, Fisba Optik AG, St. Gallen, Switzerland). A flat wave front is produced with a collimator objective and converted using a computer-generated hologram (CGH), which acts here as a measuring objective.

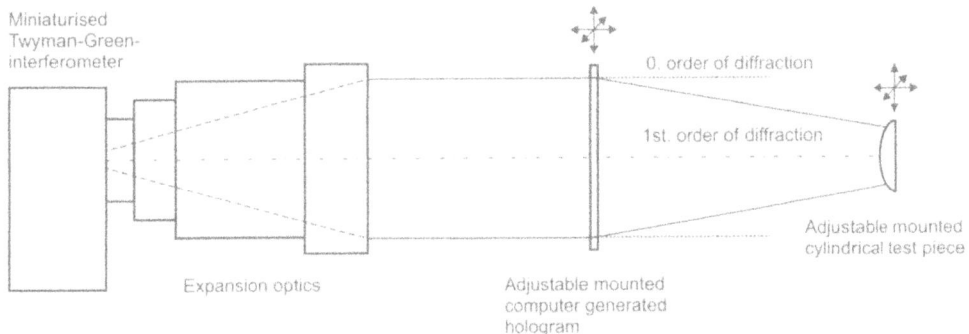

Figure 4.5-13: Interferometer with a computer-generated hologram for inspecting cylindrical surfaces

The CGH is first calculated with ray-tracing software, based on a simulated interferometer structure. Knowledge about all of the system's optics parameters and particularly, the ideal form of the inspected test piece is presupposed. After reproducing this interferogram in high resolution in an appropriate medium (e.g. electron beam lithography), it is used in the image plane of the interferometer. As the form of the test piece was explicitly specified in the calculation, it can be seen [Tu 95] that the ideal wave front is reconstructed when the reference wave is diffracted on the synthetic hologram of the first diffractive order. This meets the test piece as a reference wave and is reflected in a distorted manner, according to the deviation of the test piece from the ideal form. After passing through the CGH again, the object wave front is again collimated and interfered with the reference wave. The interference pattern, which now represents the deviation of the test piece from its ideal form, is evaluated through phase shifting, as already described in Section 4.3.3.6.

4.5.1.5 Interferometry with Diagonal Incidence on Optically Rough Surfaces

For interferometric form testing with a vertical beam of light, as described in the previous section, *optically smooth* surfaces are required with roughness below the assigned laser wavelength. It remains a prerequisite that the optical path lengths in the measurement and reference arms do not differ too strongly, as otherwise aliasing will occur and measurement evaluation is not possible.

If the inspected surfaces are, however, not polished, but matt, the incident light will be diffusely reflected. The optical path differences become too large and interferograms which can be evaluated are no longer obtainable, as either the sampling theorem has been violated, or the reflected beam can no longer be acquired by the collimator and leaves the optical system. Here, interferometry uses diagonal incidence (*Grazing Incidence Interferometry*) [Pck 92]. It is based on the known effect, that even matt surfaces become shiny if they are viewed under diagonal light. **Figure 4.5-14** shows how this effect can be used to carry out interferometric measurements on *optically rough* surfaces.

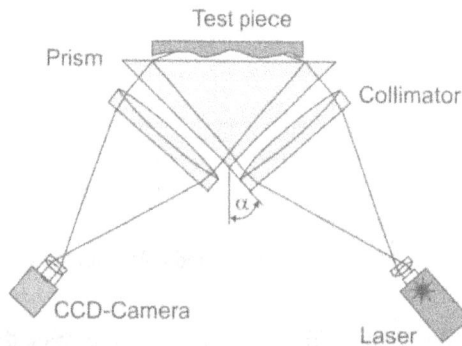

Figure 4.5-14: Interferometer with diagonal incidence for inspecting rough surfaces

After collimation, the laser light is coupled into a prism. Through reflection on the inner hypotenuse surfaces, (evenness better than 0.06 µm), a reference wave is first produced. In the actual measurement arm, there are then very flat outlet angles due to the refraction of the light as it exits through the reference surface and the measurement wave meets the test piece surface under angled incidence. Similar to the other interferometers, the measurement wave front is also deformed by the object surface and then reflected. After entering the prism again, interference takes place with the reference wave and both wave fronts are displayed on the CCD sensor in the evaluation area.

While in the case of interferometry with vertical incidence, the sensitivity, or height difference on the test piece surface which led to a light-dark stripe in the interference pattern was:

$$S_\perp = \frac{\lambda}{2} \qquad\qquad (4.5\text{-}4)$$

with diagonal incidence and an angle of incidence α, it is reduced to

$$S_\alpha = \frac{\lambda}{2\cos\alpha} \qquad\qquad (4.5\text{-}5)$$

The procedure is thus less sensitive to surface roughness and also permits form testing on flat surfaces of poorer quality. The procedure is of great interest for production metrology, particularly due to the fact that the sensitivity is variable due to the adjustable angle of incidence. However, the measurement uncertainty also falls with reduced sensitivity, while the measuring range rises. The measurement system can thus be adapted to the respective measuring function after considering the desired measurement parameters.

With commercial devices, sensitivities are adjustable between 0.5 µm and 4 µm (Topos 50, LAMTECH Lasermeßtechnik GmbH, Stuttgart). With these values, resolutions of 0.01 µm result and attainable measurement uncertainties range from ±0.1..0.5 µm. In this way, measuring ranges of up to 0.1 mm deviation from the reference are possible. The lateral local resolution is limited to 1/500 of the measuring range, due to the pixel number of the CCD sensor and the downstream image processing. **Figure 4.5-15** shows the result of a measurement of a gear wheel. Interferometry with diagonal incidence takes a position just between interferometry with vertical incidence and stripe projection procedures in terms of attainable measuring accuracy (Section 4.3.2.5).

Due to the prism used with its flat surface area, test pieces can only be investigated approximately with this metrology. The use of diffractive optical elements (DOE's) also represent an expansion [Shw 97]. A wave front which is optimally adapted to the target form of the test piece is produced with a synthetic hologram, as described in Section 4.5.1.4. Deviations of the test piece from its ideal form lead to wave front aberrations which are interferometrically measured and evaluated. The procedure permits the measurement of cylinders with free surface areas within limits. Appropriate commercial measuring systems are close to being ready for the market.

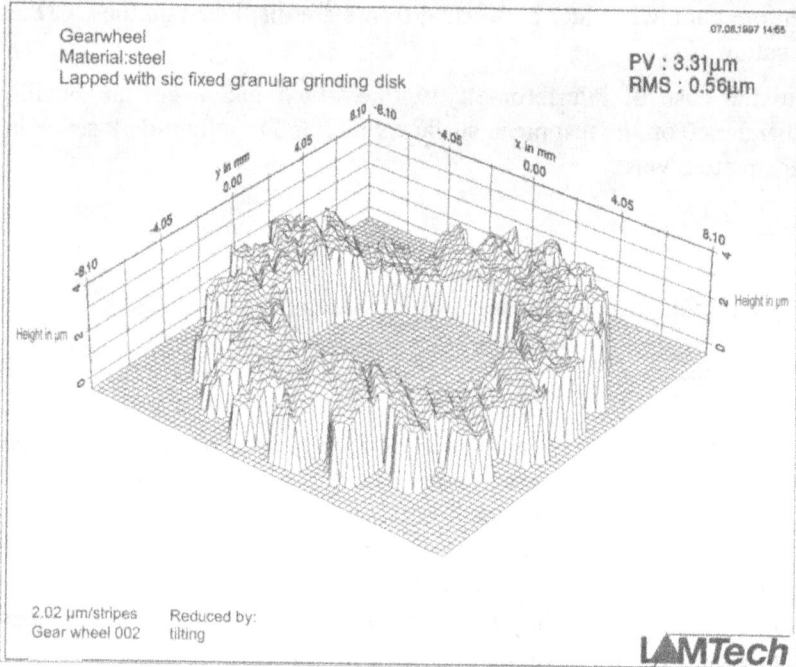

Figure 4.5-15: Measurement of the lapped face of a gear wheel with diagonal incidence

4.5.2 Surface Inspection Technology

4.5.2.1 Introduction

The function of surface inspection technology is to acquire the waviness and/or roughness of technical surfaces metrologically and to determine characteristic values for evaluating the surface quality of a workpiece from the determined measurement values.

Following the order system for structural deviations according to DIN 4760 represented in **Figure 4.5-2**, *waviness* is a structural deviation of the 2nd order. According to DIN 4774, waviness deviations predominantly occur periodically on a workpiece, the wavelengths of which are larger than the groove distances of its roughness. The relationship between wave distance and depth is generally between 1000:1 and 100:1. Structural deviations of the 3rd to 5th order (compare **Figure 4.5-2**) are called *roughness*. Roughness consists of regularly and irregularly recurring structural deviations, the distances of which amount to only a relatively small multiple of their depth [DIN 4774]. The relationship between the roughness (groove) distances and their depths generally amounts to between 150:1 and 5:1 [VDI/VDE 2601]. Waviness and roughness generally overlap, so that it is often necessary to acquire both microstructural deviations separately from one another.

4.5.2.2 Data Interpretation and Roughness Parameters

In order to determine two-dimensional surface parameters, the profile cross section is preferably positioned transverse to the direction of the processing ridges, whereby the transverse profile results (**Figure 4.5-16**).

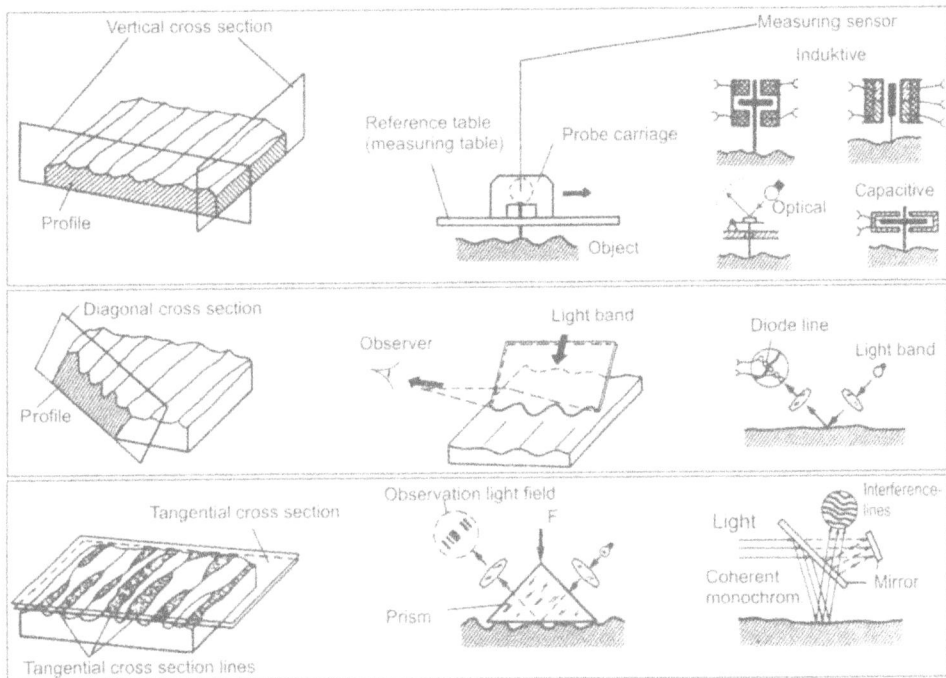

Figure 4.5-16: Different types of profile cross sections

The measured surface profiles, however, generally include the additive overlapping of waviness and roughness form deviations. To determine surface parameters, the individual deviations must be evaluated separately. This is done by using high, band or low-pass filters, which are called wave filters according to DIN 4768.

If, e.g. an unfiltered actual profile is given on the input of a wave filter, the signal proportions are suppressed or transmitted in an undistorted form, depending on the configuration of the wave filter. When applying the high-pass filter, which exclusively transmits signal proportions starting from a cut-off frequency λ_c, the roughness profile of an inspected surface can be determined. Similarly, form deviations can be extracted from a surface profile by low-pass filters and waviness through bandpass filters.

Roughness parameters are defined as single measuring segments, unless otherwise indicated. Results are indicated as average values from several measuring sections. Five single measuring segments are standard; otherwise the number of single measuring segments is indicated. The reference line for roughness evaluation is, according to DIN 4762, a middle straight line within a single measuring segment. The selection of roughness parameters and their definitions are indicated in **Figure 4.5-17**.

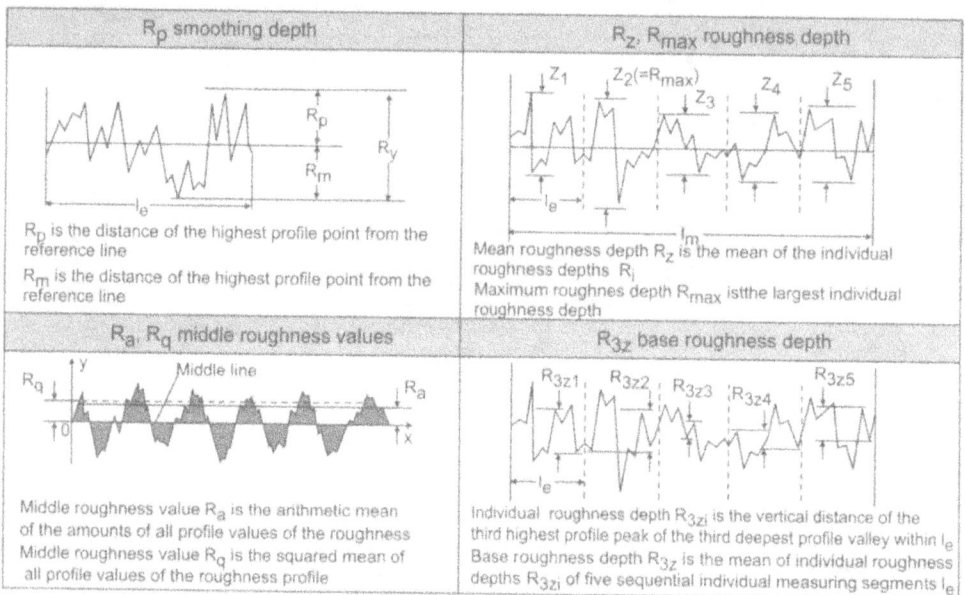

R_p smoothing depth	R_z, R_{max} roughness depth
R_p is the distance of the highest profile point from the reference line R_m is the distance of the highest profile point from the reference line	Mean roughness depth R_z is the mean of the individual roughness depths R_i Maximum roughnes depth R_{max} istthe largest individual roughness depth
R_a, R_q middle roughness values	R_{3z} base roughness depth
Middle roughness value R_a is the arithmetic mean of the amounts of all profile values of the roughness Middle roughness value R_q is the squared mean of all profile values of the roughness profile	Individual roughness depth R_{3zi} is the vertical distance of the third highest profile peak of the third deepest profile valley within l_e Base roughness depth R_{3z} is the mean of individual roughness depths R_{3zi} of five sequential individual measuring segments l_e

Figure 4.5-17: Roughness parameters

In order to quantitatively determine surface quality, in addition to profilometers, white light interferometers and even raster electron microscopes are currently used in industry. The latter two permit resolutions in the Angström range, but will not be explained here in further detail. A substantial reason for the widespread use of these surface inspection devices is the fact that the required surface parameters can be directly gained from the measured surface profile.

4.5.2.3 Profilometers

When using electrical profilometers a vertical profile cross section is probed on the workpiece surface and evaluated to determine the surface quality [War 84].

The deflections of the probe tip are taken up either inductively or laser interferometrically (**Figure 4.5-16**). As the tip touches the test piece, damage could arise in the case of soft materials.

The most simple version of an electrical profilometer, the single *skidded probing system*, is only connected with the drive unit by a joint and is supported by a skid plate on the workpiece surface. It is therefore a semi-rigid system. Through this arrangement, the probing system is primarily guided by the skid plate and to a lesser extent, according to lever conditions, through the drive unit. The drive unit must therefore be aligned parallel with the inspected surface. Single skidded probing systems require very little space on the test piece and, with an appropriately selected geometrical probe formation, permit a simple execution of the measurement, even on difficult to access workpiece areas, such as drill holes. The various geometrical versions of probes differentiate themselves in terms of the arrangement of the probe tip to the skid plate. The skid plate can be arranged in front of, behind or next to the probe tip. A great disadvantage when using single skidded probing systems is the possible falsification of measurement results during profile acquisition. The size of this falsification depends on the relationship between the distance of the probe tip-skid plate to the profile wavelength as well as on the geometry of the skid plate with the occurrence of individual profile points. Therefore, wavy profile proportions can be completely extinguished as well as being transferred with double height.

Oscillating probe systems (double skidded probing systems), as represented in **Figure 4.5-18** top, are guided over the surface of the tested object on two spherical or cylindrical skid plates, which lie behind one another in the probing direction. Due to the existence of two skid plates, these probing systems align themselves when positioned on the inspected surface. However, depending on the distance of the skid plates, the tested object needs to have a relatively long length in order to carry out a surface inspection. This has the consequence that small workpieces can not be inspected with these systems.

With *reference surface probing systems* the probe is guided over a nearly ideal geometrical reference plane which also functions as the reference for the measurement. In contrast to probing systems guided on skid plates, in this case there is only contact between the surface of the inspected object and the probe tip. Thus the surface profile is transferred in an undistorted form, apart from the influence of the probe geometry. For tactile surface inspection, reference surface probing systems represent an ideal inspection device, as it aids the acquisition of waviness and roughness of a technical surface. With sufficiently long measuring segments, the form deviation of an inspected object can also be determined.

Oscillating probe system (double skidded probing system)

Drive unit

Probe tip

Skids

Refenrence surface probing system (single skidded probing system)

Drive unit

3

2

Skid

1

1 gauge block-flat glass
2 probe tip
3 probe

Alignement

Reference surface probing system (self-aligning)

2   Drive unit

3

4

1

1 probe tip
2 probe
3 reference surface
4 support point

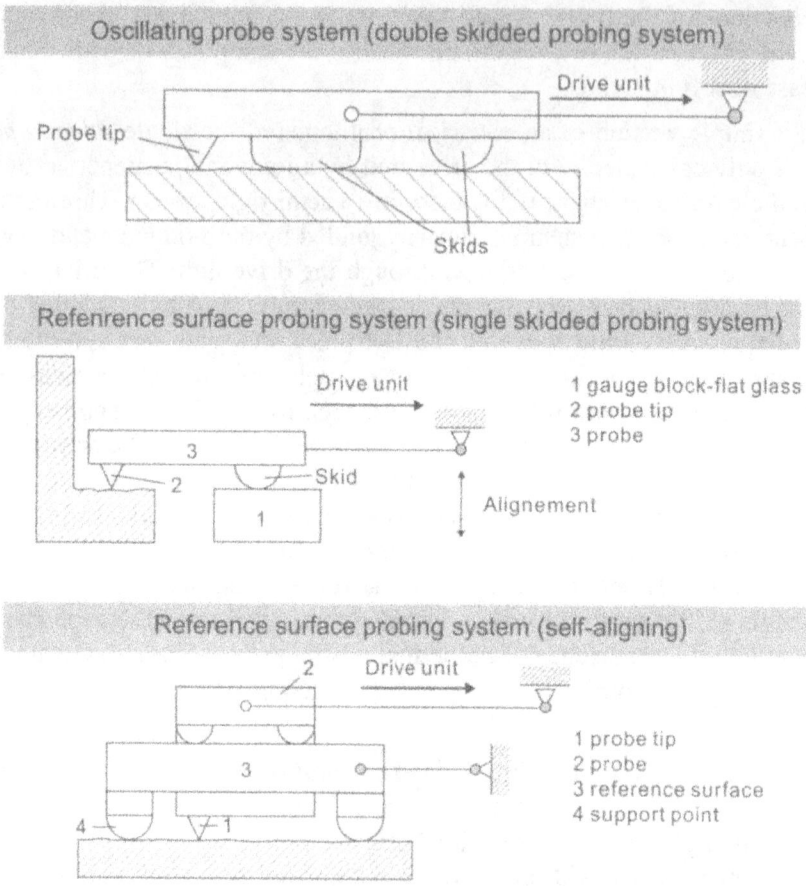

Figure 4.5-18: Examples of probing systems [Tes 1]

The simplest execution form of a reference surface system is given by using a *skidded probing system,* whereby the skid plate is guided on a gauge block or flat glass plate as a reference surface (**Figure 4.5-18** middle). This reference surface probing system has a very small space requirement on the inspected object, thereby also enabling the testing of small workpieces.

The *self-aligning reference surface system* represented in **Figure 4.5-18** bottom is characterised by simple handling and low-vibration measurement structure. With this measuring arrangement, the reference surface is directly integrated into the probing system and the probe is connected with the drive unit by a joint. The relatively high space requirement on the inspected object is seen as a disadvantage which arises due to the distance of the support points on the inspected surface.

For all of the specified profilometers, the measurement result depends strongly on the geometrical execution of the probe. The geometrical executions of probes are standardised by DIN 4772 (**Figure 4.5-19**).

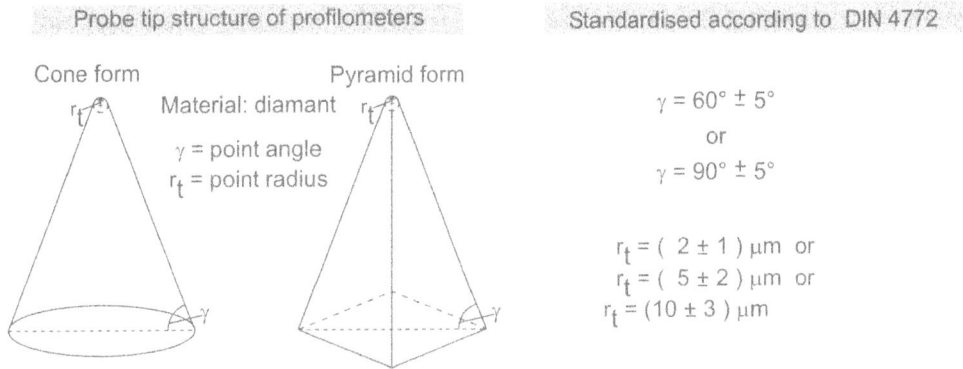

Probe tip structure of profilometers   Standardised according to DIN 4772

Cone form   Pyramid form

γ = point angle
r_t = point radius

Material: diamant

$\gamma = 60° \pm 5°$

or

$\gamma = 90° \pm 5°$

$r_t = (2 \pm 1) \mu m$ or
$r_t = (5 \pm 2) \mu m$ or
$r_t = (10 \pm 3) \mu m$

Figure 4.5-19: Execution forms of probe tips [DIN 4772]

For surface inspection, probes manufactured from diamond are generally permissible in the form of a cone or pyramid with the specified point angles and point radii. The point angle determines the limit of the detectable roughness to small wavelengths. Through the rounding of the probe tip, the measured actual profile always represents an envelope profile on the actual surface profile. A generally valid estimate of measurement error due to the probe tip geometry could, however, not be indicated so far [War 84]. The selection of an appropriate probe is thus left to the experience of the user, who, under consideration of the expected measurement results, needs to judge which geometrical form of probe tip to use for each particular application.

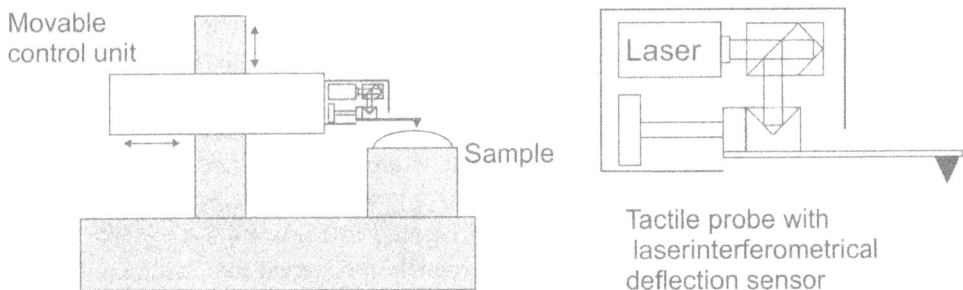

Movable control unit

Sample

Laser

Tactile probe with laserinterferometrical deflection sensor

Figure 4.5-20: Profilometer with laser interferometrical tactile probe (form Talysurf Series II, Rank Taylor Hobson)

An example of the execution forms of a reference surface system with interferometric acquisition of the tactile probe deflection is shown in **Figure 4.5-20**.

The result of a measurement with an inductively controlled profilometer system is shown in **Figure 4.5-21**. In this case a steel shaft was measured. In addition to the 2D profile, particularly the roughness analysis is of great interest in order to obtain predications about the quality of the production process.

Apart from these two-dimensional profile recordings, the three-dimensional analysis of surfaces is also of interest. Here, it is necessary to take up several parallel profile cross sections and offset them against one another. Measuring systems with appropriate hardware and software are also commercially available.

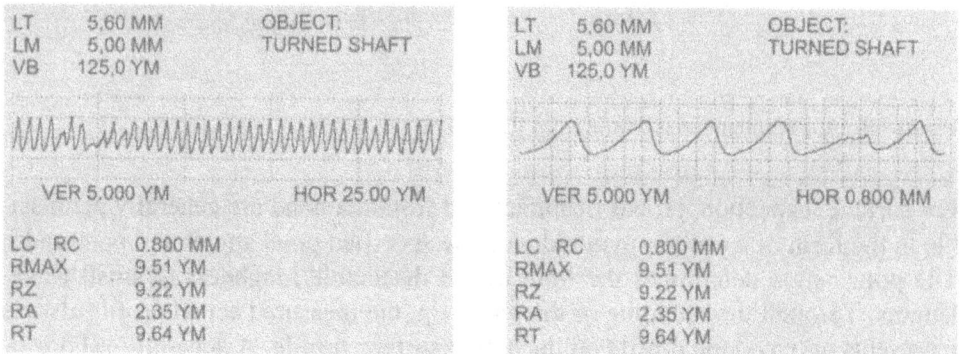

Figure 4.5-21: Measurement with an inductive profilometer (Perthometer S8P, Mahr GmbH, Göttingen)

4.5.2.4 Optical Profilometer System

In Section 4.3.3.3, the autofocus procedure was introduced as a method for non-contact distance measurement. In addition to the tactile and interferometric measuring sensors, this optical method, with resolutions in the sub-µm range, also represents an alternative to surface inspection, as long as the sensor is integrated into a suitable handling system. High demands are made on mechanics regarding positioning accuracy and linearity with sensor guidance. The execution form of such an optical profilometer system is represented in **Figure 4.5-22** (Mikrofokus, UBM Meßtechnik, Ettlingen). The sensor, which is fastened to a bridge mounting plate, can be moved over the test piece with high precision via 2 linear modules. A non-contact scanning of the sample, as it were, takes place. The working distance of the sensor is adjustable within limits, as a function of the respective collimation optics. Measuring ranges of up to 1mm are possible.

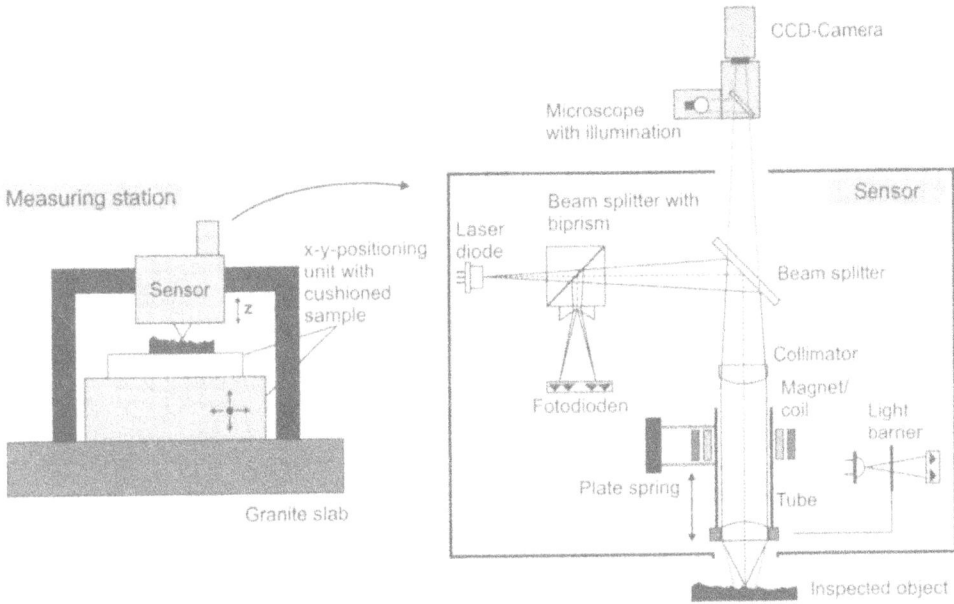

Figure 4.5-22: Optical profilometer system

It must be noted that, in place of the tactile probe with tactile systems, the laser focus mark (diameter approx. 1 µm) can be used as a non-contact scanning element for measurement with optical profilometry. This measuring technique thereby differs distinctly from tactile profilometry, where a direct interaction takes place between the tactile probe and the sample. This generally leads to a low-pass filtering of the measured surface structure, as the probe tip is not in a position to resolve all of the fine points of the surface, due to the finite expansion. In contrast, with optical scanning, the surface will modify the impacting focus mark with its microstructure, e.g. due to dispersion effects, which also has an effect on the result of an evaluation. As the distortions with tactile and optical scanning differ, it is not easily possible to compare measurement results executed with both systems on the same sample.

4.5.2.5 White Light Interferometers

A further method of non-contact measurement of rough surfaces is white light interferometry [Ca 93], with which fast, accurate and reproducible three-dimensional surface inspections can be carried out. Even height levels of up to 500 µm are measurable. Furthermore, the resolution can be increased into the nm range through the supplementary use of phase shifting methods.

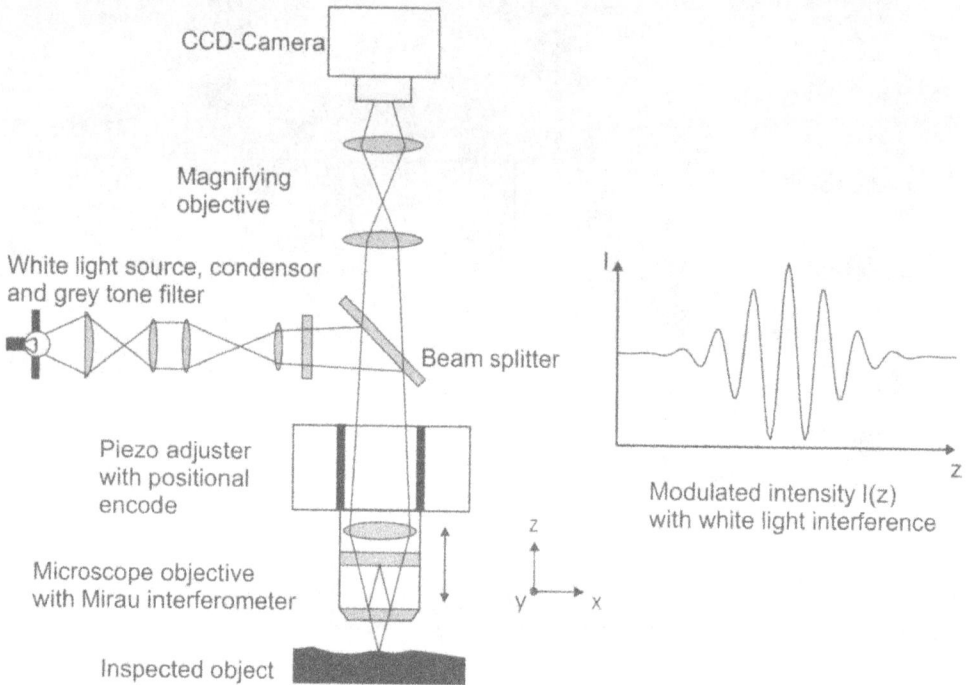

Figure 4.5-23: White light interferometer and white light interference structure

As shown in **Figure 4.5-23** left, the light from a white light source is first linked into the measuring device through a beam splitter. It then passes through a microscope objective which has a Mirau interferometer [Mal 92] built into it. If the distance of the light between the objective and the test piece is exactly the length of the light in the interferometer, white light interferences occur. As in a conventional microscope, the interferences can thereby be detected. As the height profile of the test piece is generally not constant away from the measuring aperture, the optical distance in the object arm therefore also varies. Only in areas where the height of the test piece is constant and the distances of the measurement and reference arms are the same, will the white light interferences be visible. The objective can, however, be adjusted in high resolution through piezo elements, so that the position of the white light interference can be set. The path is thereby accurately taken up by position decoders.

In contrast to coherent interferometry, with white light interferometry, the short coherence length of a white light source is directly used. The modulation of the interference signal rapidly decreases due to the short coherence length of only a few μm. **Figure 4.5-23** right shows the appropriate course of intensity for a pixel as an example.

With the adjustment of the objective and an appropriate modification of the difference in distance, the intensity in each pixel is appropriately modulated according to this intensity process. The relative measurement of the object height is carried out by extracting the modulation from the interference signal and their maximum is calculated as a function of the distance modification. The use of special hardware with digital signal processors working in parallel and the implementation of algorithms from signal processing, enable the real-time determination of the maximum modulation.

In contrast to the measurement of the phase, as carried out in coherent interferometry, with white light interferometry, the emphasis is on the determination of the phase bundle, whereby the phase reference is lost. In coherent interferometry, the measurement is thus relative to the effect that the deviation of the test piece from the used reference is determined. In white light interferometry, an interference structure is thus produced from the reference surface. This is analysed, as it were, to determine contours in the test piece. By measuring the shift of the objective over the piezo elements, quantitative elevation information can be assigned to the respective contours. Hysteresis effects of the piezos can be practically eliminated by implementing appropriate automatic controls.

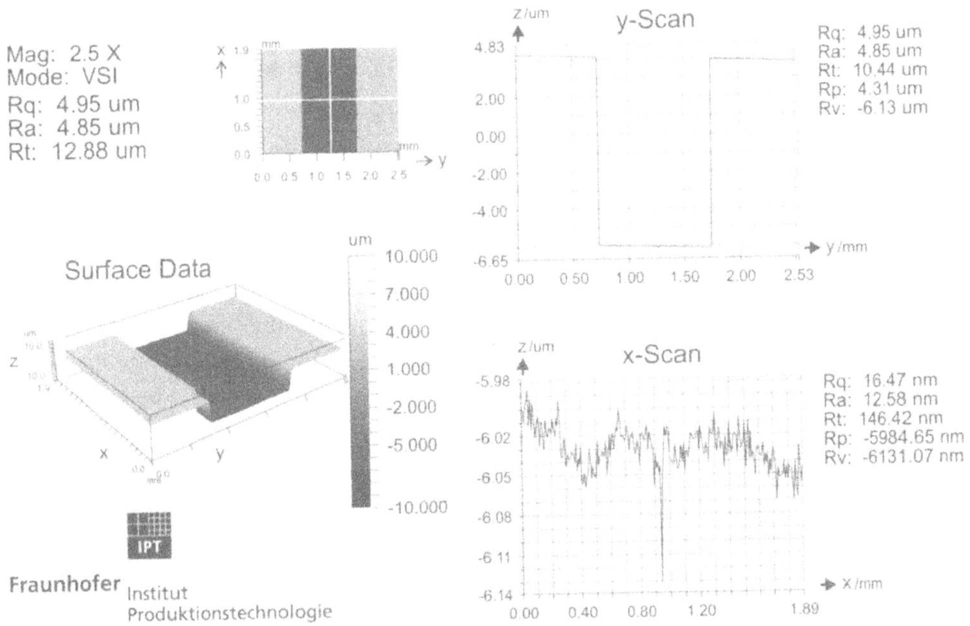

Figure 4.5-24: Measurement of a stepped structure with a white light interferometer (RSTplus Surface Profiler, WYKO Corp., Tucson, USA)

Figure 4.5-24 documents a measurement of a stepped structure. The measuring range with the stepped measurement in the horizontal range as well as the enormously high resolution (< 1 nm) in the vertical cross section are illustrated through this result. The measuring range of commercially available white light interferometers extends from a few nanometers up to 500 μm. Various microscope objectives with magnifications of 2.5x to 40x can be used. The lateral resolution is limited by the microscope objective used and the number of pixels of the CCD sensor, or frame grabbers used and are situated between 0.1 μm and 12.7 μm. The maximum measuring field is 8 mm x 6 mm with 2.5 times magnification.

4.6   Gauging Inspection

Gauging is one of the oldest methods for the measuring inspection of all types of items. It originates from the usage of models for the production of complex parts.

It is a function-orientated test, i.e. the combination of various workpiece areas is tested in such a way, as the intended application requires (e.g. a shaft must fit its bearing, independent of its mean diameter). The result of the gauging inspection makes a clear statement, i.e. the subjectiveness of a test decision is reduced, and hence the result is repeatable.

The requirements on the test personnel are low, so that untrained personnel can obtain test results after only a short introduction period. The test itself is carried out over a short period of time and the results can be used immediately, should a correction of the production process be required. Gauges can be used near the production process, as they are easy to use and of a mechanically stable build. As long as the tested workpieces and the gauge steel have a similar temperature expansion coefficient, a temperature-neutral test is possible. The gauge types used are standardised and are offered by various manufacturers as standard inspection equipment.

For decades, gauge testing has been an established practice in industry. One main area of application is the production of substitute parts, where each independently produced workpiece must be replaceable by another one. Trials to introduce modern testing methods have often failed due to their complexity and susceptibility compared with the simplicity and reliability of the gauging principle.

Gauges are used to assess simple and complex geometries, where various geometrical elements or features contribute to a function. While measuring inspection equipment checks for individually specified features, for example diameter and cylinder form, one gauging step assesses all the information necessary for the functionality. Furthermore, complicated geometries such as free form planes are extremely difficult to describe and measure, but gauging allows a quick assessment in only one step.

4.6.1 Taylor's Principle

According to Taylor's principle , a gauge always has two parts and a gauging is carried out in two steps. The function is tested on the acceptance side of a gauge with a maximum material dimension of the inspected element. The scrap side of the gauge observes the aherence of the maximum measurement with the minimum material size.

W. Taylor formulated in 1905 (**Figure 4.6-1**) as follows:

- The *"go" gauge* must be formed such that it assesses the tested form in its totality.
- The *"not go" gauge*, on the other hand, should only test individual determining features of the geometrical form of the workpiece.

This requirement assures that the tolerance limits are not exceeded by form deviations of the workpiece. In cases where the form tolerance exceeds the measurement tolerance, Taylor's principle does not apply. Most gauges do not apply to Taylor's principle as they are designed for either a one step gauging process or do not test the envelope of the geometrical elements, but a two-point dimension (e.g. gap gauge) [Lot 78].

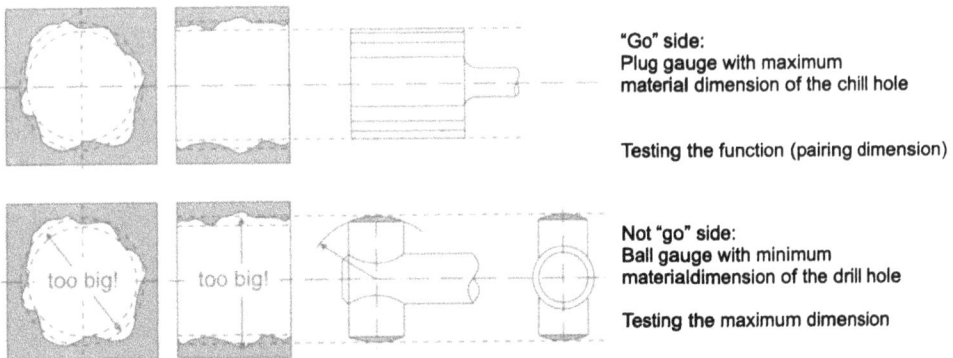

"Go" side:
Plug gauge with maximum material dimension of the chill hole

Testing the function (pairing dimension)

Not "go" side:
Ball gauge with minimum materialdimension of the drill hole

Testing the maximum dimension

Figure 4.6-1: Taylor's principle

During the course of the 20th century, various versions and generalisations of Taylor's principle have been formulated. One extension states that it is also applicable to gauges for which a certain position to the workpiece is specified in a different regulation. For example, if a drill hole is subject to a right angle requirement with regard to a workpiece plane, the bolt must be vertical to this plane. In such a case, the extension of Taylor's principle is sensible according to Weinhold [Lot 80]: "The acceptance gauge must measure all elements at once, which are part of a coupling. During gauging, the workpiece and acceptance gauge must assume relative positions, which can occur between the workpiece and its counter part or those which are defined by additional position requirements."

4.6.2 Types of Gauging Inspections

Gauging inspections can be differentiated according to various aspects. When differentiating with regard to its applications, there are the following gauge types:

- *workshop gauges* to test workpieces
- *standard gauges* as counter gauges to test shop gauges and
- *inspection gauges* to test again those workpieces, which had to be reworked after the initial shop gauging.

Another possibility lies in differentiation according to their principal geometrical and functional features. **Figure 4.6-2** shows a list and typical representatives or characteristics of the respective gauge classes.

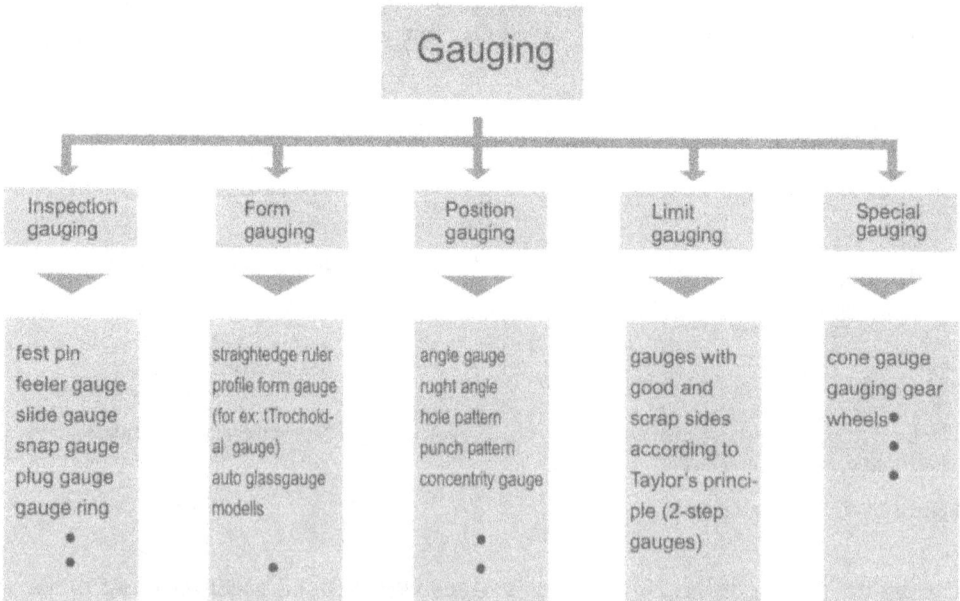

Figure 4.6-2:   Classification of the gauging inspection

The division into measuring, form and positional gauges is orientated toward tolerancing in technical drawings in accordance with the standardised tolerancing principles (Section 2.5). Gauges normally measure more than one dimensional, positional or form feature. A respective division is therefore interlinked.

Limit gauges are a further group of gauges with which the maintaing of tolerance can be tested. These have always two parts or are two-sided (acceptance and scrap gauges) and are applied in accordance with Taylor's principle.

Special gauges are all those inspection instruments used for tests with a gauging character which do not fit into any of the categories listed above.

4.6.2.1   Inspection Gauging

Inspection gauging is used when either a minimum or a maximum value must not be exceeded. This is the case when a single critical dimension is given.

For example, for a washer, the drill hole must be of at least a minimum dimension so that the intended screw fits. Even if it is only marginally too small, it prohibits its functional use. A small excess of the diameter is not critical. Another example is an hexagonal bolt, which is unusable if the width across the flat is exceeded by a small amount, whereas if it is smaller, the dimensional difference can be treated more tolerantly. **(Figure 4.6-3)**

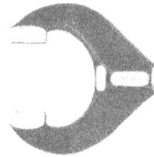

Feeler gauge   One-jaw snap gauge   Gauge ring

Measuring pin   Ball gauge (front as ball section)

Figure 4.6-3: Inspection gauges

Inspection gauges can only make one of two statements: either 'fits' or 'does not fit'. An assessment with regard to maintaining the tolerance is not possible unless two independent inspection gauges are used as acceptance and scrap gauges in accordance with Taylor's principle.

Traditional dimensional gauges are feeler gauges and test pins. The slide gauge should not be included in this category. When using it as a vernier calliper, it is not a gauge as such, but it can be fixed to a defined value to carry out dimensional gauging. Other gauge types, such as thread gauges and gap gauges can be adjustable, allowing for a more flexible application. Ring gauges and snap gauges are, when looked at separately, dimensional gauges, but if they are adjusted to one another they form a limit gauge in accordance with Taylor's principle (Section 4.6.2.4).

4.6.2.2   Form Gauging

The tasks of form gauges include monitoring of contours, profiles and models. Standard geometrical forms (straightness, flatness, circle and cone) cannot be measured suitably by gauging. For a long time, free-formed edges and planes were only assessed by gauging. For this the original models were used, which were also the templates for the manufacture of the tools. Glass gauges are a well-known example for form gauges, which were, until recently, used in the automotive industry. Until a few years ago, car companies made a so-called original gauge for each window in a car, which was, in principal, a model of the original part of the car. From this, templates were made to enable the cutting of flat glass. For these flat glass gauges contact points are mounted, which represent the car body. During inspection, the glass must lie within the closed conture; the distances of the edges to the support point must not be too great. The assessment of the finished glass is similar. The matching up of the edges is checked with an original gauge. **(Figure 4.6-4)**

The problems of the form gauging become clear with the example of a straightedge ruler. The linearity of an edge or plane can be judged with the aid of a straightedge ruler. When positioning the ruler on an uneven edge, either a stable dual point or a unstable one-point position is reached. Dual point positions can also exist more than once. For assessment, the gap is measured, which exists between ruler and tested object. This can be best ascertained in counter light (high contrast). The assessment is questionable and reliant on the subjectivity of the tester. Such a gauge is therefore only used if target geometries only need to be kept as a tendency, i.e. there is no substantial tolerance.

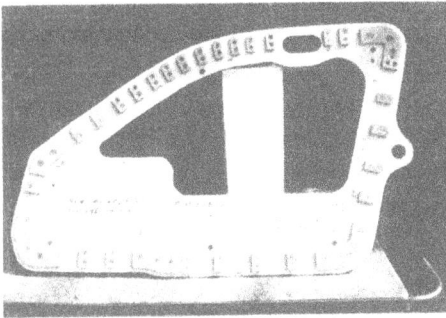

Flat glass gauge   Glass gauge
                                                       (side window BMW)

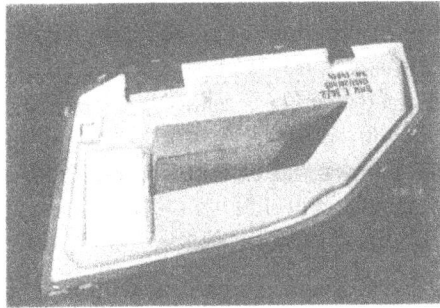

Figure 4.6-4:   Various form gauges (source glass gauges: Sekurit Saint Gobain GmbH)

4.6.2.3   Positional Gauging

Positional gauging is characterised by a high complexity due to multiple fits. As the possible position relationships of geometric elements in a component are very diverse, the positional gauge must be specially manufactured for the respective type of workpiece. Due to multiple fits, the sufficiently precise manufacture is quite problematic and therefore expensive. Right-angled or angled gauges, for which the same basics apply as for the ruler, are excluded from this. Therefore, the complicated positional gauges concentrate on position and concentricity (**Figure 4.6-5**) .

(Workpiece)

Right-angle gauge   Concentricity gauge

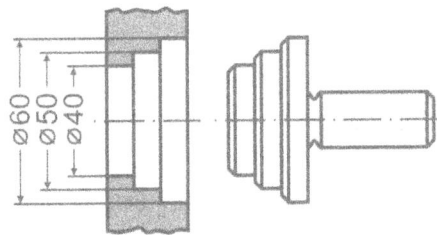

Figure 4.6-5:   Various positional gauges

One of the tasks of positional gauges is the checking of functional connections of various geometric elements as well as the existence of features (templates). A concentricity gauge is made of a multi level bolt, which is coupled with the workpiece. It is equivalent to the acceptance side of a limit gauge, as it tests the interaction of the geometrical elements concerned. According to Taylor's principle and in order to have a complete gauging, it must be determined if the individual diameters exceed the maximum dimension.

The maximum material principle, which is either impossible or difficult to handle for many measuring devices, is elegantly and quickly realised by a positional gauge . The extension of an individual tolerance is permitted in case that other tolerances contributing to the component function are not fully used (Section 2.5.2.4).

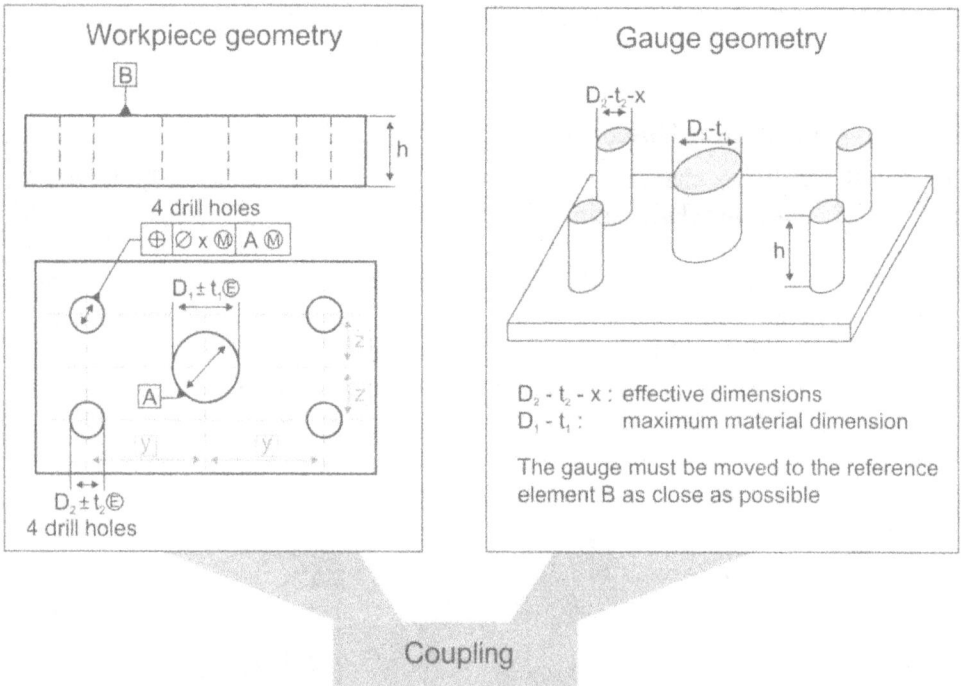

Figure 4.6-6: Example of a positional gauge

Figure 4.6-6 shows a corresponding example: a plate with five drill holes, their positions to one another are defined to one another by theoretically exact dimensions. The diameter of each of the five drill holes is toleranced. Furthermore, a positional tolerance of the four small drill holes is given with respect to a larger drill hole. This is achieved by applying the maximum material principle. In the case that one of the small drill holes deviates from its theoretically exact position but still lies within the upper limit of the tolerance field, the connection with the gauge or the functional counterpart can be established. The same applies for the central reference drill hole.

The gauge is made up of five pivots, which are vertically fixed onto a base. Their mean axis is at theoretically exact positions of the respective drill holes. Their length equals the thickness of the inspected base, as the whole length of the drill holes needs to be assessed. The central pivot's diameter is the maximum material dimension of the mean drill hole as the reference shows a circled "M" (symbol of the application of the maximum material principle MMP). The small pivot's diameter is the "effective dimension", which is derived from the maximum material dimension of the drill hole minus the positional tolerance.

4.6.2.4   Limit gauging

Task of limit gauging is to assess the coupling capability of defined single component features. The limit gauge is a shop gauge. It is made of two gauges, one one gauge „go" and one „not go". The assessment is related to the dimension, the form and/or the position of a workpiece element [Lun 90], [Mah 28].

Today it is principally required of the acceptance side of a limit gauge that, according to Taylor's principle, it is a perfect counterpart. Through a function test it guarantees the joining capabilities of the respective tested object with the according workpieces. This rule can be deviated from if the production process excludes various error probabilities and this permits the simplification of the acceptance gauge. An example is the utilisation of snap gauges instead of gauge rings when testing shafts, which are produced in rotation. Due to this production process it can be assumed that the shafts will be of a sufficient roundness. It is therefore not necessary to test the diameter with a gauge ring over its overall dimension. A two-point test with a gap gauge will permit a predication for the overall dimension. **(Figure 4.6-7)**

According to Taylor's principle, screwplug gauges have some thread pitches on their acceptance side of the gauge, so that the functionality can be tested over the overall length of the thread. On their scrap side, gauges only have one complete thread pitch. Another type of limit gauge is the limit plug gauge, which is also standardised in accordance with the standard fit system. Limit snap gauges exist in one-jaw or two-jaw form. The one-jaw form has stepped measuring surfaces on one side, so that one inspection process is sufficient to carry out acceptance and scrap gauging.

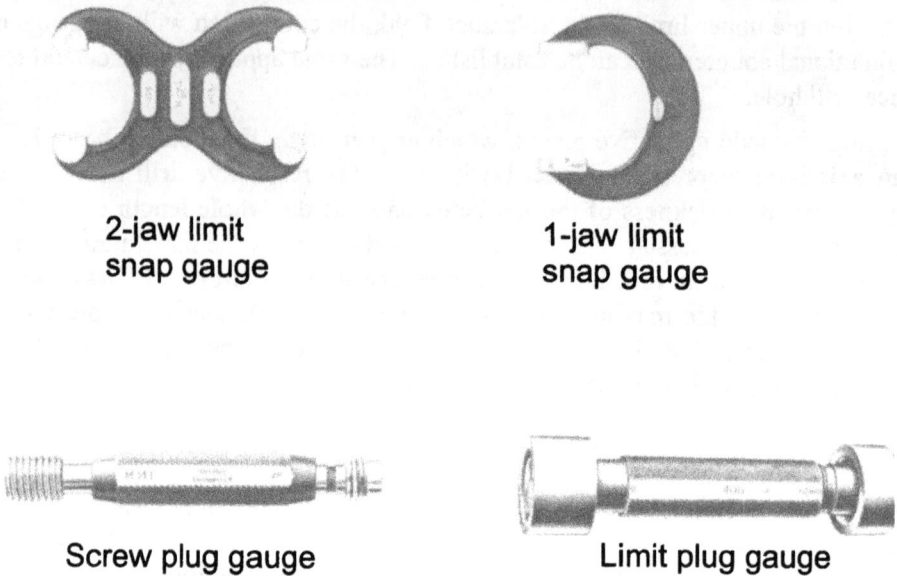

2-jaw limit
snap gauge

1-jaw limit
snap gauge

Screw plug gauge   Limit plug gauge

Figure 4.6-7: Types of limit gauges

4.6.2.5   Special Gauging

Special gauging refers to those gauging inspection procedures which do not normally fit within the normal classification. A special gauging is often carried out in connection with display measurement devices.

An example for a special gauge is tangential composite of gears. For this procedure, gearing gauges with defined features are used [VDI 2608].

The gear wheel gauge and the tool gear wheel are rolled together under a predetermined axial distance and either the right tooth side or the left tooth side are constantly in use. The deviations from the faultless even movement transfer are measured. The tangential composite of gears of the assessed wheel is the deviation of its revolving position from its target position. It can be measured as an angle or distance along the circumference of the circle (roller path).

4.6.3 Norms for Gauging Inspection

Gauges are widely represented in the German standards (103 DIN standards) [DIN 87]. Although most of the "current" versions date from the 1970's, this diversification also shows the importance of the gauges in industrial practice. **Figure 4.6-8** provides an overview.

103 DIN-Norms with regard to gauging

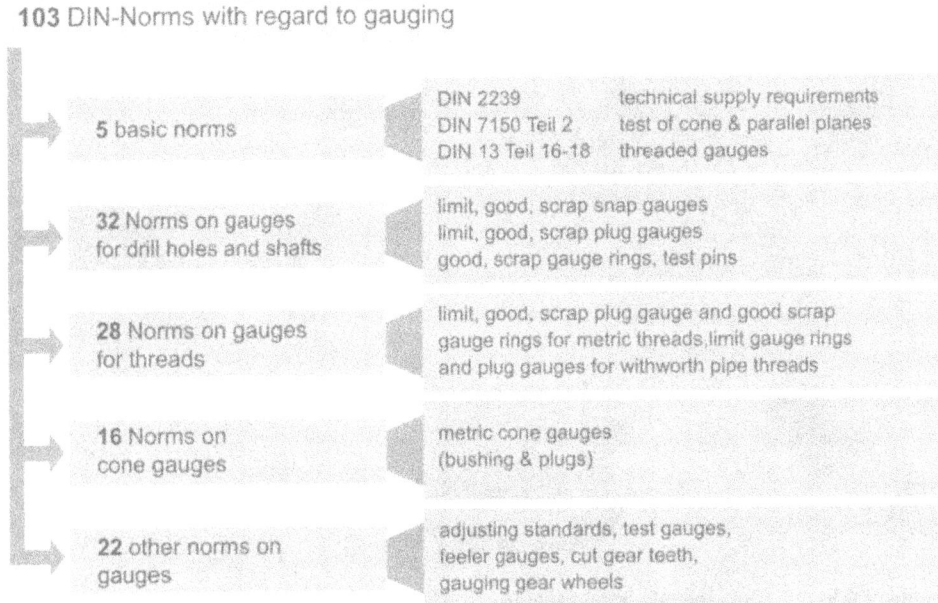

5 basic norms	DIN 2239 technical supply requirements DIN 7150 Teil 2 test of cone & parallel planes DIN 13 Teil 16-18 threaded gauges
32 Norms on gauges for drill holes and shafts	limit, good, scrap snap gauges limit, good, scrap plug gauges good, scrap gauge rings, test pins
28 Norms on gauges for threads	limit, good, scrap plug gauge and good scrap gauge rings for metric threads,limit gauge rings and plug gauges for withworth pipe threads
16 Norms on cone gauges	metric cone gauges (bushing & plugs)
22 other norms on gauges	adjusting standards, test gauges, feeler gauges, cut gear teeth, gauging gear wheels

Figure 4.6-8:   Gauging standards

These standards are a binding basis between the gauge manufacturers and the customer. They are basically limited to gauges, which are used as mass inspection equipment for standardised fits and thread levels. Further information can be found in [DIN 87].

4.6.4 Principle of Virtual Gauging

Contrary to the traditional physical gauging, virtual gauging has the potential to make differentiated statements about the gauging process rather than just the "good/bad" - assessment [Pf 94]. Hence it is possible to find trends in the form of functional features or feature groups and to interfere with the production process in a corrective manner rather than condemn products to be rejected. Furthermore, a statistical observation of the production results can be carried out, which permits a reduction in the test effort and speeds it up. Even in the case of "non-joining - capability" an analysis of the gauging result is possible. This generates information for possible corrective work.

Workpiece   Digitalisation

z
y
x

n points $P_i(x,y,z)$
on the inner
cone plan

Virtual
gauging

Gauging   Modelling

Conventional statement:
good, scrap, re-work

Additional statement:
minimum diameter,
maximum diameter,
form, orientation

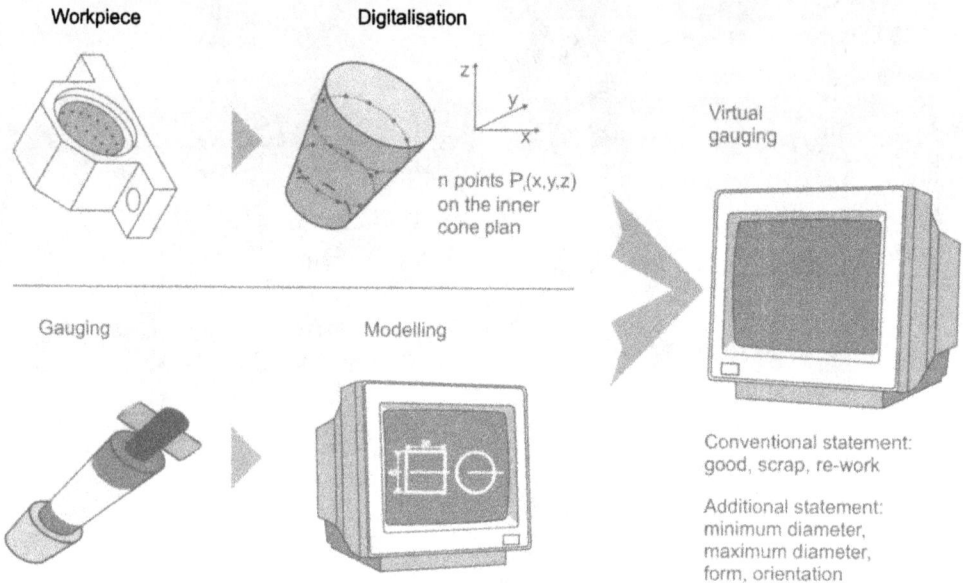

Figure 4.6-9:   Principle of virtual gauging

Figure 4.6-9 shows the principle of virtual gauging. The idea is to simulate the gauging process on a computer. For this the geometry of a gauge is derived from the CAD data of the tested elements represented within the computer. The surfaces of the related real component elements are scanned by a coordinate measuring machine and digitised. The computer-internal representation of the gauge as well as the digital model of the real elements permit the numerical running of the gauging process with the aid of a computer programme. The result not only shows the "good/bad" - assessment but also additional information, which is important for the control of the production process.

The clear advantage of virtual gauging lies in the higher information value of its test results. It provides indications for controlling the production process. For example, with gauging, which tests various elements, such as positional gauging, virtual gauging indicates which of the elements have been produced with a fault. This is not achieved with physical gauging; it only informs whether the totality of the checked features is faultless without pointing to separate features.

The disadvantage of virtual gauging is the high cost which the principle requires. A scanning coordinate measuring machine or a form measuring device and computers with respective software are needed for virtual gauging. The requirements for the tester are much higher than for conventional gauging. Therefore, this principle cannot be used as closely to the production process as conventional gauging.

4.7   Integration of inspection devices into automated measuring applications

A major aim of industrial automation is to free humans from routine tasks. This also naturally applies to measuring and testing methods. A routine task would, for example, be the testing of the many parts of a production lot with the aid of manual inspection equipment. Without automation

- the measuring devices and the workpieces would need to be guided manually,
- the measurement displays would need to be read after each measurement,
- the measurement results would need to be recorded manually,
- the measurement records would have to be evaluated and archived by staff.

Thankfully, there are a number of methods available today with which various levels of automation can be reached.

Automation within production metrology is primarily the computer aided processing of measured values, apart from the automated guidance of measuring devices or workpieces. For this, the measuring device must have suitable electronic interfaces, i.e. purely mechanical measuring devices are not suited for the integration into automated measuring applications.

Section 4.7.1 on electronically guided measuring devices at a computer-aided measurement stations shows the first step toward automated test data collection. Measuring aids and workpieces are still guided manually, but the electronic application of measured values has no limits.

A higher level of automation can be reached if the guidance can be automated in addition to the data-technical integration of the measuring devices. This is the case, for example, with coordinate measuring machines, which were introduced in Section 4.4. These automatically scan geometrical features with the aid of a measurement programme. The measurement results can be provided by data interfaces in the machine control or control computer for further usage, or they can be read off the machine directly. The same principle applies to the workpiece measurement on machine tools, if a probe, which is fixed like a tool, is used to automatically scan the features with the aid of an NC program. In this case, the tool machine temporarily becomes a coordinate measuring machine.

The next step of the degree of automation is the automatic handling of measuring device and workpiece. This is often the case due to integration of measuring and test technology into the material flow of production applications and especially needed if various workpieces and production features must be measured in a short period of time. Representative for these applications are robot-aided measuring devices and multi-point measuring devices, which will be briefly described in Sections 4.7.2 and 4.7.3.

4.7.1 Electronic Handheld Measuring Devices at the Computer-Aided Measuring Station

Due to the advances, which have been achieved over the last decades in the field of electronic and data technology, a number of measuring aids and computer accessories are available at a reasonable price, which allow the integration of manual measuring devices at a computer-supported measuring station. Apart from purely mechanical manual measuring aids, electronic and digital types are increasingly used, which are interfaced with external evaluation units. Evaluation units can be specialised equipment, which add certain functions to manual measuring devices or computer systems based on industrial standards (PC systems, VMEbus computer, etc) with programs to evaluate measured values and utilise these **(Figure 4.7-1)**. At the moment, apart from the manufacturer's specific interfaces, RS-232 or V.24 are used, which offer an easy, star-shaped connection of measuring devices for a practical solution at a measuring station site computer.

Interfaces RS-232 or V.24 are the most widely used data interfaces. In terms of equipment, it is very easy and cost effective and exists in every computer system, from a single chip micro control system to a workstation.

Normally, only one application is connected to an RS-232 or V.24-interface, i.e. only one interface in the computer system is needed per measuring device. Standard RS-232 was specified by the American "Electronic Industries Association" (EIA). The correct name of the currently used third revised version of the RS-232 norm is EIA RS-232-C. The international version is CCITT-recommendation V.24 of the CCITT (Comité Consultatif International de Télégraphique et Téléphonique), which is equivalent to the EIA RS-232-C, apart from a few, hardly used differences.

The RS-232 norm or the V.24-standard do not define the data format or the mechanisms for the data exchange. The way in which the RS-232 interface exchanges data is specific to the manufacturer and the equipment.

As long as measuring devices, computer components and software are bought from one manufacturer, all individual components should be compatible. However, the compatibility of products from various manufacturers cannot be assured by industry standards.

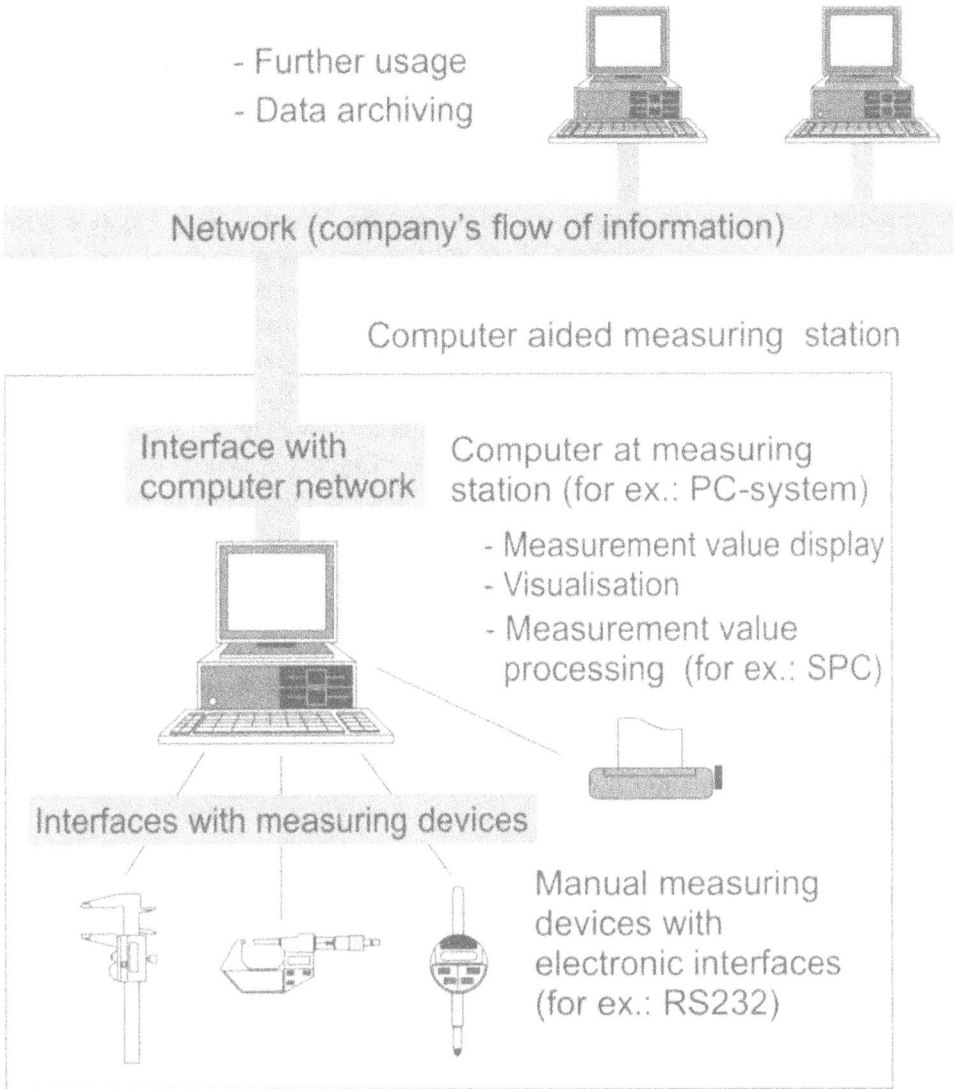

Figure 4.7-1: Computer-supported measuring station with hand-guided measuring devices

Once the measuring devices are plugged into a computer system, a wide range of industrial techniques for data use and data communication is available, to automatically conduct measurement acquisition, recording, evaluation and archiving. Usually an evaluation and visualisation is carried out at the measuring site. For this CAQ programs are available, for example, which offer a statistical evaluation of test data and support a direct measurement adoption from manual measuring devices of the same manufacturer.

The consolidated test results can, after the first evaluation at the test site, be transported to other departments or computers, e.g. into the archive, monitoring or further processing.

For data transfer between computer systems, various standards and products exist. However, those techniques prevailed, which profited from mass production and the price slump for computer components. This was the Ethernet [Heg 92] as a standard transmitter medium, linking technology and retrieval procedures and, on the other hand, TCP/IP (Transmission Control Protocol / Internet Protocol) as a communication protocol based there upon [Tan 89]. The Ethernet was standardised by the American "Institute of Electrical and Electronical Engineers" (IEEE), amongst other communication systems for the computer network [Sta 93]. TCP/IP and additional protocols as for example the "File Transfer Protocol" (FTP) were developed from the ARPANET, a network initiated by the "Defence Advance Research Projects Agency" of the US Ministry of Defence, and the Internet [Tan 89]. Ethernet cards can be used on nearly every module computer system, TCP/IP and additional software is available for nearly every system and for all Ethernet interfaces. The Ethernet and TCP/IP are a part of the basic equipment of nearly every computer system.

However, even with the TCP/IP specification, no definition of the data formats or interpretation of the transported data is made. Based on TCP/IP are a number of software products, for example databank systems and retrieval functions over the network, which support the construction of networked computer applications for test data acquisition and data usage.

4.7.2 Robot-Aided Measurement Devices

To carry out fully automated measurements, the technical integration of the measurement devices is needed as much as the fully automated handling of measuring devices and workpieces.

This usually requires the setting up of special measuring applications, which are adapted to the work piece spectrum and therefore differ considerably in their constructive set up. However, the used acquisition principles and the principles of the data technical integration are less different. This is not surprising if one takes into consideration that about 90% of all features of mechanically produced workpieces are related to length and length conditions [Dut 96]. Automated measuring equipment for testing geometrical features are therefore mainly position and angle detectors with electronic interfaces.

Industrial robots are mainly used for assembly and mounting applications. If various parts, which are for example worked on in various tolerance-afflicted steps need to be built into component groups, the problem often arises that before the assembly process a joining selection of matching workpieces must be carried out in order to reach an optimum assembly quality.

A test cell which is flexibly automated through the application of robots, can be made available during assembly to acquire the data needed for the workpiece selection. It can control the material flow depending on the selection and provide ideally paired workpieces for the assembly process.

Furthermore, the use of a robot application enables flexible measurement and test equipment, which can be reconfigured by the robot and therefore adapted to a class of workpieces, e.g. variations in the highly varied small and medium sized production batches.

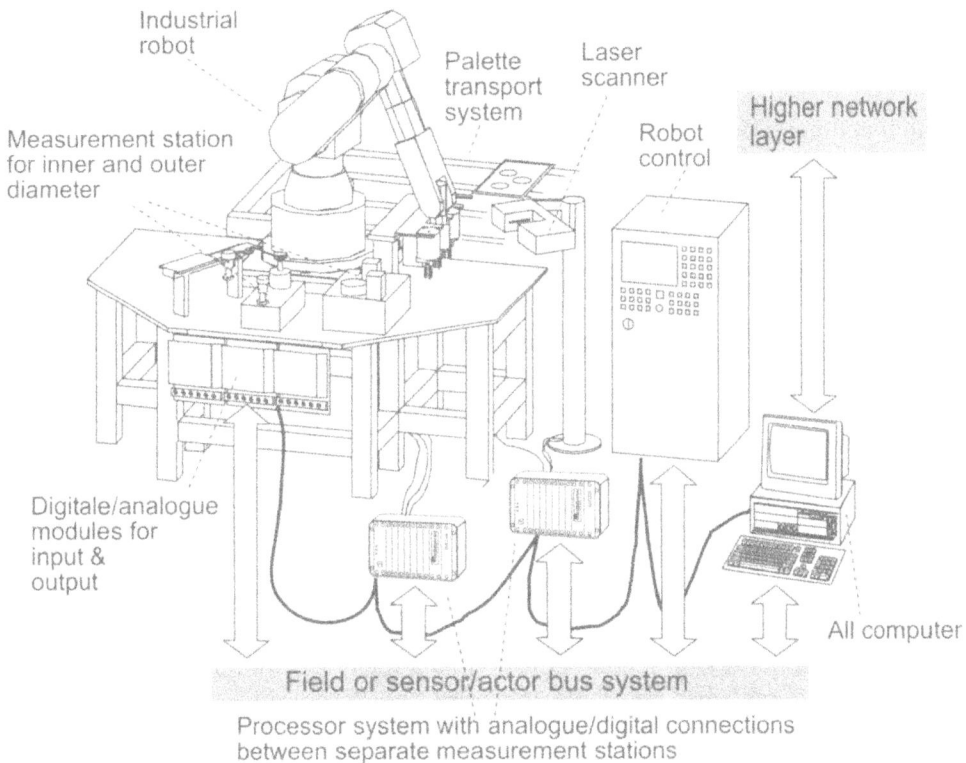

Figure 4.7-2: Robot-aided test cell for the automated testing of different workpieces and features

The example in **Figure 4.7-2** shows the main component of a clutch release bearing. The single parts are transported via a palette transport system to the test cell where the quality-relevant features and those controlling the material flow are automatically collected with various inspection devices (tactile position sensor, laser scanner).

For the controlling of the measurement station and for acquiring the test data as well as for controlling the material flow within a test cell, a sensor-/actuatorbus system is used. All applied inspection equipment is connected directly with the input/output module joined with a bus interface, or they are connected with the processor systems for the measurement control and data handling, which in turn has bus interfaces. The robot control, switches and actuators of the palette transport system as well as the cell computer are connected with the bus system with respective interfaces.

The cell calculator has an interface to the cell-internal bus system as well as to the overlying network. It is therefore integrated into the process information flow. Through this interface, control information is exchanged and test data is transported for further usage and archiving to other computer systems.

From a communication technological point of view, this interface to a higher network level can be realised with the aid of an Ethernet card and TCP/IP software, as introduced in Section 4.7.1. The choice of the suitable bus system is not so straight forward as many different standards have been established. Therefore, many products are offered on the market. The European market is led by PROFIBUS-DP, INTERBUS [DIN 19258], CAN [ISO-IS 11898], [Ets 94] and FIP [Let 93] in the field of standardised field and sensor-/actuatorbus systems. On an international level, a global field bus norm has been worked on for years by the IEC (International Electro-technical Commission) [IEC 95]. By the time this has been passed - the norm must comply with many company requirements as well as technical and organisational points - national and European norms will have been established. Specifications by PROFIBUS-FMS and -DP (Germany), FIP (France) and P-NET (Denmark) have already been standardised as European norm EN 50170 [Böt 96].

These facts alone show that, for the near future, industrial users and manufacturers of automated components will have to deal with a number of field and sensor-/actuatorbus systems and handle quite complex interfaces and user regulations. It has now been established that there cannot be one universal bus system for all applications.

The concrete conditions which led to the selection of a bus system for the application in **Figure 4.7-2,** will not be explained here as they require further information. Specialised literature is recommended for more detailed information.

4.7.3 Multi-Point Measuring Devices

Multi-point measuring devices are measuring devices, which measure one-dimensional features with many contact elements at the same point in time at various positions on the workpiece **(Figure 4.7-3)**.

These measuring devices usually work with a modular system, i.e. they are adjusted to their measuring task by combining different elements. The modularly structured multi-point measuring device is quickly available, easy to modify, maintain and separated into its basic modules after use. However, even though they can be put together in various ways, these devices are only suitable for a certain workpiece spectrum and can only be adapted for the use on a limited number of workpieces.

The variety of the realised solutions makes it more difficult to show an overall diagram. Multi- point measuring devices for shaft parts with a horizontal workpiece recorder are primarily used in industry and with a vertical workpiece recorder for disk-shaped parts. The assembly parts for both are similar:

- base, possibly assembled from various separate components
- callipers and components to take up measurement value recorders
- measuring inserts, measuring heads, deflection heads, buffers
- prisms, crests and feeds for the acceptance of the measured object

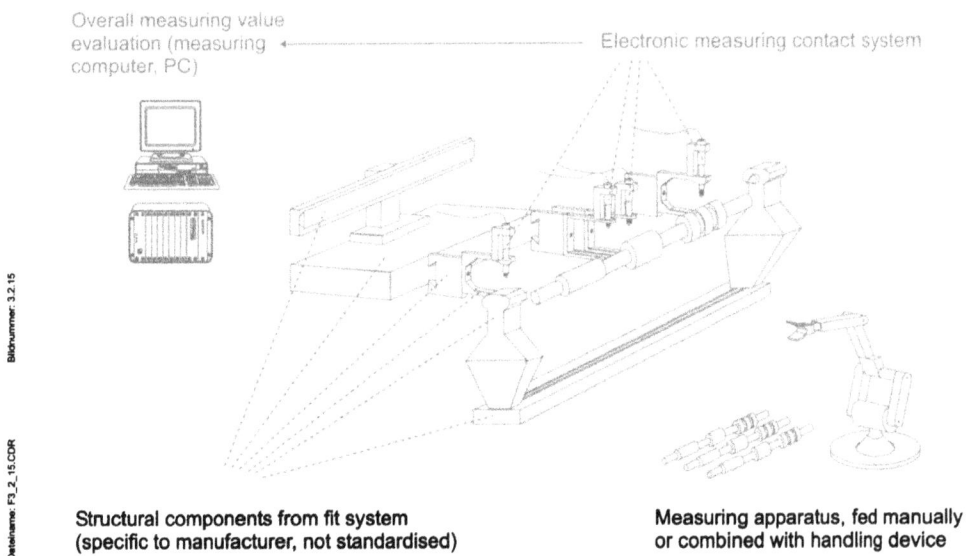

Overall measuring value evaluation (measuring computer, PC)

Electronic measuring contact system

Bildnummer: 3.2.15

Dateiname: F3_2_15.CDR

Structural components from fit system (specific to manufacturer, not standardised)

Measuring apparatus, fed manually or combined with handling device

Figure 4.7-3:   Components of a typical multi-point measuring device

These elements are not standardised and the modular design systems offered by the various manufacturers are usually not compatible with one another.

Inductive or digital measuring sensors elements are mainly used as measurement value recorders. The evaluation of the measurement values and linking of various measuring points as well as the visualisation is carried out with external evaluation units and connected computer systems. Due to the modularity of the modern computer systems, such as PC systems or VME-bus systems, the combination of suitable interface cards for the connection of the inductive or digital touch probes is possible.

Multi-point measuring devices are normally calibrated with a standard (master piece, workpiece with known dimensions or gauge block). However, multi point measuring devices are subject to wear and tear, which can only be counteracted by calibration to a certain extent. It is therefore necessary, to monitor these inspection devices in regular intervals as described in Chapter 6.

Another feature of multi-point measuring devices is that when they cannot be fed manually, they are often combined with a handling device and integrated into automated production devices.

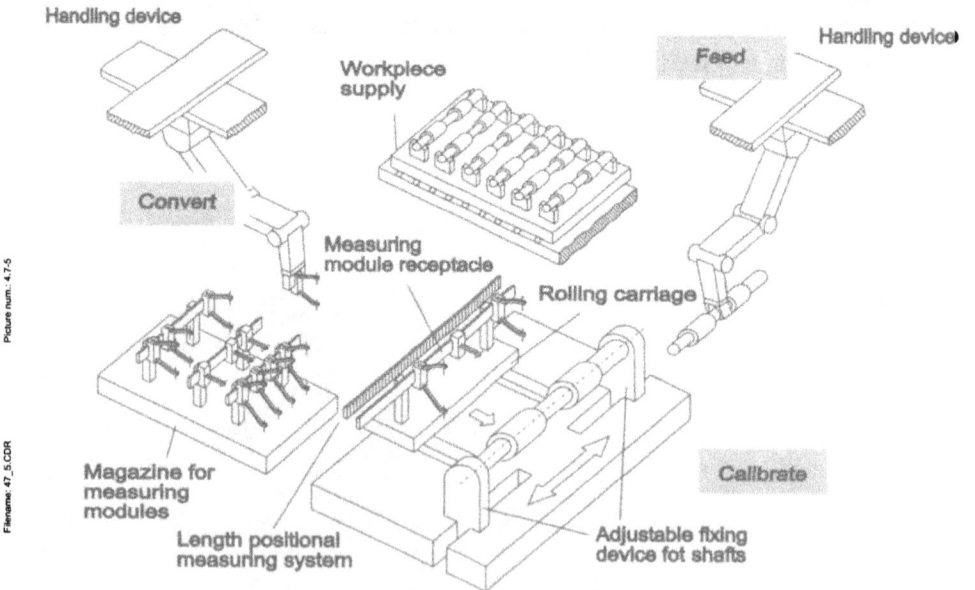

Figure 4.7-4: Flexibly configurable multi-point measuring apparatus with tactile diameter inspection

Figure 4.7-4 shows a typical application of a robot-aided multi-point measuring device for the acquisition of diameter and length dimensions on shaft like workpieces. As shown in the application in **Figure 4.7-3**, the measuring and device components are handled with the aid of an industrial robot. The basis is a conventional apparatus kit, which has been modified for the automated robotic operation. The appliance mechanics contains a calliper receptacle as well as the required number of callipers, which have been fitted with measuring contact elements. The calliper receptacle is fixed to a pneumatic carriage, which is driven toward the shaft during the measuring process. The callipers, which are pre-set to the defined diameter ranges, are arranged in magazine. They are taken by the robot from there when needed and mounted onto the calliper base automatically. The overall facility also contains gripping device change apparatus and a palette-orientated workpiece feed system.

The industrial robot not only configures the number and position of the measurement modules but also modifies the shaft recorder flexibly. This is the recording between crests, which can be either changed or adapted with the aid of the robot.

After set-up, the measuring device is automatically calibrated. Afterwards, the industrial robot handles the parts and the feeding of the measuring device.

Figure 4.7-5 shows the non-contact inspection of features of a car door as a further example for the application of a multi-point measuring device [GFM 97].

An important quality feature of car doors is their flexion. Deviations of 0.5 mm can already lead to optical faults and wind noise in the end product. To acquire these features fully automatically during production, a multi-point measuring device was installed. Contrary to the tactile acquisition of test data, this is done with the aid of laser distance sensors.

At the end of a production cell, every eighth front door is inserted into the measuring device by the robot. Within a few seconds, 12 laser distance sensors determine the actual data of the flexion of the outer contour of the door. Afterwards, the robot takes the door and places it on the assembly line for assembling.

The measurement results are acquired automatically and are conveyed to an evaluation programme for statistical process control (SPC).

This optical multi-point measuring device also needs calibration at regular intervals, which is also done automatically. For this, a master door is used, which can be loaded during shift changes.

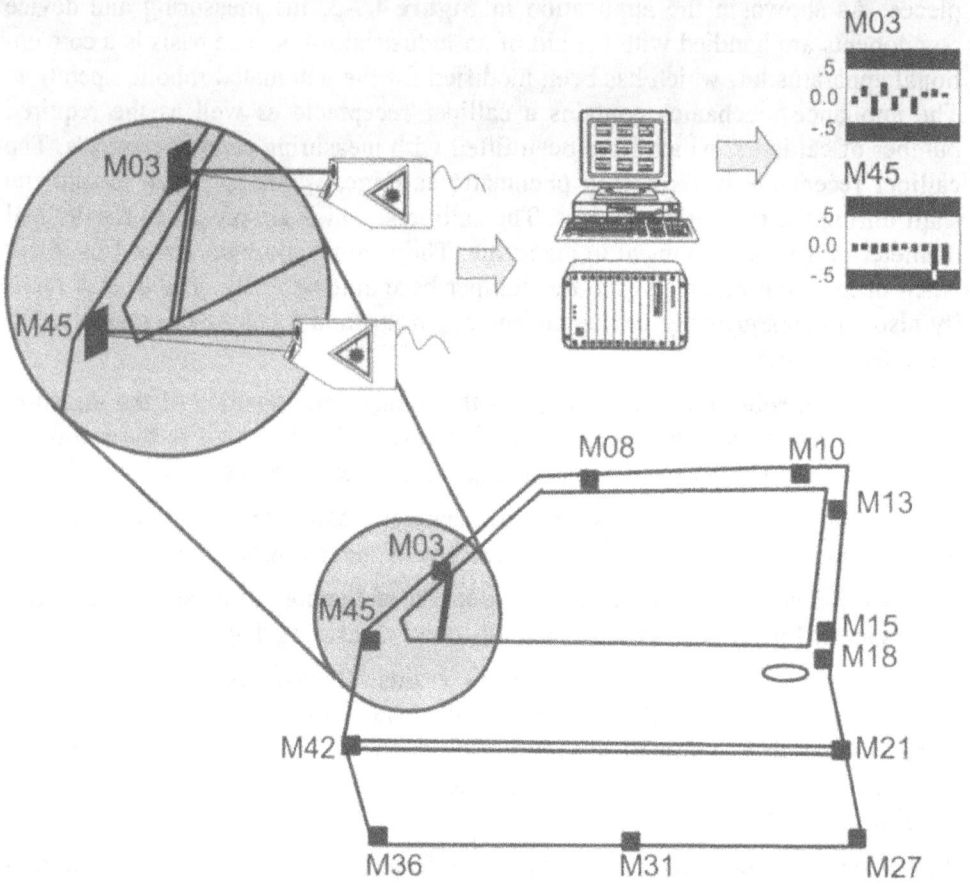

Figure 4.7-5:   Multi-point measuring device with triangulation sensors

Further readings

[Ah 91]   Ahlers, R. J.: Industrielle Bildverarbeitung, Addison-Wesley, 1991

[App 77]   Appold, H. et al.: Technologie Metall für maschinentechnische Berufe. 9. Aufl., Verlag Handwerk und Technik, Hamburg 1977

[Bar 97a]   Bartelt, R.: Formtester oder 3D-Koordinatenmeßgerät? Werkstatt und Betrieb Vol. 130, München: Carl Hanser Verlag, 1997

[Bar 97b]   Bartelt, R.: Oberflächenrauheit richtig messen, Teil1: Beschreibung der gebräuchlisten Kenngrößen. F&M Vol. 105, München: Carl Hanser Verlag, 1997

[Bäs 89]   Bässmann, H; Besslich, P.W.: Konturorientierte Verfahren in der digitalen Bildverarbeitung. Springer-Verlag Berlin, Heidelberg, New York, London, Paris, Tokyo, 1989

[Ber 93]   Berthold, G.: Potentiometrische Sensoren als Weggeber und Stellungsmelder, Sensoren in der Praxis, S28-37, München, Franzis-Verlag GmbH: 1993

[Böt 96]   Böttcher, J.: EN 50170 - Die europäische Feldbusnorm. Entstehung, Spezifikationen und Folgen für Hersteller und Anwender, in: Elektronik, Heft 12/96 vom 11.Juni '96, Franzis-Verlag, München 1996

[Bre 93]   Breuckmann, B.: Bildverarbeitung und optische Meßtechnik in der industriellen Praxis. München, Franzis-Verlag, 1993

[Brü 96]   Brüggeman, C.; Kross, J.: Charakterisierung von CCD-Kameras. Feinwerktechnik und Meßtechnik F&M 104, 9/1996

[Ca 93]   Caber, P.J., Martinek, S.J., Niemann, R.J.: A new interferometric profiler for smooth and rough surfaces, Proc. SPIE 2088, Laser Dimensional Metrology, Photonex'93, October 1993

[Chi 95]   De Chiffre, L.; Hansen, H.N.: Metrological limitations of optical probing techniques for dimensional measurements. Annals of the CIRP Vol.44/1995

[Chr 91]   Christoph, R., Wiegel, E.: In drei Dimensionen optoelektronisch Messen. Sonderdruck Kontrolle Juli/August 1991, Konradin Verlag

[DIN 87]   N.N.: Längenprüftechnik 2, Lehren. Hrsg.: DIN, Deutsches Institut für Normung e.V, Beuth-Verlag, Berlin/Köln, 1987

[Don 93]   Donges, A.; Noll, R.: Lasermeßtechnik, Grundlagen und Anwendungen. Heidelberg: Hüthig-Verlag GmbH, 1993.

[Dut 96]   W. Dutschke: Fertigungsmeßtechnk, Teubner Verlag Stuttgart 1996, 3. Auflage

[Edl 66]   Edlén, B.: The Refractive Index of Air. Metrologia 2, S. 71-80, 1966.

[Ets 94]   Etschberger, K. (Hrsg.): CAN - Controller area network, Hanser, München Wien, 1994

[GFM 97]   N.N.: Laser-Vielstellenmeßsystem IST-3 prüft Kontur von PKW Türen in der Fertigung; Produktinformation der Gesellschaft für Meßtechnik mbH, Hirzenrott 2, 52076 Aachen

[Hab 91]   Haberäcker, P.: Digitale Bildverarbeitung Grundlagen und Anwendungen, Carl Hanser Verlag München Wien, 4. Auflage, 1991

[Hec 87]   Hecht, E.: Optics. Reading, Masachusetts: Addison-Wesley Publishing Company,
                1987.

[Heg 92]   Hegering, H.-G.; Läpple, A.: Ethernet - Basis für Kommunikationsstrukturen,
                DATACOM Buchverlag, Bergheim, 1992

[Hei 91]   Heime, K.: Elektronische Bauelemente. Vorlesungsskript, RWTH Aachen, 1991

[Her 89]   Hering, E., Martin, R., Stohrer, M.: Physik für Ingenieure. 2. Aufl., VDI-Verlag
                Düsseldorf, 1989

[Her 92]   Hering, E.; Martin, R.; Stohrer, M.: Physik für Ingenieure. VDI-Verlag GmbH,
                Düsseldorf 1992

[Hmd 89]   Hemd, A. vom: Standardauswertung in der Koordinatenmeßtechnik Dissertation
                RWTH Aachen 1989

[Hou 62]   Hough, P.V.C.: Methods and Means for Recognizing Complex Patterns, U.S Pat-
                ent 3069654, 1962

[IEC 95]   IEC1158 (Teile 2 ...7 im Entwurf), Feldbus für industrielle Leitsysteme, Beuth-
                Verlag Berlin 1992 ... 1995

[Jäh 96]   Jähne, B.; Massen. R.; Scharfenberg, H.: Technische Bildverarbeitung - Maschi-
                nelles Sehen. Springer-Verlag Berlin Heidelberg, 1996

[Jäh 97]   Jähne, B., Digitale Bildverarbeitung, Springer Verlag, Berlin, 4. Auflag, 1997

[Kle 88]   Klein, M. V.; Furtak, T. E.: Optik. Berlin, Heidelberg: Springer-Verlag, 1988.

[Koc 91]   Koch, A.: Streckenneutrale und Bus-fähige faseroptische Sensoren für die Weg-
                messung mittels Weißlicht-Interferometrie, VDI-Verlag GmbH Düsseldorf 1991

[Kör 95]   Körner, K.; Nyarsik, L.; Fritz, H.: Schnelle Planitätsmessung von großflächigen
                Objekten. Messen, Steuern und Regeln MSR, 11-12/1995

[Kr 86]   Krautkrämer, J.: Werkstoffprüfung mit Ultraschall. Springer-Verlag, Berlin Hei-
                delberg New York London Paris Tokyo 1986

[Krm 86]   Krumholz, H.-J.: Optimierte Istgeometrie-Berechnung in der Koordinatenmeß-
                technik Dissertation RWTH Aachen 1986

[Ku 88]   Kuttruf, H.: Physik und Technik des Ultraschalls. Hirzel Verlag, Stuttgart 1988

[Ler 96]   Lerner, E.J. Charge-coupled devices capture image information in Laser Focus
                World, August 1996, S. 103–116

[Let 93]   Leterrier, P.: The FIP Protocol, Centre de Compétence FIP, Nancy, 1993

[Lez 90]   Lenz, R.: Grundlagen der Videometrie, angewandt auf eine ultra-hochauflösende
                Kamera. Technisches Messen tm 57, 10/1990

[Lic 94]   Lichtman, J.W.: Konfokale Mikroskopie. Spektrum der Wissenschaft, Oktober
                1994, S. 78ff.

[Lip 97]   Lipson, S. G.; Lipson, H. S.; Tannhauser, D. S.: Optik, Springer Verlag, Berlin,
                Heidelberg, 1997

[Lot 96]   Lotze, W.: Das Gelenkarmmeßgerät-Ein neues Koordinatenmeßgerät VDI-
                Berichte Nr. 1258, VDI Verlag GmbH Düsseldorf 1996

[Lot 78]   Lotze, W; Glaubitz, W: Taylorscher Grundsatz - Grundlage für das Prüfen im
                Austauschbau. Feingerätetechnik Berlin 29 (1980) 2, S.51-55

[Lük 85]   Lüke, H.D.: Signalübertragung, Grundlagen der digitalen und analogen Nachrichtenübertragungssy-steme, Springer-Verlag Berlin Heidelberg, 1985

[Lun 90]   Lunze, U.: Beschreibung und Prüfung der Paarungsgeometrie prismatischer Werkstücke. Dissertation, Dresden, 1990

[Lv 91]   Leavers, V.F.: Shape Detection in Computer Vision using the Hough Transform, Springer Verlag, Berlin, Heidelberg, New York, London, Paris, Tokyo, 1992

[Mah 28]   Mahr, C.: Die Grenzlehre. Carl Mahr, Esslingen, 1928

[Mal 92]   Malacara, D.: Optical Shop Testing. New York, Chichester, Brisbne, Toronto, Singapore, John Wiley & Sons, 2nd Edition 1992

[Mal 92]   Malacara, D.: Optical Shop Testing. New York, Chichester, Brisbne, Toronto, Singapore, John Wiley & Sons, 2nd Edition 1992

[Mar 80]   Marr, D.; Hildreth E.: Theory of Edge Detection. Proc. R. Soc. London ser. B. Vol. 207, 1980

[Moh 90]   Möhrke, G.: Mehrdimensionale Geometrieerfassung mit optoelektronischen Triangulationssensor - Verfahren, Meßunsicherheit, Anwendungsbeispiele. Dissertation RWTH Aachen 1992

[Neu 88]   Neumann, H.J. (Hrsg.): CNC-Koordinatenmeßtechnik (Kontakt und Studium, Bd. 172) Expert-Verlag Ehningen 1988

[Neu 93]   Neumann, H.J. (Hrsg.): Koordinatenmeßtechnik (Kontakt und Studium, Bd. 426) Expert-Verlag Ehningen 1993

[NN 91]   Digitale Längen- und Winkelmeßtechnik, Verlag Moderne Industrie, 1991

[NN 95a]   Optische Werke G. Rodenstock München: Produktioninformation Foto-Optik

[NN 95b]   Carl Zeiss Jena GmbH - Industrial Optics and Lasers: Produktinformation Telezentrische Objektive - Reihe VisionMess TVM

[NN 97]   Pressekonferenz 97: Sicher, schnell und wirtschaftlich produzieren mit absoluten Meßsystemen, Heidenhain 13.06.1997

[Pc 97]   Pietschmann, C.: Merkmalorientierte Fertigungsintegration von Koordinatenmeßgeräten Shaker Verlag Aachen, 1997

[Pck 92]   Packroß, B., Pfister, B., Schmidt, G.: Interferometer mit schrägem Lichteinfall. Kontrolle 1992, November

[Pf 72]   Pfeifer, T.: Neuere Meßverfahren zur Beurteilung der Arbeitsgenauigkeit von Werkzeugmaschinen. Habilitationsschrift RWTH Aachen, 1972

[Pf 90]   Pfeifer, T.; Czuka, F.-J.: Meßsystem zur 100%-Kontrolle von Stanzteilen im Produktionstakt. Technisches Messen tm 57, 2/1990

[Pf 92a]   Pfeifer, T. et al.: Optoelektronische Verfahren zur Messung geometrischer Größen in der Fertigung. Kontakt&Studium, Bd. 405, Hrsg. Technische Akademie Esslingen, Expert-Verlag 1992

[Pf 92b]   Pfeifer, T. (Hrsg.): Koordinatenmeßtechnik für die Qualitätssicherung VDI Verlag 1992

[Pf 94]   Pfeifer, T.; Pietschmann C.: Numerische Paarungslehrung mit Koordinatenmeßgeräten In: Innovative Qualitätssicherung in der Produktion. Beuth-Verlag, Berlin, 1994

[Pf 96a]   Pfeifer, T.; Rümenapp, S.; Feldhoff, J: Ultraschall zur Bestimmung von Faserori-
               entierungen in Verbundkunststoffen, Kunststoffberater 7/8 1996

[Pf 96b]   Pfeifer, T.; Beyer, W.; Freudenberg, R.; Meyer, S.: Tascspecific Mechanical
               Standards e.g. Measuring of Bevel Gears on Coordinate Measuring Machines
               VDI-Berichte Nr. 1230, VDI Verlag GmbH Düsseldorf 1996

[Pf 97]   Pfeifer, T., Mischo, H., Evertz, J., Manekeller, S.: An Approach to Model Based
               Interferometry, in: Kunzmann, Waldele, Wilkening, Corbett, McKeown, Weck,
               Hümmler (Eds.): Progress in Precision Engineering and Nanotechnology, 1997,
               Volume 1

[Pra 91]   Pratt, William K.: Digital image processing 2. ed. Wiley Verlag New York 1991

[Pre 97]   Pressel, H.-G.: Genau messen mit Koordinatenmeßgeräten Expert-Verlag, Ren-
               ningen-Malmsheim, 1997

[Pro 92]   Profos, P.; Pfeifer, T.: Handbuch der industriellen Meßtechnik. R. Oldenbourg
               Verlag München Wien 1992

[Rob 93]   Robinson, D.W.; Reid, G.T.: Interferogram Analysis, IOP Publishing Ltd 1993,
               Bristol, Philadelphia

[Rod 1]   N.N.: RM 600, Optisches Oberflächenprüfgerät, Firmenschrift der Firma Optische
               Werke Rodenstock, München

[Ru 96]   Rümenapp, S: Automatisierte Ultraschallprüfung von Faserverbundkunststoffen.
               Dissertation RWTH Aachen, Shaker Verlag, Aachen 1996

[Sch 54]   Schmidt, H.: Lehren. Springer-Verlag, Berlin/Göttingen/Heidelberg, 1954

[Sei 95]   Seitner, R.: Si-Positions-Detektoren. Firmenschrift, 1995

[Sha 96]   Scharsich, P., Pfeifer, T.: Abnahme und Überwachung optischer Koordinaten-
               meßtechnik - Die neuen Blätter 6.0, 6.1, 6.2 der Richtlinie VDI/VDE 2617.VDI-
               Berichte 1258 'Koordinatenmeßtechnik', VDI-Verlag, April 1996

[Sha 97]   Scharsich, P., Pfeifer, T.: Kalibrierung und Anwendungsmöglichkeiten aktiver
               Photogrammetriemeßsysteme. GMA-Bericht Nr. 30 zur DGZfP-VDI/VDE-GMA
               Fachtagung 'Optische Formerfassung'

[Shw 76]   Schwider, J.;Burov, R.: Testing of Aspherics by means of Roational-Symmetric
               Synthetic Holograms. Optica Applicata, Vol.VI, 1976, p. 83-88

[Shw 97]   Schwider, J.: DOE-Based Interferometry, in: Jüptner, Osten: FRINGE97, Berlin:
               Akademie Verlag, 1997, S. 205-212

[Sne 78]   Schneider, C. A.: Entwicklung eines Laser-Geradheits-Meßsystems zur Durchfüh-
               rung geometrischer Prüfungen im Maschinenbau. Dissertation RWTH Aachen,
               1978

[Srö 90]   Schröder, G.: Technische Optik. 7. Aufl., Vogel Fachbuch Würzburg, 1990

[St 88]   Steeb, S.: Zerstörungsfreie Werkstück- und Werkstoffprüfung, Expert Verlag
               1988

[St 93]   Steinbrecher, R.: Bildverarbeitung in der Praxis, R. Oldenbourg Verlag, 1993

[Sta 93]   Stallings, W.: Networking Standards, Addison-Wesley, Reading (Massachusetts),
               1993

[Swa 97]   Schwab, O.; Lorscheider, H.; Scheuvens, B.: Oberflächenstrukturen mit der zei-
              lenkamera erfaßt. Feinwerktechnik und Meßtechnik F&M 105, 1-2/1996

[Tan 89]   Tanenbaum, A.S.: Computer Networks, Prentice-Hall, Englewood Cliffs (New
              York), 1989

[Tes 1]   N.N.: Elektronisches Meß- und Steuersystem für Werkzeugmaschinen, Fir-
              menschriftder Fa. Tesa S.A., Renens/Schweiz

[Thi 93]   Thiel, J.: Entwicklung eines wellenlängenstabilisierten Halbleiterlaser-Interfero-
              meters zur relativen Längen- sowie absoluten Abstandsmessung. Fortschr.-Ber.
              VDI Reihe 8 Nr. 354, Düsseldorf: VDI-Verlag 1993

[Thi 95]   Thiel, J.; Pfeifer. T.; Hartmann, M.: Interferometric Measurement of absolute dis-
              tances of up to 40m. Measurement 16 (1995), S.1-6.

[Tiz 88]   Tiziani, H.J.; Packroß, B.; Schmidt, G.: Testing of aspheric surfaces with com-
              puter generated holograms. In: Weck, M.; Hartel, R. (eds.): Ultraprecision in
              Manufacturing Engineering. Springer-Verlag, Berlin, Heidelberg, New York,
              London, Paris, Tokyo, 1988, p. 335-342

[Tiz 91a]   Tiziani, H.J.: Kohärent-optische Verfahren in der Oberflächenmeßtechnik. Tech-
              nisches Messen (tm), Jg. 58, 1991, S.228-234

[Tiz 91b]   Tiziani, H. J.: Optische Meßtechnik und Meßverfahren. Vorlesungsskript, 1991,
              Stuttgart

[Tr 89]   Tränkler, H.-R.: Taschenbuch der Meßtechnik, R. Oldenbourg Verlag München
              Wien 1989

[Tra 96]   Trapet,E.; Wäldele,F.: Rückführbarkeit der Meßergebnisse von Koordinatenmeß-
              gerätenVDI-Berichte Nr. 1258, VDI Verlag GmbH Düsseldorf 1996

[Tra 82]   Trapet, E.: Ein Beitrag zur Verringerung der Meßunsicherheit von Fluchtungs-
              meßsystemen auf Laser-Basis. Dissertation RWTH Aachen, 1982

[Tu 95]   Tutsch, R.: Formprüfung allgemeiner asphärischer Oberflächen durch Interfero-
              metrie mit.... Aachen: Verlag Shaker, Band 29/94

[Wa 84]   Warnecke, H.-J.; Dutschke, W.: Fertigungsmeßtechnik Handbuch für Industrie
              und Wissenschaft, Springer-Verlag Berlin Heidelberg New York Tokyo, 1984

[War 84]   Warnecke, H.J. ; Dutschke, W. (Hrsg.): Fertigungsmeßtechnik - Handbuch für
              Industrie und Wissenschaft Springer Verlag Berlin Heidelberg New York Tokyo
              1984

[Wk 95]   Weck, M: Werkzeugmaschinen Band 3.2 Automatisierungs und Steuerungstech-
              nik 2, VDI-Verlag 4. Auflage 1995

[Wk 96]   Weck, M.: Werkzeugmaschinen Band 4, Meßtechnische Untersuchung und Be-
              urteilung, VDI-Verlag 5. Auflage 1996

[Wol 75]   Wolf, H.: Ausgleichsrechnung I - Formeln zur praktischen Anwendung Dümmler-
              Verlag, Bonn 1975

[Wol 84]   Wollersheim, H.-R.: Theorie und Lösung ausgewählter Probleme der Form- und
              Lageprüfung auf Koordinatenmeßgeräten Dissertation RWTH Aachen 1984

[Zam 91]   Zamperoni, P: Methoden der digitalen Bildverarbeitung, Wiesbaden Vieweg-
              Verlag, 1991

Norms and guidelines

DIN 862	DIN 862: Meßschieber, Anforderungen Prüfung. Beuth-Verlag GmbH, Berlin 1988
DIN 863	DIN 863: Meßschrauben, Teil 1, Bügelmeßschrauben, Normalausführung, Begriffe, Anforderungen, Prüfung. Beuth-Verlag GmbH, Berlin 1983
DIN 878	DIN 878: Meßuhren. Beuth-Verlag GmbH, Berlin 1983
DIN 879	DIN 879: Feinzeiger, Teil 1, Feinzeiger mit mechanischer Anzeige. Beuth-Verlag GmbH, Berlin 1983
DIN 2258	Normung graphischer Symbole für Zeichnungseintragungen
DIN 2270	DIN 2270: Fühlhebelmeßgeräte. Beuth-Verlag GmbH, Berlin 1985
DIN 4760	Gestaltabweichungen; Begriffe, Ordnungssystem
DIN 4760	N.N.: Gestaltabweichungen - Begriffe Ordnungssystem Beuth Verlag Berlin 1982
DIN 4762	Oberflächenrauheit; Begriffe
DIN 4768	Ermittlung der Rauheitsmeßgrößen Ra, Rz, Rmax mit elektrischen Taschnittgeräten; Grundlagen
DIN 4769	Oberflächenvergleichsmuster; Technische Lieferbedingungen, Anwendung
DIN 4772	Elektrische Tastschnittgeräte zur Messung der Oberflächenrauheit nach dem Tastschnittver-fahren
DIN 4774	Messung der Wellentiefe mit elektrischen Tastschnittgeräten
DIN 19245	N.N.: DIN 19245, PROFIBUS, Process Field Bus, Beuth Verlag, Berlin, 1991 (Teil 1 und 2), 1993 (Entwurf Teil 3)
DIN 19258	N.N. DIN 19258 (Entwurf), INTERBUS-S, Sensor-/Aktornetzwerk für industrielle Steuerungssysteme, Beuth-Verlag Berlin 1994
DIN 66301	N.N.: VDAFS Format zum Austausch geometrischer Informationen Version 1.0 Beuth Verlag Berlin Köln 1986
DIN ISO 1101	Technische Zeichnungen; Form- und Lage-tolerierung; Form-, Richtungs-, Orts-, und Lauftoleranzen; Allgemeines, Definitionen, Symbole, Zeichnungseintragungen, 1985
DIN EN ISO 10360-2	N.N.: DIN EN ISO 10360-2 Coordinate metrology Part 2: Performance assessment of coordinate measuring machines Referencenumber ISO 10360-2 :1994
DMIS 3.0d	N.N.: DMIS - Dimensional Measuring Interface Standard Draft Revision 3.0d CAM-I, Inc. Arlington Texas USA 1994
ISO -IS 11898	N.N.: ISO-IS 11898, Road vehicles - Interchange of digital information - Controller Area Network (CAN) for high speed communication, 1993
VDI/VDE 2601	VDI/VDE 2601: VDI/VDE - Richtlinien 2601, Blatt1, VDI/VDE-Handbuch der Meßtechnik, August 1977

VDI/VDE 2617 1-6	N.N.: VDI/VDE 2617 Blatt 1-6 Genauigkeit von Koordinaten-meßgeräten -Kenngrößen und deren Prüfung, Beuth-Verlag Berlin und Köln 1983-1997
VDI 2617 Blatt 6	VDI/VDE 2617 Blatt 6.0, 6.1, 6.2: Koordinatenmeßgeräte mit optischer Antastung. Beuth-Verlag, 1994, 1995, 1996
VDI/VDE/DGQ 2618	VDI/VDE/DGQ 2618 Blatt 12: Prüfanweisung für Winkelmesser. VDI-Verlag, Düsseldorf 1991
VDI/VDE-GMR	Dokumentation Laserinterferometrie in der Längenmeßtechnik.VDI-Berichte 548, Düsseldorf: VDI-Verlag 1985.

5   Test Data Evaluation

In moving away from a quality assurance strategy, which simply carries out an inspection, and the increased implementation of error avoidance, a fundamental change has taken place in the field of test data evaluation. Today, test data is no longer used only to demonstrate that manufactured products conform to specifications, but is also increasingly implemented as an indicator for early detection of process errors **(Figure 5-1)**.

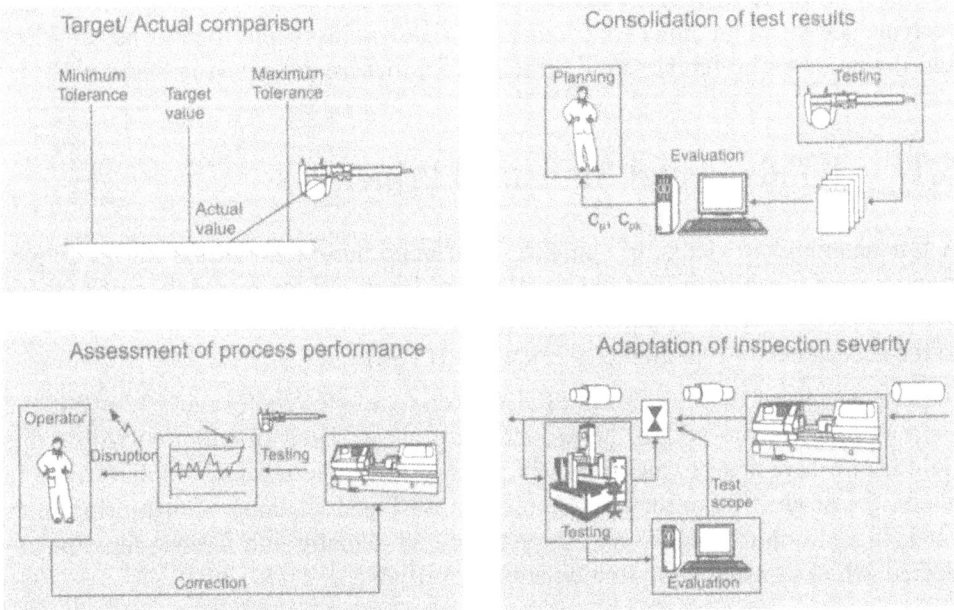

Figure 5-1:   Functions of test data evaluation

Traditionally, a comparison is made between the actual and the nominal values of a feature, based on information gained through test data acquisition. With the aid of expanded test data evaluation, control loops are increasingly set up in operational areas. Within these control loops, the machine operator receives a decision basis for possible necessary corrective intervention in the process, with the aid of statistical data evaluation. The test data also reaches the planning areas in a consolidated form through multilevel control loops. They facilitate an appropriate re-action (i.e. basically adapted to the situation) to the actual qualitative abilities of the individual processes on the operational level. On the basis of analysing test data, the degree of testing can be dynamically adapted to the current process per-formance. Alternatively, a critical comparison can be made of tolerances set by the construction area with the capability values for the processing of tolerated features on the production lines.

In the sense described, today's test data evaluation forms an important pillar of quality management. It constitutes a prerequisite for fulfilling the demand on mod-ern quality management systems, following continuous operational performance process improvement [QS-9000], [VDA 4 Part 1].

It is only possible to develop the potential of test data comprehensively if the un-derlying statistical methods are used appropriately with regard to their possibilities and limitations. The fundamental principles for this are described in Section 5.1.

5.1   Fundamental Statistical Principles

A fundamental knowledge of statistics is an important precondition for the imple-mentation of meaningful test data evaluation. Based on the parameters of individ-ual workpiece criteria, production processes can be assessed or decisions made as to the acceptance or rejection of workpieces [Pf 96].

Experience shows that the values of test criteria are scattered irregularly within the field of values given within the production process. Their distribution within this field is unknown prior to measurement. This is due to the amount of interference to which the production process is exposed as well as deviations of measuring tools used. In addition to coincidental interference, systematic interference must be ex-pected, which is usually not immediately identifiable as such.

In the interest of reducing costs and time, it is often necessary to carry out tests on only a sample inspection (random sample) of the workpieces produced. The ran-dom sample represents the entire mass, or population to be evaluated. For the ran-dom sample to reflect a highly representative picture of the population, a suffi-ciently wide sample scope ($n \geq 5$) and a sufficiently short time interval between the samples must be selected [Pf 96].

Within a statistical framework, the workpieces are described as feature media. The features (e.g. dimensions) receive qualitative or quantitative parameters through the production process, which can be registered or measured and are described here as feature values (feature parameters). All feature values of a set to be evalu-ated form the population of the feature. With the aid of descriptive statistical methods, the comprehensive data sets, which accumulate as random samples, as well as 100% tests, can be characterised by a few, albeit indicative, feature values (e.g. mean $\bar{x}$, variance s^2).

When using descriptive statistics alone, especially with random samples, the quality of workpieces cannot be described sufficiently, as the feature values gained from random samples cannot be directly transferred to the feature values of untested feature media. In this case, inductive statistical methods are used to apply the required conclusions gained from random samples to the population. For this, the probability calculation procedure and methods based on theoretical distribution models are available, the most important of which will be introduced here.

5.1.1 Descriptive Statistics

Descriptive statistical methods make it possible to present a large amount of individual data through a few feature values, whereby the diversification of the feature values being measured must be taken into consideration **(Figure 5.1-1)**. Different types of scale are used, depending on feature parameters.

With a *nominal scale*, parameter values can only be classified under the criteria of same or different, without a sequence. It is only established whether the parameter value of a feature medium is the same as or different from a defined feature parameter. When using an ordinal scale, parameter values can also be put into a rising or falling sequence.

When using a *cardinal*, or metric, scale, the parameters of the investigated feature cannot only be put into a sequence but intervals and relationships between parameter values can also be established. The parameter values are real numbers, which indicate dimension.

Features on a nominal and ordinal scale are described as qualitative features. Both scales have no intervals and the parameters are qualitative, rather than numerical, even though they are represented by numbers and rankings.

A feature on a cardinal scale is also described as a quantitative feature. It is discrete if the number of its parameters can be counted (e.g. counting faulty parts). It is described as being continuous if every real number of a given area can be assumed as a given parameter (e.g. arbitrary exact measuring process) [Bam 96], [Rin 95].

Type of feature	Qualitative features		Quantitative features
	nominal	ordinal	continuous or discrete
Type of scale	Nominal scale	Ordinal scale	Cardinal scale metric scale
Examples			
Feature media	Workpiece (bearing support)	Surface of a workpiece	Workpiece (bearing support)
Feature	Maintaining a drilling measurement tolerance of 40^{H8}	Surface features	Distance between main drilling and drillings in the flange
Feature value/ parameter	ok, not ok	smooth, rough, very rough	Measured value, quantity
Examples of statistical values	Frequency, modal value	Frequency, modal value, quartile, median	Frequency, modal value, quartile, median standard deviation, mean

Figure 5.1-1: Types of features and scales

The next section introduces evaluation methods for quantitative features. Regarding the evaluation of qualitative features, please refer to [ISO 2859], [DIN 53804], [DIN 55350], [Rin 95].

5.1.1.1 Describing a set of data through a frequency distribution

The listing of frequencies of various feature values in a set of data provides a simple overview. **Figure 5.1-2** shows such a frequency distribution for a *discrete* feature. It is based on an inspection of faulty workpieces with 10 random samples each comprising 16 example workpieces (bearing supports). The faulty workpieces are characterised by not maintaining the drill measurement 40^{H8}, which can, for example, be tested by gauging. The number of faulty workpieces within this sample portray a quantitative feature, which can assume discrete values. The test result is reflected in the table and in the bar chart. In addition to absolute frequency, the relative frequency is shown and is calculated by relating the absolute frequency to the total number of feature values.

Testing of n=10 random samples of 16 workpieces each
on unusable pieces with regard to the measurement b=40^{H8}

Feature values: 1) per workpiece:
ok, n.ok (qualitative, nominal)

2) per random sample:
number of parts n.i.o.
(quantitative, discrete)

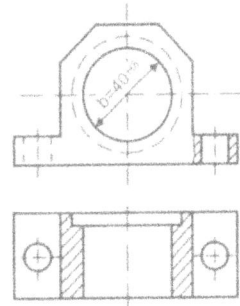

relative
frequency:

$$h_i = \frac{n_i}{n}$$

Legend:
ok = Workpiece ok
n.ok = Workpiece not ok

Bar chart:

Num. of work-pieces per sample n.i.o. j	Absolute frequency of samples with j workpieces n.i.o. n_j	Relative frequency of samples with j workpieces n.i.o. h_j
0	4	0,40
1	4	0,40
2	1	0,10
3	1	0,10
4	0	0,00
5	0	0,00
⋮		⋮
16	0	0,00

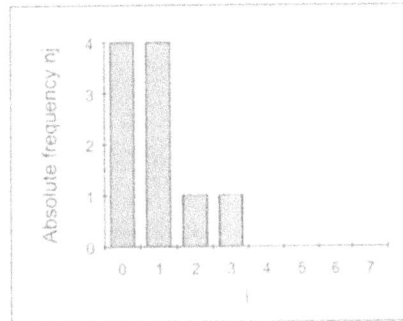

Figure 5.1-2:   Example of a discrete frequency distribution

In order to portray a frequency distribution of a *continuous* feature, the range of values must be divided into intervals or so-called classes. For this, it is generally recommended to choose $k = \sqrt{n}$ classes for n measurement values, where the class width should always be greater than the uncertainty of an individual measurement value [Dut 84].

Figure 5.1-3 shows the frequency distribution of a continuous feature. This example is based on measurements of intervals between two drilled holes on 75 workpieces (bearing supports). The values scattered around the reference measurement of 40mm are categorised into 8 equal class ranges. In addition to the determined absolute frequency, the relative frequency is shown. The graph shows the absolute frequency distribution. It can no longer be described as a bar chart, as a frequency is assigned to every parameter value in a bar chart. In this case, the frequency of various parameter values are grouped into classes. The construction of a bar chart is not advisable for a continuous feature, as all parameter values generally differ.

Determining the measurement a= 40 $\pm$ 0,1 mm on n=75 workpieces

Feature values:measurement values x (quantitative, continuous)

Number of classes:

$\sqrt{n} = 8,66$

→ k = 8

Relative frequency:

$h_j = \dfrac{n_j}{n}$

a=40±0.1

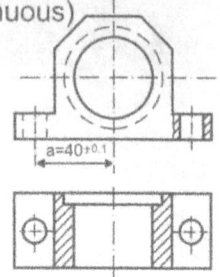

Class no. j	Class values xi [mm]	Absolute frequency nj	Relative frequency hj
1	39,88≤xi<39,91	2	0,026
2	39,91≤xi<39,94	6	0,080
3	39,94≤xi<39,97	12	0,160
4	39,97≤xi<40,00	14	0,187
5	40,00≤xi<40,03	17	0,227
6	40,03≤xi<40,06	11	0,147
7	40,06≤xi<40,09	9	0,120
8	40,09≤xi<40,12	4	0,053

Frequency distribution:

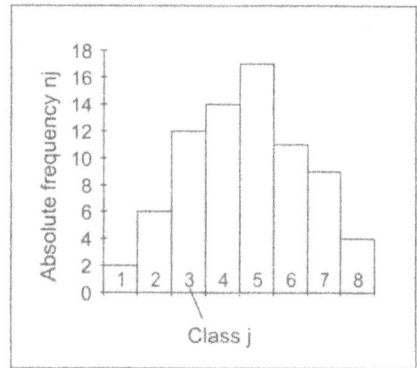

Measurement between 39,88 mm and 40,12 mm
→ class width:

$\Delta x_j = \dfrac{40,12mm\text{-}39,88mm}{8} = 0,03mm$

Figure 5.1-3: Example of a continuous frequency distribution

The upper diagram in **Figure 5.1-4** shows the relative frequency distribution for this example. For continuous features, a histogram (middle diagram) is a favourable method of presentation. In a histogram, rectangles are drawn in across the class widths, where the surfaces are proportional to relative frequencies. The width of the rectangles relate to the class width and the level of frequency density. The frequency density is calculated from the proportion of relative frequency to class width. The relative frequencies are added together successively to yield the sum frequency. The bottom diagram in **Figure 5.1-4** shows the sum frequency distribution, which is also described as the cumulative frequency distribution. The representation of relative frequency and sum frequency are also advisable for discrete features, while a frequency density is not ascertainable for discrete features, as there is no class width.

Determining the measurement a= 40±0,1 mm on n=75 workpieces

Feature values: measurement values x (quantitative, continuous)

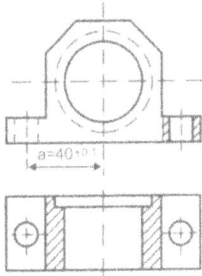

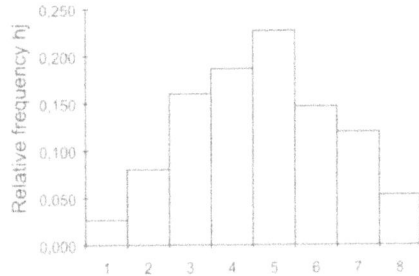

Frequency density:

$$f_i = \frac{h_i}{\Delta x_i}$$

Frequency density f
(bar surface)

Relative frequency
h (bar surface)

Frequency density f
(bar height)

$\Delta x=0,03$ mm

Class Nr. j	Relative frequency h_j	Frequency density f_j	Cumulative frequency
1	0,026	0,866	0,026
2	0,080	2,667	0,106
3	0,160	5,333	0,266
4	0,187	6,233	0,453
5	0,227	7,567	0,680
6	0,147	4,900	0,827
7	0,120	4,000	0,947
8	0,053	1,767	1,000

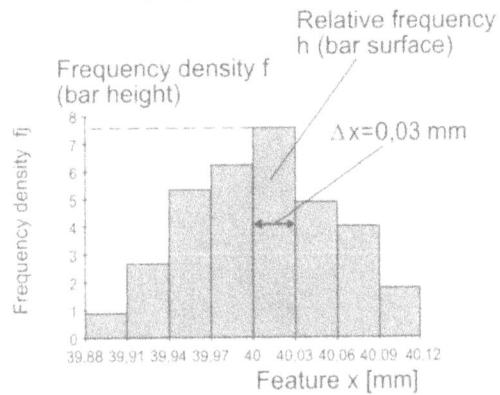

Cumulative frequency:

jx := class containing feature value x

$$F_{(x)} = \sum_{j=1}^{j_x} f_i \cdot \Delta x = \sum_{j=1}^{j_x} h_i$$

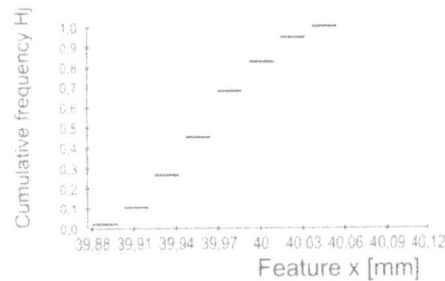

Figure 5.1-4:   Frequency density and cumulative frequency

5.1.1.2 Describing a set of data with positional and dispersion parameters

In **Figure 5.1-5** a fitted curve is drawn in over the frequency density distribution diagram, which will be explained below. In addition to this, the most important positional and dispersion parameters are shown. With these elements, the summarising of data is taken one step further than with a representation through a frequency distribution.

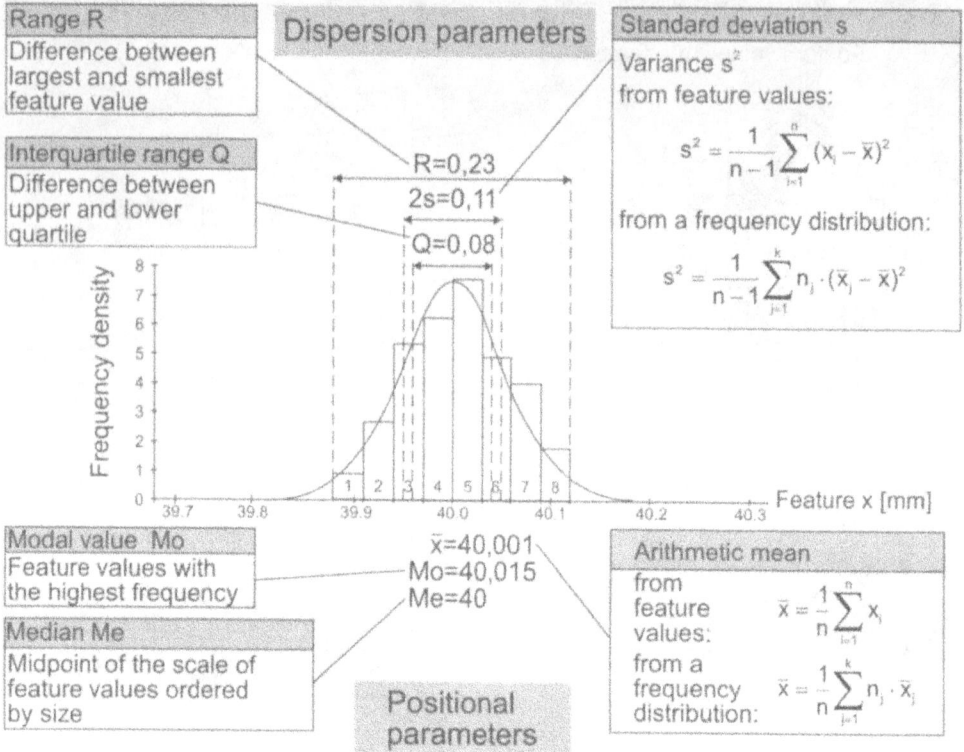

Figure 5.1-5:   Positional and dispersion parameters

Positional parameters

Positional parameters localise the set of data. The best-known positional parameter is the *arithmetic mean,* referred to below as the mean. It is often also described as the average value. With growth processes, a geometric mean is used in place of the arithmetic mean. Please refer to [Sta 70]. The *modal value(s)* identify the feature value(s), which have the highest frequency. When data is divided into classes of equal width, the mid-point of the class with the highest frequency is defined as the modal value [Har 95]. The *median* is characterised by the fact that at least 50% of the values are less than or equal to or at least 50% of the values are greater than or equal to the value of the median. The median therefore lies exactly "in the middle" of the feature values arranged by size. If there are an odd number of values, the median can be determined with certainty. With an even number of values, the two middle values fulfil the conditions of the median. In this case, the average of these two values is frequently used as the median [Bam 96].

Dispersion parameters

From the specification of positional parameters, it cannot be determined whether the feature values are mainly situated near these positional parameters, or lie further away from them. For the complete description of a set of data, the positional parameters must be supplemented by dispersion parameters, which describe these circumstances. The *span* represents a simple dispersion measure. It is the difference between the largest and smallest feature value. The *interquartile range* indicates the size of the range between the *upper* and *lower quartile* (quarter). It encompasses the area of the middle 50% of the feature values arranged by size. If the feature values cannot be divided into 4 groups due to their number (e.g. 75 feature values), an interquartile range must be chosen such that at least 50% of the values lie within the interquartile range [Har 95]. The *variance* is calculated using the formula shown in **Figure 5.1-5**. The standard deviation is the square root of the variance. This also identifies a particular range of feature values, as long as the values follow a particular distribution function. This is described in further detail in Section 5.1.2

The *quantiles* of a distribution represent a further important dispersion measurement. After specifying a proportion of the distribution, quantiles indicate the maximum, or minimum, feature value that still belongs to this proportion. The α-quantile of the standardised normal distribution (detailed in Section 5.1.2.2) in **Figure 5.1-6** marks those boundaries u_α of the standardised feature value u, where all values less than or equal to u_α occur with a probability, or frequency, α. Conversely, the 1-α-quantile marks the boundary $u_{1-\alpha}$. All feature values to the left of this boundary occur together with a probability of 1-α. If the existing distribution is truncated simultaneously on both sides, one speaks of a $\alpha/2$-quantile (right boundary) and 1-$\alpha/2$-quantile (left boundary). Quantiles are often also defined as percentile or fractiles and are listed in tables for various distributions e.g. in [Har 95] or [Rin 95]. As **Figure 5.1-6** concerns a theoretical continuous distribution, the probability density function φ is used in place of the relative frequency density. The distribution function Φ in place of the cumulative frequency, is calculated through the integration of the density function. The quantiles therefore represent the function value of the reverse function of a distribution.

There are additional feature values to those mentioned in **Figures 5.1-5** and **5.1-6**. In this context, it is worth mentioning the *skew*, which expresses the symmetry of a distribution and the *curvature*, which measures the steepness of a distribution. Further details can be found, for example, in [Har 95] and [Sta 70].

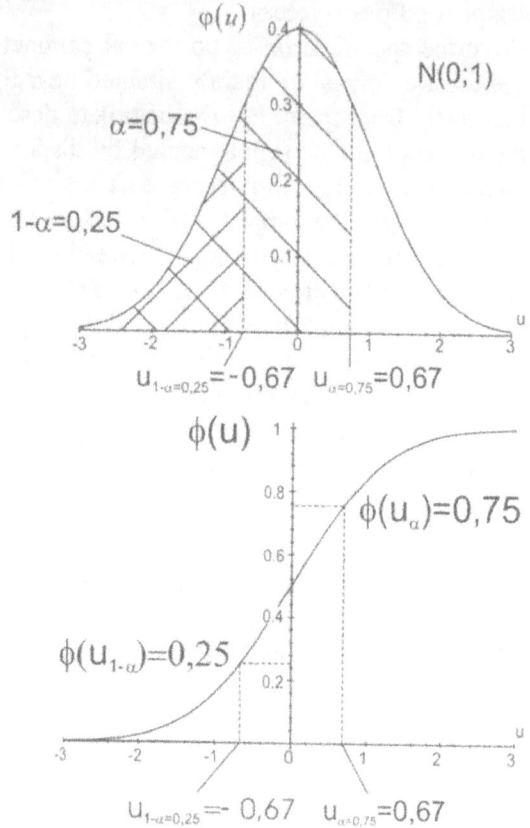

α-quantile $u_\alpha = \Phi^{-1}_{(\alpha)}$

e.g.: $\alpha=0,75 \rightarrow u=0,67$

$(1-\alpha)$-quantile $u_{1-\alpha} = \Phi^{-1}_{(1-\alpha)}$

e.g..: $1-\alpha=0,25 \rightarrow u=-0,67$

Figure 5.1-6:   Quantiles (also percentiles or fractiles)

Outliers

If feature values differ only slightly within the investigated group and an individual value deviates extremely from the rest, this value is described as an outlier. If the reason for the extreme deviation is known, such that it unequivocally fails to represent the population being investigated, the outlier can be eliminated. Where at least 10 feature values are being investigated, a value can be rejected if it lies outside the area $\bar{x} - 4s \leq x \leq \bar{x} + 4s$, where the mean $\bar{x}$ and standard deviation s are calculated without the suspected outlier value [Dut 84]. In addition to this "rule of thumb", there is literature available on numerous testing procedures for determining outliers, e.g. [Dut 84] and [Har 95].

Such outliers have different effects on the individual positional and dispersion parameters. The range is very easy to calculate, but very sensitive to outliers. In addition to influencing the range, outliers also influence the standard deviation and the mean. In contrast, the modal value, median and interquartile ranges are not influenced; they are "outlier proof".

5.1.1.3 Multivariate quantities of data

When two or more features are captured on the same medium, the procedure specified above for representation and compression of data can be applied to these values, by treating the features independently. The typical question about the relationship between the features cannot, however, be analysed in this way. Such a relationship can only be described by one statistic (coefficient of correlation) or by a functional connection (regression analysis). This is used below for two features.

The (Pearson) coefficient of correlation r **(Figure 5.1-7)** is a measure of the linear relationship between two features, referred to here as x and y. The values of the correlation coefficients are between -1 and 1. A correlation near +1 means that high x-values mostly occur with high y-values and low x-values occur with low y-values. A negative correlation occurs if high y-values occur with low x-values and vice versa. The extreme cases of r=+1, or r=-1 occur exactly when the feature values are situated on a line with a positive or negative gradient. The correlation coefficient is visibly reflected in the proportional length of the axes of the ellipse, which enclose the feature values. In the case of r=0, the two feature values are referred to as uncorrelated; the values then form an "irregular cluster of points" in the diagram, which is enclosed by a circle **(Figure 5.1-7)**.

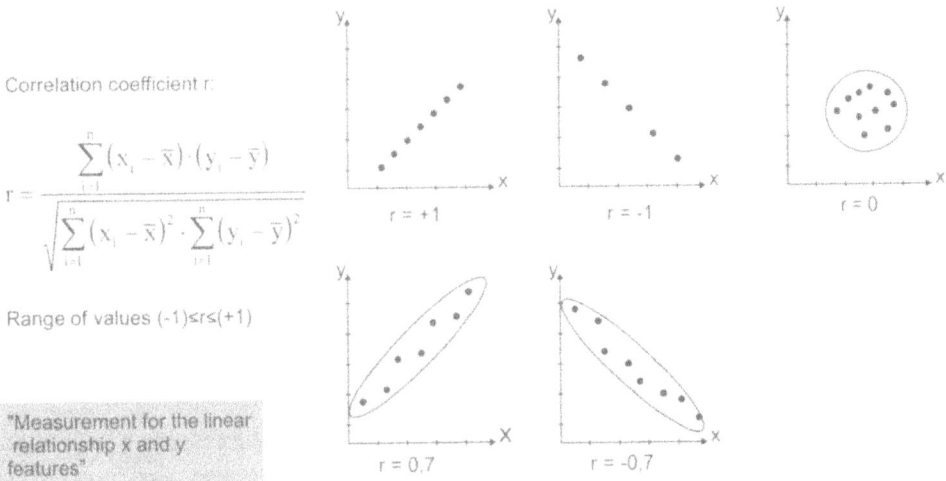

Correlation coefficient r:

$$r = \frac{\sum_{i=1}^{n}(x_i - \bar{x}) \cdot (y_i - \bar{y})}{\sqrt{\sum_{i=1}^{n}(x_i - \bar{x})^2 \cdot \sum_{i=1}^{n}(y_i - \bar{y})^2}}$$

Range of values $(-1) \leq r \leq (+1)$

"Measurement for the linear relationship x and y features"

$r = +1$ $r = -1$ $r = 0$

$r = 0.7$ $r = -0.7$

Figure 5.1-7:   Coefficient of correlation

If the value of the correlation coefficient allows the assumption of a linear relationship this connection can be determined more precisely with a linear regression analysis. The linear relationship between x and y is described by the equation y=a+bx. The coefficients a and b are determined with the formulas shown in **Figure 5.1-8**. The sum of the squared distances between the straight line and the feature values in y-direction is thereby minimised. The residual dispersion is determined in order to measure the quality of fit of the straight lines to the feature values. In the example the linear relationship between the carriage positions of a coordinate measuring device and the length measurement deviation was examined. More than 60% of the dispersion of the length measurement deviation in this example is explained by the regression. This linear relationship is typical for coordinate measuring devices (Section 6).

Feature values:		Equation for lines of regression:	Regression from y to x:	Residual dispersion:

i	x	y
1	x_1	y_1
2	x_2	y_2
.	.	.
n	x_n	y_n

$$y=a+bx$$

$$\text{with } b= \frac{\sum_{i=1}^{n}(x_i - \bar{x})\cdot(y_i - \bar{y})}{\sum_{i=1}^{n}(x_i - \bar{x})^2}$$

$$a = \bar{y} - b\cdot\bar{x}$$

"Proportion of the dispersion not explained by the regression"

$$rsd = s_y \cdot \sqrt{1 - r^2}$$

(witht r=correlation coefficient)

Example:

Dependecy of the length measurement deviation of a coordinate measuring device on the carriage position.

y: length measurement deviation [μm]
x: carriage position [mm]

Nr.	x	y_{Mess}	y_{regr}
1	0	0,0	0,8
2	83	3,5	2,7
3	166	4,0	4,5
4	249	8,0	6,4
5	332	7,0	8,2
Mean standard deviation.	166	4,5 3,16	4,52 2,61

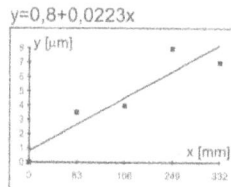

$y=0,8+0,0223x$

Residual dispersion

rsd=1,20216

"38 % of the standard deviation of y is not explained by the regression"

Figure 5.1-8: Linear regression

If the correlation coefficient is not equal to +1 or −1, two different lines of regression can be determined from the data. These correspond to the regression from y to x (minimisation of distances in y-direction) and the regression from x to y (minimisation of distances in x-direction). If there is a causal direction, the choice of regression lines is obvious. The distances in direction of dependent variables (regressand) must then be minimised. In the example of the carriage position of a coordinate measuring device, the length measurement deviation is dependent on the carriage position, so that a line of regression can be clearly determined. If there is no causal relationship, no best-fit line can be determined using this method [Ehr 86].

The minimisation of the perpendicular distance between the feature values and the lines offers a solution to this problem. This method of so-called orthogonal regression is, however, not appropriate for investigating feature values with differing units of measurement. If there is reason to assume that the relationship between the feature values is not linear, a relationship can be determined using non-linear regression calculation methods. These methods are described for example in [Har 95].

5.1.2 Distributions

Observed frequency distributions can be approximated with mathematical formulas. These formulas are constructed on the basis of idealised random events (random models). They describe the probability of an event occurring, by which the occurrence of a feature value is understood. The parameters for a formula, which describes the observed data, are determined from the observed frequency. With a random sample, the formula describes the probability of expecting parameter values, as long as the data follows the underlying model. By enlarging the random sample, the accuracy of predictions about the population can be increased accordingly (Section 5.1.3). In this sense, the relative frequency converges against the probability [Bam 96].

The probability calculation forms the basis for describing data with the aid of distribution models. With the applications discussed here, the probability can be interpreted as relative frequency with respect to the arithmetical processing of observed data.

5.1.2.1 Distributions of discrete features

Both distributions of discrete features introduced below require the events (occurrence of a feature value) to occur independently from one another, randomly and with a constant average relative frequency, or probability [Ehr 86]. This is generally the case for questions arising in production metrology (e.g. number of faulty parts in a random sample).

The description of data from **Figure 5.1-2** with the aid of a *Poisson distribution* is shown in **Figure 5.1-9.** It is used to describe rare events. The mean relative frequency, or probability, p, which is calculated from the quotient of the number of events (total number of faulty parts) and the total number of feature media (number of samples a • sample size n), is a measure to determine if there is a distribution of rare events.

f(x)

Probability function

$$f(x) = \frac{\lambda^x}{x!} \cdot e^{-\lambda}$$

■ calculated value
□ measured value

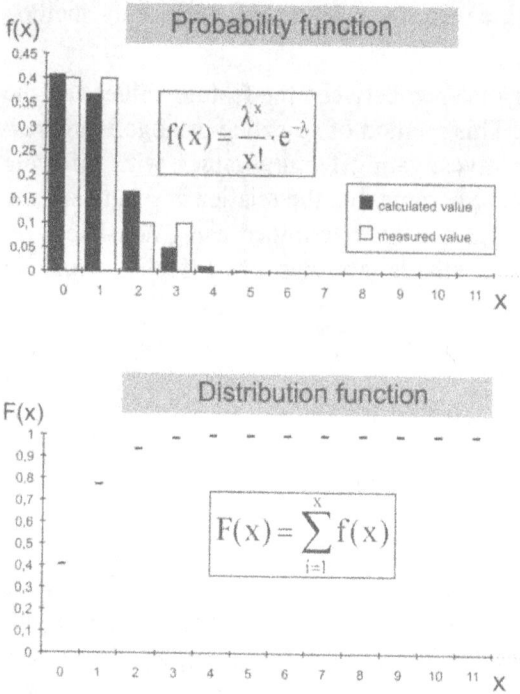

Random sample Nr.	-Faulty parts T
1	0
2	1
3	2
4	1
5	0
6	1
7	0
8	1
9	3
10	0
total	9
mean expectation value	0.9

F(x)

Distribution function

$$F(x) = \sum_{i=1}^{x} f(x)$$

λ = expectation value = variance

$$p = \frac{\sum T}{n \cdot a} = 0.05625$$

$n = 16$ (sample size)

$a = 10$ (number of samples)

Figure 5.1-9:   Poisson distribution

A description using the Poisson distribution is suitable for $p \leq 0.1$ [Bam 96], [Dut 84]. It represents an approximation possibility of the *binomial distribution* (see below), where the mean should be approximately equal to the variance. The probability function, which can also be called the frequency function, describes the course of probability, or relative frequency, dependent on the feature values x. The parameter λ is calculated from the mean (also called the expectation), or the data variance. With $\lambda = 0.9$, the mean of the data in this example, theoretical frequency values are derived, which approximately correspond with the observed frequency values. The distribution function describes the course of the cumulative frequency, or the probability that feature values occur, which are less than or equal to a value x. It is determined by consecutive addition of the probabilities, or relative frequencies.

If the observed events occur more frequently, so that the probability of occurrence for a given set of data is $p > 0.1$, then the data is examined with the aid of a binomial distribution. This, however, does not rule out the possibility that the data with $p \leq 0.1$ can also be described with the binomial distribution. In **Figure 5.1-10**, the probability, or frequency, function of a binomial distribution is shown for the example in **Figure 5.1-2**. The function has two parameters: the sample size n and the mean probability, or frequency, p.

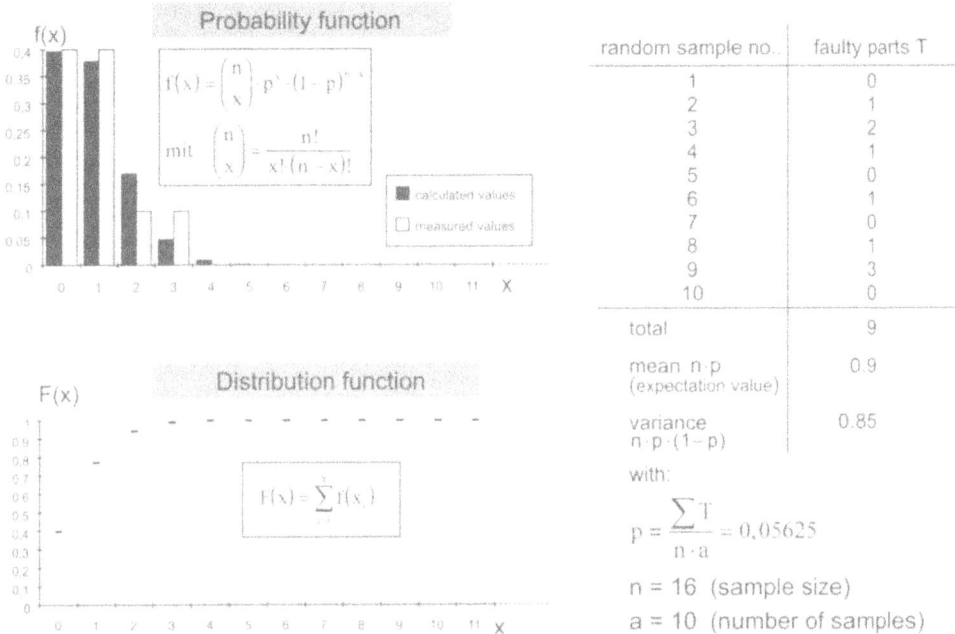

Probability function

$$f(x) = \binom{n}{x} p^x \cdot (1-p)^{n-x}$$

$$\text{mit} \quad \binom{n}{x} = \frac{n!}{x! \, (n-x)!}$$

□ calculated values
□ measured values

Distribution function

$$F(x) = \sum_{i=1}^{x} f(x_i)$$

random sample no.	faulty parts T
1	0
2	1
3	2
4	1
5	0
6	1
7	0
8	1
9	3
10	0
total	9
mean $n \cdot p$ (expectation value)	0.9
variance $n \cdot p \cdot (1-p)$	0.85

with:

$$p = \frac{\sum T}{n \cdot a} = 0.05625$$

n = 16 (sample size)

a = 10 (number of samples)

Figure 5.1-10: Binomial distribution

It produces nearly the same picture as with the Poisson distribution. This clearly shows that the Poisson distribution is a good approximation of the binomial distribution, when the mean p is less than the given limit. This approximation is often used, as the Poisson distribution requires fewer calculations. The mean, or expectation value and variance of the binomial distribution function are derived as described for the Poisson distribution. Both distributions can be adapted through the appropriate choice of corresponding parameters from different empirical frequency distributions.

5.1.2.2 Distributions of continuous features

For numerous practical cases, the normal distribution (often also called the Gauss distribution) is an appropriate statistical model for the evaluation of observed continuous features [Sta 70]. **Figure 5.1-11** represents the probability, or frequency density, function and the distribution function of the data from **Figure 5.1-3**.

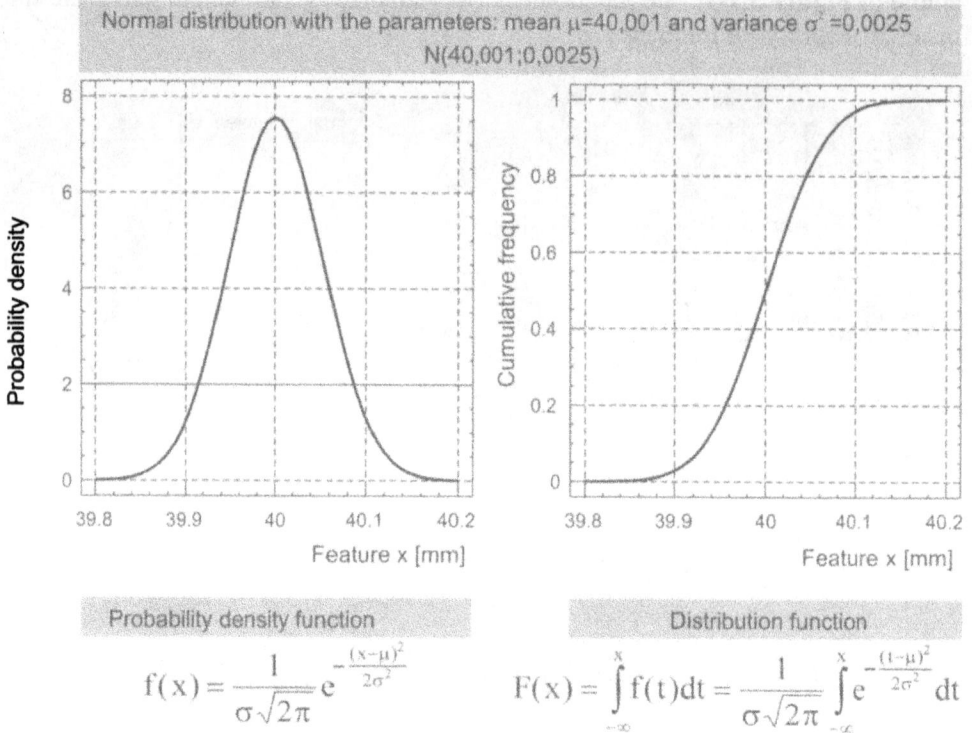

Figure 5.1-11: Normal distribution

The normal distribution is characterised by two parameters; the mean, or expectation value, and the variance (for the population μ and σ^2, for random samples $\bar{x}$ and s^2). The shorthand for such a distribution is usually $N(\mu, \sigma^2)$. The values $\bar{x}$ and s^2 of the sample give approximation values for the parameters μ and σ^2 of the population when the data from only one random sample is known.

The probability density function corresponds to the relative frequency densities of the observed frequency distribution, where the surface under the density function represents a measure of the probability or relative frequency. The indication of a probability for a parameter value, with discrete features is not meaningful, as a nil value would always result for a parameter value. With a continuous feature, a probability can only be given for an interval. The distribution function Φ is determined though the integration of the density function. As this integral can only be determined numerically, the function values are calculated with the aid of computers or from tables for the standardised normal distribution, into which every normal distribution can be transformed.

The corresponding transformation regulation with which the normal distribution can be standardised is shown in **Figure 5.1-12**. The parameters μ and σ^2 arise to $\mu=0$ and $\sigma^2=1$ for the standardised normal distribution. Accordingly, the standardised normal distribution is abbreviated as N(0,1). The values of the standardised normal distribution can be determined from tables, e.g. in [Bam 96], [Dut 84] or [Har 95]. With the aid of the described transformation, the function values of every N(μ, σ^2)-distribution can be determined.

In order to determine the probability P, with which the parameter value of a feature x is found within an interval [a, b], the interval boundaries are first transformed following the same calculation specification. Subsequently, the values of the distribution function at the transformed interval a' and b' can be determined from tables. These tables frequently contain only function values for positive values of u, as the distribution function is point symmetrical to the point $\Phi (u=0)=0,5$ and therefore $\Phi (-u)=1-\Phi (u)$ applies. The probability values given in the tables indicate the probability of a parameter value being greater than or equal to the function value. Accordingly, these values can also be regarded as square measure numbers of the surfaces under the density function. The probability for feature values within the interval [a, b] or [a', b'] can be calculated by subtracting the probability of the upper interval boundary from the probability of the lower interval boundary. **Figure 5.1-12** shows this process for the interval [-1, 1] of the standardised normal distribution. This indicates the area $\pm 1\sigma$ as $\sigma=1$ for the standardised normal distribution. The turning points of the density function are on the level of the standard deviation. 68.3% of the feature values are to be expected in the area between the turning points,

The normal distribution and standard normal distribution are especially important due to the **central limit theorem of statistics**. This generally valid theorem states that the distribution of a variable formed from a summation of n random sample sizes, increasingly approaches a normal distribution as n increases. The reasoning behind it is that, with the summation of two random variables, the convolution is undertaken by their distribution densities. The convolution of two functions, or a function with itself, results from the reflection of one function at the ordinate and a shift around the offset t. The value of the convolution/convolution integral at t corresponds to the product of function 1 and the reflected function 2, integrated within the range of 0 to t. The central limit theorem of statistics applies, even if the summed random variables are not normally distributed [Gim 91], [Har 95].

Probability/frequency density function

$$\varphi(u) = \frac{1}{\sqrt{2\pi}} e^{-\frac{u^2}{2}}$$

with the transformation:

$$u = \frac{x - \mu}{\sigma}$$

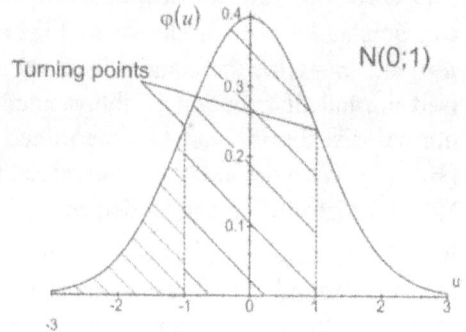

Turning points

$\varphi(u)$

N(0;1)

Distribution function

$$\Phi(u) = \int_{-\infty}^{u} \varphi(t)\,dt = \frac{1}{\sqrt{2\pi}} \int_{-\infty}^{u} e^{-\frac{t^2}{2}}\,dt$$

$\Phi(u)$

$$\Phi(-u) = 1 - \Phi(u)$$

$\Phi(1)$

Probability/frequency

$$P_{(a \leq u \leq b)} = \Phi(b) - \Phi(a)$$

$\Phi(-1)$

e.g.: $P_{(-1 \leq u \leq 1)} = \Phi(1) - \Phi(-1) = 2\Phi(1) - 1 = 0{,}683$

(i. e. 68,3 % of all feature values
are situated within this interval)

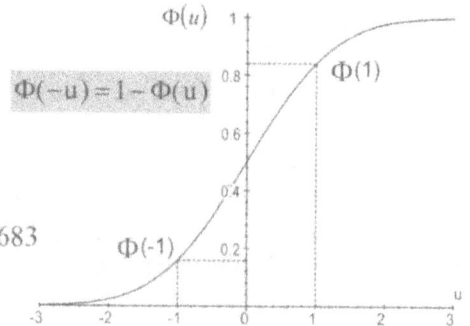

Figure 5.1-12: Standardised normal distribution

This normal distribution model can only be used if the feature values are normally distributed, or the deviation from the normal distribution is negligible. This should be checked with a statistical test if necessary (Section 5.1.3). In production metrology, there are many continuous features, which are not normally distributed. For features, which describe deviations in form or position, there are amount values. For example, with a roundness test, (**Figure 5.1-13**), the deviation from an ideal set is determined, without signs and thus as an amount. A frequency distribution would result in a steep left distribution with a mean close to zero. The normal distribution is not a suitable model for this form of distribution.

The acquisition of data by amount – e.g. for the roundness measurement – for normally distributed measurement values can, however, be deduced from the normal distribution. The probability density function of the normal distribution is convoluted at the zero point and a so-called type-I amount distribution emerges. The convolution at the mean of the normal distribution (zero point) described here, represents a special form of type-I amount distribution. Basically, the normal distribution can be convoluted at any point $x \leq \mu$, so that a type-I distribution develops by superimposing the values to the left of the convolution point with those to the right of the convolution point [Die 95].

Features have a "steep left distribution"

Transformation through
the measurement : y=|x|

(Recording of amounts of prefixed feature values which were originally normally distributed)

Frequency density: $f(x) = \dfrac{1}{\sigma\sqrt{2\pi}}\left(e^{-\frac{(x-\mu)^2}{2\sigma^2}} + e^{-\frac{(x+\mu)^2}{2\sigma^2}} \right)$

Example: roundness measurement

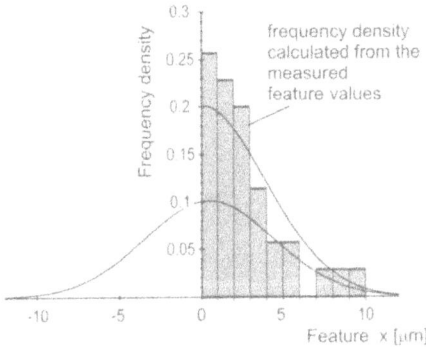

frequency density
calculated from the
measured
feature values

Expectation value μ , and variance $\sigma_{\hat{i}}^2$ of the type I absolute normal distribution can be calculated from the expectation μ value and the variance σ^2 of the normal distribution:

$$\mu_{\hat{i}} = \mu\left[\phi\left(\frac{\mu}{\sigma}\right) - \phi\left(-\frac{\mu}{\sigma}\right)\right] + 2\frac{\sigma}{\sqrt{2\pi}}\exp\left(-\frac{1}{2}\left(\frac{\mu}{\sigma}\right)^2\right)$$

$$\sigma_{\hat{i}}^2 = \sigma^2 + \mu^2 - \mu_{\hat{i}}^2$$

Figure 5.1-13: Amount distribution

Figure 5.1-13 represents the analysis of an exemplary roundness measurement with the help of a type-I amount distribution. By specifying the process of the density function of the amount distribution and the density function of the normally distributed data, the convolution process becomes clear. By standardising, the average value of the mean here is zero, thus achieving this form of amount distribution. Apart from the type-I amount distribution, there is a type-II amount distribution, which is suitable for the description of vectorial sizes (e.g. coaxiality). Regarding the calculation fundamentals for type-I and type-II amount distributions, please refer to [Ang 92].

In addition to the distributions introduced here, there are many other forms of distributions. However, with the distributions explained here, it is already possible to describe a multitude of features arising in production metrology.

Figure 5.1-14 shows distributions for the evaluation of continuous production features. When selecting an appropriate distribution model, attention must be given to using a statistical model, which takes into account the technological origins of the features.

Feature		Evaluation procedure
Length measurements		N
Form tolerances		
Symbol	Toleranced feature	
—	straightness	B1
▱	evenness	B1
○	roundness	B1
⌀	cylinder form	B1
⌒	line form	B1
⌓	surface form	B1
roughness		N

Feature		Evaluation procedure
Positional tolerance		
Symbol	Toleranced feature	
∥	parallelism	B1
⊥	trueness	B1
∠	inclination (angularity)	B1
⊕	position	B1
◎	coaxiality, concentriaty	B2
⩶	symmetry	B1
↗	cyclic running	B1/B2 form dev./ posit. dev.
	plan run	B1

Legend: N: normal distribution, B1: absolute normal distribution (type I), B2: absolute normal distribution (type II)

Figure 5.1-14: Continuous distributions for the evaluation of production features (Source: [Ang 92])

5.1.3   Inductive Statistics

Since exhaustive data acquisition is not possible in many applications, it is often necessary to draw conclusions about the population from data, which has been gained from a random sample. Frequently, the goal is to determine an interval for feature values gained from the sample (e.g. mean and standard deviation) in which the feature values of the population and the sample are evaluated as corresponding statistically. As feature values acquired from random samples can only be indicated with a statistical uncertainty, these confidence defined intervals must be specified only with particular care. A further goal of inductive statistics is the examination of a basic assumption regarding the population, based on sample results (e.g. "the feature values are normally distributed").

5.1.3.1   Confidence Intervals

When estimating feature values of the population from sample data, few real samples are usually taken whereas, theoretically, many further samples could also be taken. Both the determined and theoretical feature values follow a distribution model, the selection of which must depend upon the value itself. In this distribution model, an area is now selected with the aid of quantiles (Section 5.1.1.2), in which the value of the real sample must lie within a given probability, so that the result of the sample can become "familiar". The probability that the feature value determined from the random sample lies within this confidence interval is called the confidence level and is described using the term 1-α. The probability of the converse case is α and is described as the significance level. The confidence level and the significance level can be presented graphically as surfaces under the density function.

The confidence intervals form the connection between the sample feature values and the population, where the sample values provide the estimation values for the population parameters, which are supplemented by a confidence interval. These are limits, within which deviations of the determined parameter are tolerated. In addition to the procedures introduced here to determine the confidence interval of sample feature values, there is literature on numerous other procedures, e.g. in [Bam 96], [Har 95] and [Sta 70], which take other conditions into account. .

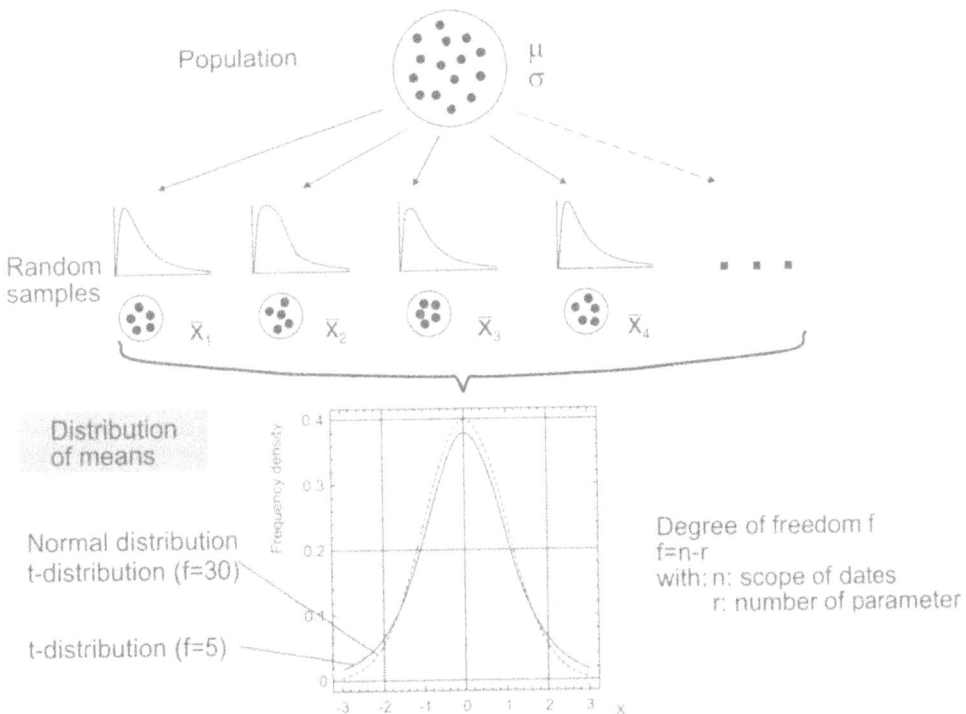

Figure 5.1-15: Distribution of means and the effect of the central limit theorem of statistics

The procedure is clarified in **Figure 5.1-15** through the distribution of the mean parameter. With a small sample size, the means follow the t-distribution (also Student's distribution), a test distribution, the standardised form of which is portrayed here. The expectation value of the distribution of the means is the same as that of the population. The form of the t-distribution is determined by a degree of freedom parameter f, which results from the difference between the amount of data (sample size n) and the number of parameters calculated from the data (a parameter: mean → r=1). For large sample sizes (n>30) the means are almost normally distributed. This is a consequence of the central limit theorem of statistics (Section 5.1.2.2). In **Figure 5.1-15** it can be identified by the fact that the graph of the t-distribution for f=30 is no longer distinguishable from the normal distribution. If the population is normally distributed, the means are always normally distributed, irrespective of the sample size. However, the t-distribution should also be used to describe the distribution of the means in this case (the standard distribution is not known and the sample size is smaller than 30).

$$\bar{x} - t_f \frac{s}{\sqrt{n}} \le \mu \le \bar{x} + t_f \frac{s}{\sqrt{n}}$$

t-distribution (f=74)
with $t_{(1-\alpha)} = -t_{(\alpha)}$

Example:

n=75   (sample size)
x̄=40,001 mm (mean)
s=0,052 mm (standard deviation)

Number of degrees of freedom:
f=n-1=74

Selection of confidence level
1) for α=0,1
   t (for f=74 and $1-\frac{\alpha}{2}$=0,95)= 1,66

   39,991 mm ≤ x̄ ≤ 40,011 mm

2) for α=0,05
   t (for f=74 and $1-\frac{\alpha}{2}$=0,975)= 1,99

   39,989 mm ≤ x̄ ≤ 40,013 mm

t=-1,99   t=1,99

confidence interval
(for α=0,05)

for large sample size (n>30):
normal distribution ≈ t-distribution

for α=31,74% t=1 (normal distribution) $\sigma_x = \frac{s}{\sqrt{n}}$
→ standard deviation of x̄

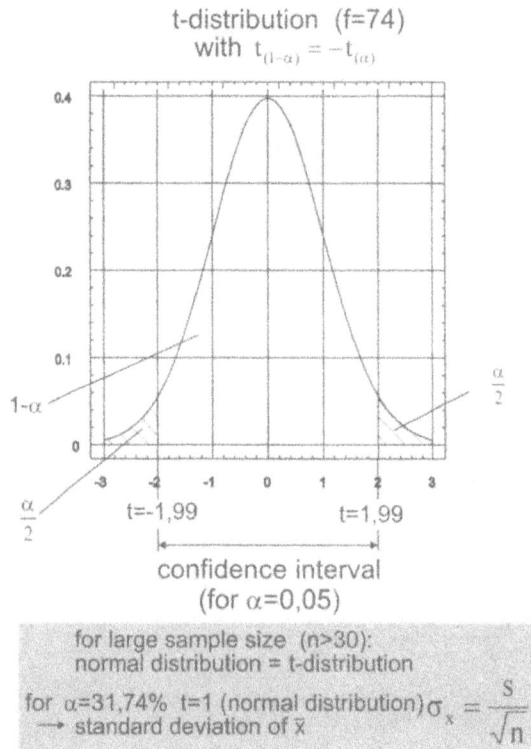

Figure 5.1-16: Confidence interval of the mean

In **Figure 5.1-16** the confidence interval of the mean is formulated as an inequality, which is derived from the standardisation of the distribution [Bam 96]. Mean $\bar{x}$ and standard deviation s are determined from the sample data with the sample size n. This is to be used if the standard deviation of the population is known. The factor t_f is a quantile of the standardised t-distribution and thus it is that t, which limits the confidence interval with a pre-determined significance level α or confidence interval 1-α. It is thus determined as a quantile of 1-α/2 from tables, e.g. in [Har 95], [Bam 96] or [Rin 95]. As the t-distribution is symmetrical, $t(1-\alpha)=-t(\alpha)$ applies. The values for the upper and lower interval limit are the same amount and differ only with respect to the sign. In **Figure 5.1-16**, the confidence interval of the mean is determined for two different confidence levels. Here it is clearly recognisable that the confidence interval becomes larger for a higher confidence level and thus for a smaller significance level.

Frequently, a so-called standard deviation of the mean is indicated. This represents a special case of the relationship described for the confidence interval. In order to determine the factor t_f, the quantiles of the normal distribution are used here, instead of the quantiles of the t-distribution, which is appropriate for samples with more than 30 parts.

A significance level of α=31.74% is selected, which corresponds to a confidence level of 1-α=68.26%. The (1-α/2)-quantile to be determined is exactly at the height of the standard deviation in this case, so that the confidence interval corresponds to the $\pm1\sigma$-area. With the specification of the standard deviation of the mean, it must be considered that the confidence interval is very narrow, due to the high significance levels compared with the confidence intervals with the common levels of confidence of α=0.1 or α=0.05.

In the case of a normally distributed population, calculations with quantiles of the standard normal distribution can also be carried out for a smaller sample size (n<30). However, in this case the standard deviation of the population should be known. If there is only an estimate available for the standard deviation of the population in the form of the standard deviation of the sample, the quantiles of the t-distribution should be used to calculate the confidence intervals.

In a similar way, a confidence interval can be determined for the standard deviation, for which the interval is indicated as an inequation in **Figure 5.1-17**, which is derived from the standardisation of the distribution [Bam 96]. The standard deviation of the samples follows an χ^2-distribution, which, like the t-distribution, is a test distribution. Their form also depends on parameter f (degree of freedom). Only one parameter calculated from the data is considered, the standard deviation, so that in the case of the sample size the same number of degrees of freedom result as with the determination of the confidence interval of the mean.

$$\sqrt{\frac{f \cdot s^2}{\chi^2_{(1-\frac{\alpha}{2})}}} \leq \sigma \leq \sqrt{\frac{f \cdot s^2}{\chi^2_{(\frac{\alpha}{2})}}}$$

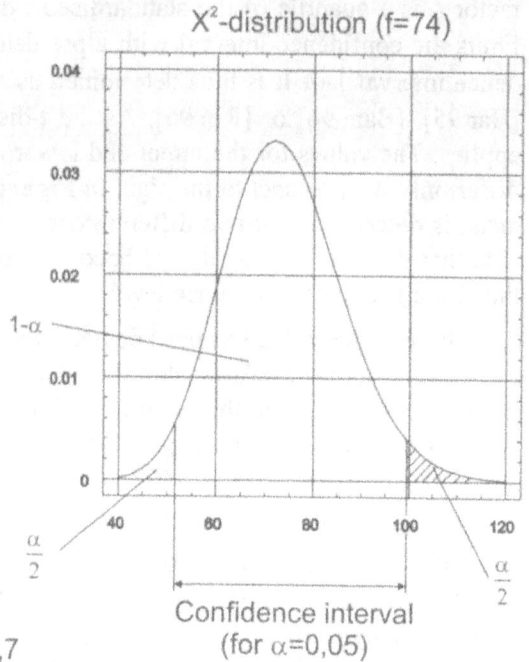

χ^2-distribution (f=74)

Example:

n=75   (sample size)
x̄=40,001 mm (mean)
s=0,052 mm   (standard deviation)

Number of degrees of freedom:
f=n-1=74

Selection of the confidence level
1) for α=0,1

χ^2 (for f=74 and $1-\frac{\alpha}{2}$ =0,95)=95,1

χ^2 (for f=74 and $\frac{\alpha}{2}$ =0,05)=55,2

$0,046 \text{ mm} \leq \sigma \leq 0,060 \text{ mm}$

2) for α=0,05

χ^2 (for f=74 and $1-\frac{\alpha}{2}$ =0,975)= 99,7

χ^2 (for f=74 and $\frac{\alpha}{2}$ =0,025)= 52,1

$0,045 \text{ mm} \leq \sigma \leq 0,062 \text{ mm}$

Confidence interval
(for α=0,05)

Figure 5.1-17: Confidence interval of the standard deviation

In addition to the standard deviation s of the sample, the number of degrees of freedom is directly entered into the formula for determining the interval limits. The quantiles of the χ^2-distribution are determined from tables, e.g. from [Har 95], [Bam 96] or [Rin 95], for a selected confidence level or significance level, depending on the available degrees of freedom. It must be considered here that the distribution is not symmetrical; therefore the quantiles must be individually determined for both the upper and lower interval limits. From a sample size of n>30 the χ^2-distribution can be approximated by a normal distribution. The two examples in **Figure 5.1-17** clearly show that the confidence interval is reduced by a larger significance level.

5.1.2.3   Statistical tests

Certain preconceptions frequently exist regarding the features of a population. In this case, a so-called hypothesis exists about the population in which experiences, assumptions or theoretical considerations are expressed. This hypothesis can be checked on the basis of sample data with the aid of statistical tests.

The basic methodology for such a test is described in **Figure 5.1-18**. Firstly, the so-called null hypothesis H_0 is formulated with regard to a feature of the population, e.g. the specification of the expectation value μ. In connection with this, a so-called alternative hypothesis H_1 is often set up. This is omitted, if the alternative hypothesis represents a denial of the null hypothesis. After setting up the null hypothesis H_0, a test statistic is determined from the sample data. With this size, it can be either a sample feature value (e.g. mean $\bar{x}$) or a specially designed size. The test statistic follows a specific test distribution, irrespective of whether or not the hypothesis is correct. Following the definition of the test statistic, an area,

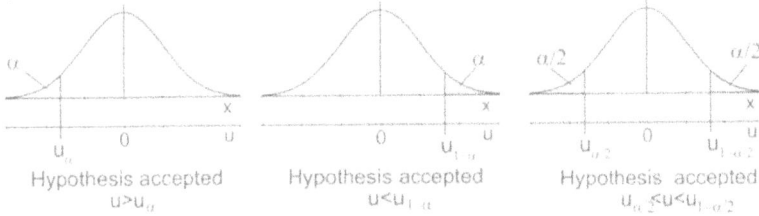

1) Hypothesis H_0 about a feature of the population   e.g..: μ

2) Determining a test statistic u out of the sample data   e.g.: $\bar{x}$

3) Determining limits $u_a, u_{1-a}, u_{a/2}, u_{1-a/2}$ for a level   e.g.: determining a confidence
of significance α interval

4) Comparison of limits and test statistics:

	Hypothesis accepted $u > u_a$	Hypothesis accepted $u < u_{1-a}$	Hypothesis accepted $u_{a/2} \leq u \leq u_{1-a/2}$

Errors:

Test decision	Actual situation	
	H_0 true	H_0 false
H_0 is regected	type I error (α-error)	correct decision
H_0 is accepted	correct decision	type II error (β-error)

Figure 5.1-18: Statistical test

the so-called acceptance region, is determined for the significance level, within which the expected test statistic is certain with the probability $1-\alpha$. The usual value for this is 0.01 or 0.05. The majority of values are expected to be within this acceptance region. Depending on the situation, the area is indicated by a boundary on one side (one-sided test) or on two sides (bilateral test). These boundaries are quantiles of the appropriate test distribution.

Statistical tests assume that the hypothesis is correct if the sample lies within the acceptance region. Conversely, if the test statistic lies outside the area, the hypothesis is rejected. In this case a defining (significant) difference is assumed between the sample and the population. Here, a hypothesis can be rejected, although it is actually correct, if a sample is pulled which has a test statistic lying outside the acceptance region. The probability of such a type I error, also described as an α-error, is equal to α, as this value indicates the probability of the distortion area. For this reason, α is also described as the probability of error. If the probability of error, or significance level, is reduced, then accordingly the acceptance region becomes larger. This increases the danger of a test value of a sample lying within an acceptance range, although the hypothesis is false. Such an error is defined as a type-II or β-error. It cannot be directly derived from the significance level. The quality function provides a complete description of a test regarding the possibility of type-I and type-II errors and is described, for example, in [Bam 96] or [Har 95].

When executing statistical tests, the selection of the correct significance level is problematic. Frequently, two significance levels are used. If the test statistic has a significance level of 0.05 within the acceptance region, the hypothesis is accepted. If it is situated in the distortion area, it must be checked whether it is still in the distortion area with a significance level of 0.01. The hypothesis is rejected if this is the case. Should the test statistic, however, be within the acceptance region at this level, the difference is neither coincidental nor significant. In this case, a larger sample must be tested, as the amount of information in the sample is not sufficient [Dut 84], [Slö 94].

Testing for a particular distribution
The graphic test on a normal distribution with the aid of a probability network differs in procedure from that described above, as no test statistic is used in this case. The cumulative frequency is plotted against the feature values in the probability network. The axis of the cumulative frequency is thereby divided in such a manner in the probability network – in contrast to a diagram with linear divided axes – that a straight line results if the underlying data is normally distributed. **Figure 5.1-19** shows the probability network of the data from **Figure 5.1-3** created with the aid of software for computer-supported test data evaluation.

The vertical axis of the representation corresponds to the feature axis, the horizontal axis of the non-linearly divided axis corresponds to the cumulative frequency. With a cumulative frequency of 50%, the mean can be read off with the aid of lines of equilibrium on the feature axis. The feature values of the lines of equilibrium, which can be read off with 1s, 2s and 3s, are quantiles of the single, double and three-way value of the standard deviation. On the right, next to the probability network, the distribution of the absolute frequencies is shown. The small deviation between the cumulative frequency values and the lines of equilibrium confirms the hypothesis that the data is normally distributed. A more detailed description of the methodology for data analysis with a probability network for normally distributed feature values can be seen, for example, in [DGQ 172b] or [ISO 5479].

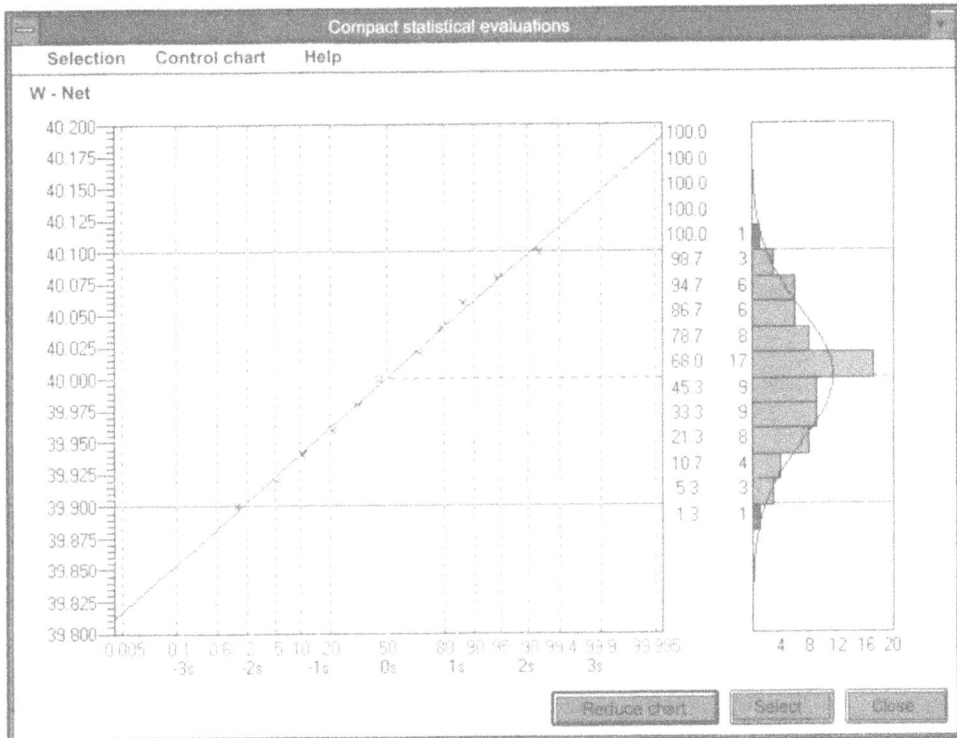

Figure 5.1-19: Test for normal distribution: probability network

Whether or not the cumulative frequency values in a probability network acknowledge the hypothesis about the distribution form frequently depends on the subjective impression. With the aid of a computational test, a more objective evaluation of data is possible with respect to its form of distribution.

The χ^2-test can be used for a computational check of whether a distribution determined from measured feature values corresponds to a theoretically expected distribution. With this test, the hypothesis is a statement about the type of distribution model. In order to carry out this test, the data must be classified in such a way, that the absolute frequency in each class is larger than four [Bam 96]. As a test statistic,· the squared sum of the differences between the frequencies of observed data and the frequencies calculated from the theoretical distribution model are used. This test is therefore not limited to the normal distribution, but can also be used for any other forms of distribution. The test statistic follows the model of a χ^2-distribution, which was already used to determine the confidence interval of the standard deviation. Its form depends on the parameter f, the degree of freedom. In the case of an adjustment test, the degree of freedom results in f=k-r-l. Here k defines the number of classes and r the number of distribution parameters. The adjustment test is a one-sided test. Only if the test statistic, and thus the difference between the observed distribution and the theoretical distribution, becomes very large, the hypothesis about the distribution form is to be rejected. Conversely, a very small test statistic value speaks for the acceptance of a hypothesis. The boundary is determined according to the definition of a significance level α from the α-quantile of the χ^2-distribution. When comparing the boundary with the test statistic, it is checked accordingly, whether the test value is smaller than the boundary. In this case, the hypothesis can be accepted, otherwise it is to be rejected.

Hypothesis: The feature values are normally distributed.

Test statistic: $\chi^2 = \sum\limits_{j=1}^{k} \dfrac{\left(n_j - n \cdot P_j\right)^2}{n \cdot P_j}$

Example: - number of feature values n=75
 - number of classes k=6
 (class size > 4)

Absolute frequency		Test value
theoretical	observed	$\dfrac{\left(n_j - n \cdot P_j\right)^2}{n \cdot P_j}$
$n^* P_j$	n_j	
7,47	8	0,03760375
11,67	12	0,00933162
16,47	14	0,37042502
16,65	17	0,00735736
12,07	11	0,09485501
7,98	13	3,15794486
		$\Sigma 3,67751762$

Determining the limits:
Number of estimated parameters r=2
(mean and standard deviation)

$\longrightarrow$ number of degrees of freedom
 f=k-r-1=3

Example: f=3, significance level $\alpha=0,05 \rightarrow \chi^2_{limit} = \chi^2_{1-\alpha} = 7,8$

Comparison of limits and test statistic:

$\chi^2 = 3,7 < \chi^2_{limit} = 7,8$

$\longrightarrow$ **Hypothesis accepted**

χ^2-distribution with f=3

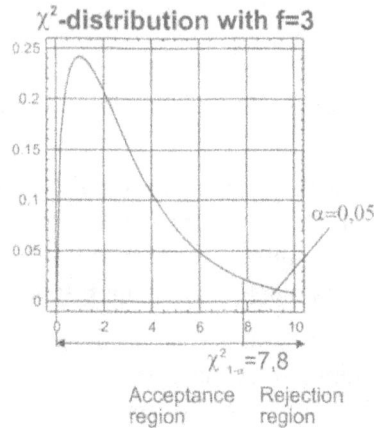

$\chi^2_{1-\alpha} = 7,8$

Acceptance region   Rejection region

Figure 5.1-20: Adjustment test: χ^2-Test

Figure 5.1-20 represents the methodology for a χ^2-test using the data from **Figure 5.1-3**. Here classes 1 and 2, as well as classes 7 and 8, are summarised in order to give a class occupation number greater than four. The appropriate hypothesis states: the data is distributed normally. The mean and standard deviation were determined from the measurement data. These values are parameters of the theoretical distribution model of the normal distribution. The values for the absolute frequency density can be calculated, or determined from tables. The frequency density can then be determined by dividing standardised values by the standard deviation. Through multiplication with the class width, the relative frequency, or probability P_i, arises, from which, through multiplication with the total number of feature values n, theoretical absolute frequency is calculated [Sta 70]. The test statistic is calculated in accordance with the formula indicated. In order to describe the normal distribution, r=2 parameters (mean and standard deviation) are required. With the number of classes k=6, the number of degrees of freedom are therefore calculated as f=3. The boundary results, in accordance with the selected significance level α=0.05 as (1-α)-quantile of the χ^2- distribution to $\chi^2_{1-\alpha}$=7.8. The hypothesis can be accepted, as the test statistic is smaller than the boundary

With a second sample, which is generally independent of the first sample, additional information can be gained about the population. In this case, it is necessary to check whether the features have remained the same, so that the second sample actually does expand the information base. The test described below can only be used if the population is distributed normally, or at least approaches a normal distribution.

Mean testing
The test introduced in **Figure 5.1-18** is a random sample Gauss test. Concerning the population, the hypothesis exists that the mean has the value μ. The test statistic is then the mean of a sample. As long as the population is normally distributed and the mean μ and its standard deviation σ are known, a confidence interval for the mean can be established with the aid of quantiles of the normal distribution. This confidence interval forms the acceptance region of the statistical test. Now a check is done on whether or not the mean of the sample is situated within the confidence interval. If it is situated within this area, the hypothesis is accepted otherwise it is rejected. The described random sample Gauss test forms the basis for statistical process control (SPC) with the aid of charts containing mean values (Section 5.2).

The so-called t-test enables a comparison of two sample means. The hypothesis states that when belonging to a common population, the means of the samples do not differ significantly. This test is conditional upon the standard deviations of both samples not differing significantly. The test statistic is calculated with the formula indicated in **Figure 5.1-21**. It follows the model of a t-distribution, which was already used to determine the confidence interval of the mean. The degree of freedom parameter is determined from the total of the sample sizes (n_1+n_2), reduced by two (number of means to be examined). The boundaries result from the quantiles of the t-distribution, as a function of the selected significance level. As the t-distribution is symmetrical, it is sufficient to determine the amount of the test statistic and only compare it with the positive boundary, the (1-α/2)-quantile.

Hypothesis: Means $\bar{x}_1$ and $\bar{x}_2$ from 2 samples (size n_1 and n_2, standard deviation $s_1 \approx s_2$*) belong to a common population

Test statistic: $t = \dfrac{|\bar{x}_1 - \bar{x}_2|}{s_d}$ mit $s_d^2 = \dfrac{(n_1-1)s_1^2 + (n_2-1)s_2^2}{n_1+n_2-2} \cdot \dfrac{n_1+n_2}{n_1 \cdot n_2}$

Example: two samples with $\bar{x}_1=40,001$ mm, $\bar{x}_2=39,99$ mm
                        $s_1=0,052$ mm, $s_2=0,058$ mm
                      $n_1=75$, $n_2=50$

Determining the limits: estimated parameters $(\bar{x}_1, \bar{x}_2)$

Example: f=123, significance level α =0,05
   $\rightarrow t_{limit}=t_{1-\frac{\alpha}{2}}=-t_{\frac{\alpha}{2}}=1,98$

Comparison of limits and test statistics:

$t=1,11 < t_{limit}=1,98$

→ **Hypothesis accepted**

(As the distribution is symetrical and only the differential between the means is being compared, it is sufficient to carry out a comparison with the upper limit)

*(Check with the F-Test)

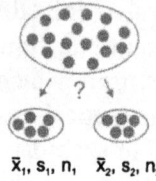

→ $s_d=0,00994$ and t=1,11

→ Number of degrees of freedom:
   $f=n_1+n_2-2$

t-distribution for f=123 $(t_\alpha=-t_{1-\alpha})$

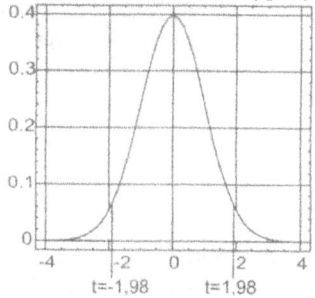

Figure 5.1-21:   t-test: comparison of two means

In the example represented here, with 123 degrees of freedom, the quantiles of the t-distribution can be approximated by the quantiles of the normal distribution. The t-test should not be confused with the random sample Gauss test, which presupposes knowledge about the mean and standard deviation of the population. Often the t-test is also called the two sample t-test.

Dispersion Test

The F-Test can be used to check whether two variances or standard deviations differ significantly, as this is a pre-requisite for carrying out the t-test. The hypothesis for the F -Test is formulated in such a way that the variances of the random samples match statistically, i.e. they do not differ significantly. The F-distribution test feature F is calculated from the quotient of the two variances, as shown in **Figure 5.1-22**.

The F-distribution is a test distribution, as are χ^2- and t-distributions. With the aid of the degrees of freedom f_1 and f_2, which are defined by the size of the tested random samples, the F-distribution is characterised by two parameters. After fixing a level of significance α the limits of the acceptance area can be defined from the quantiles of the F-distribution. It is important that the quantile can be arrived at when exchanging the sequences of the degrees of freedom from the reciprocal value of the original quantile, i.e. $F(f_1,f_2)_\alpha = 1/F(f_2,f_1)_{1-\alpha}$. When evaluating the hypothesis, the test statistic must be checked to see whether it lies within the range of rejection or acceptance. **Figure 5.1-22** shows the F-Test for the sample in **Figure 5.1-21**. In this case the hypothesis is accepted and therefore the result of the t-test confirmed.

Hypothesis: Variance s_1^2 and s_2^2 of 2 random samples (range n_1 and n_2) statistically agree.

Test statistic: $F = \dfrac{s_1^2}{s_2^2}$

Example: two random samples with s_1=0,052 mm, s_2=0,058 mm
n_1=75, n_2=50   → F=0,804

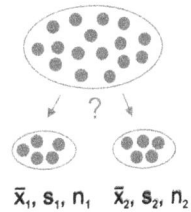

$\bar{x}_1, s_1, n_1$ $\bar{x}_2, s_2, n_2$

Determination of limits: number of degrees of freedom: $f_1 = n_1-1$, $f_2 = n_2-1$

Example: f_1=74 and f_2=49, level of significance α=0,05 → $F_{limit\,(\alpha/2)}$=0,595; $F_{limit\,(1-\alpha/2)}$=1,72

Comparison of limits and test statistics:

$F_{limit(\alpha/2)}$=0,595<0,804<$F_{limit(1-\alpha/2)}$=1,72

→ **Hypothesis accepted**

F-distribution with f_1=74 und f_2=49

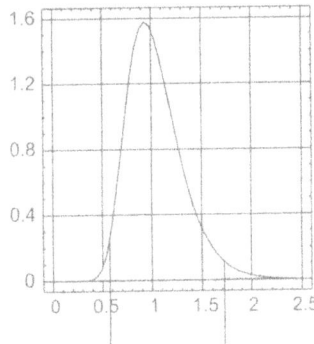

$F_{f1,f2(\alpha)} = \dfrac{1}{F_{f1(1-\alpha),f2}}$

Figure 5.1-22: F-Test: Comparison of two variances

If three or more random samples are accessible, variance analysis can be used for the evaluation. When carrying out variance analysis, the total variance is divided into single variances. The so-called simple variance analysis separates the total variance into two variances, the variance within the random sample and the variance between random samples. For the variance analysis to be carried out the feature values have to be distributed in a normal or near-normal form. With k random samples, which are numbered from j=1 to j=k and have the statistic n_j, the squared deviations within the random sample or between the samples are calculated using the formulas in **Figure 5.1-23**.

Variance analysis: Separation of the total variance into single variances

simple variance analysis: Division of the variance in the variance between the random samples and the variance within the random samples

k random samples (j=1 to j=k), each random sample n_j-value $x_{i,j}$ (i=1 til n_j), measured value $x_{i,j}$ normally distributed

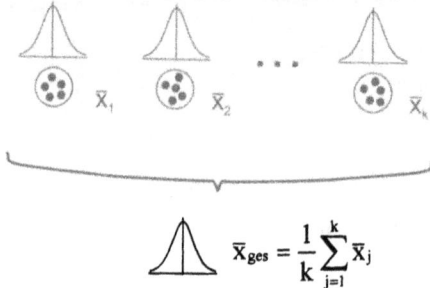

Squares of deviation

$$Q_i = \sum_{j=1}^{k} \sum_{i=1}^{n_j} (x_{i,j} - \overline{x}_j)^2 \qquad \text{within the random samples}$$

$$\longrightarrow \quad \overline{Q}_i = \frac{Q_i}{n-k}$$

$$\overline{x}_{ges} = \frac{1}{k} \sum_{j=1}^{k} \overline{x}_j \qquad Q_z = \sum_{j=1}^{k} n_j (\overline{x}_j - \overline{x}_{ges})^2 \qquad \text{between the random samples}$$

$$\longrightarrow \quad \overline{Q}_z = \frac{Q_z}{k-1}$$

Statistical test:

Hypothesis: all random samples have the same expectation value.

Test statistic: $F = \dfrac{\overline{Q}_z}{\overline{Q}_i}$

Determination of limits: number of degrees of freedom $f_1 = k-1$, $f_2 = n-k$, $\alpha \longrightarrow F_{limit\,(\,a)}$

Comparison of the limit and the test statistic: Hypothesis is accepted if $F < F_{limit\,(\,a)}$

Figure 5.1-23: Variance Analysis

The mean squared deviations $\overline{Q}$ are calculated by dividing the squared deviations within the random samples by the difference between the overall number of feature values of all random samples and the number of random samples and the squared deviations between the random samples by the number of random samples k minus one. With the aid of the mean squared deviations, the variances are assessed.

It seems reasonable to assume that if the dispersion between the random samples is much greater than within the samples, the expected values of the samples differ or the samples do not belong to the same population. This can be assessed with a statistical test. The hypothesis argues that all random samples should have the same expectation value. For test statistic F the quotient of the mean squared deviations is used within the random samples and the mean squared deviations between the random samples. This test statistic follows the model of the F-distribution, which has been described within the F-Test. Both parameters, or degrees of freedom, of the distribution are defined by the formulas in **Figure 5.1-23**. The acceptance area has only an upward limit. A very small test statistic value, i.e. the dispersion within the random sample is much larger than between the random samples indicates that the hypothesis is right. At a certain level of significance the limit equals α the $(1-\alpha)$-quantile of the F-distribution. To be able to establish whether the hypothesis is right or wrong, the test statistic is analysed to see whether it is smaller or greater than the limit.

Variance analysis is also used to analyse the efficient links between numerous variables and objective factors of a production process. In this context it is a statistical experimental technique, which aids the analysis and optimisation of production processes [Gim 91].

Apart from the processes and methods introduced here, there are a number of other possibilities to statistically analyse data in production metrology. Please refer to the further reading section. When choosing the processes and methods a sensible method of description must be found, which acknowledges the technological origins of the data.

5.2   Statistical Process Control

The statistical fundamentals described in the previous section are a basis for the overall understanding and application of the methods described at the beginning to carry out comprehensive and meaningful test data analysis. This not only guarantees quality but also allows the possibility of intervention to rectify the production process. The following explains several statistical process control (SPC) procedures in more detail.

5.2.1 Random Sample Testing Schedules

As early as the 1940s random sample testing schedules were developed. Their aim was to reduce the existing testing effort to fit into the continual flow of production units [Rin 95]. The basis for this draft was the so-called *Continuous Sample Plan 1* (CSP-1) from which all later random sample testing schedules have been derived as modifications **(Figure 5.2-1)**.

A precondition for the application of CSP-1 is a reject rate, which is approximately known. In principal, there is the differentiation between 100%-testing and random sample testing. The 100%-test is always the first step. The random sample test only starts after i tested units have been classified as fault free. In this phase, one unit out of the units subsequent to k are tested and evaluated. Should a faulty unit be found during these tests, the 100%-test starts immediately.

Pre-requisite	process waste rate approximatelly known
Aim	optimising of costs

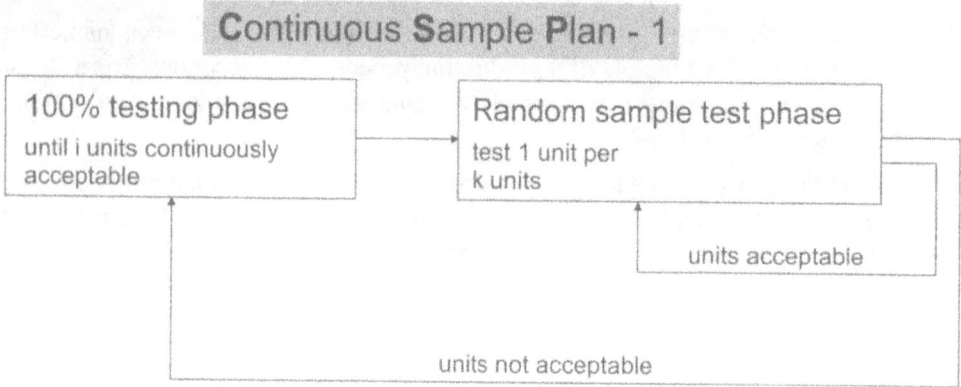

Continuous Sample Plan - 1

100% testing phase	Random sample test phase
until i units continuously acceptable	test 1 unit per k units

units acceptable

units not acceptable

Figure 5.2-1: Construction of continuous random sample test schedules

The random sample test schedule is defined by the following parameters:

- Reject rate P
- Number i of the required fault-free units to start the random sample test phase
- Number k of units, of which one is tested during the random sample test phase

With these values some feature values can be determined such as mean length of the test phases $u(P|i)$ or $v(P|k)$ as well as the mean non-detection of faulty units $AOQ(P|i;k)$, which in turn allow insight into the expected course of the test schedule **(Figure 5.2-2)**. The numerical examples with various parameter combinations in this figure also show that extreme care has to be taken when laying down random sample test schedules. On the one hand unnecessarily high test costs have to be avoided. On the other hand the non-detection of faults has to be as low as possible. Principally an optimum between cost and fault detection has to be found. Parameters, which purely operate on cost optimisation, normally result in a low fault detection ratio.

5.2.2 Structure, Design and Application of Shewart-Quality Control Charts

Statistical Process Control describes the course of a process as accurately as possible with the aid of testing and analysis of samples and corrects deviations immediately. As the machine operator usually introduces these corrections, the test results have to be presented in an appropriate simple, visual form. For this purpose quality control charts (QCC) are used. (**Figure 5.2-3**) [Kir 94]. The American W.A. Shewart introduced this control chart method in the 1940s.

	Waste rate P = 1% i = 100, k = 50	Waste rate P = 5% i = 100, k = 10
Mean length of 100% testing $u(P|i) = \dfrac{1-(1-P)}{P(1-P)}$	173 units	3358 units
Mean length of the random sample test phase $v(P|k) = \dfrac{k}{P}$	5000 units	200 units
Mean slippage of faulty units	0,94%	0,25%

Figure 5.2-2: Feature and number samples for continuous random sample test schedules

Random sample test results from single tests are either immediately used again or logged onto a quality control chart as a random sample test function g(x) in relation to time or random sample test number. The quality control chart takes the form of either a manual data sheet or a spreadsheet on a computer. The defined coordinate system usually has a centre line M, which shows the target value of the tested feature on the ordinate in the form of an abscissa-parallel line. A special emphasis is placed on the set control limits (UCL, LCL) and warning limits (UWL, LWL). While the warning limits are not imperative, the control limits, which determine when the process will be manually corrected, have to be defined before the first random sample is taken.

When using quality control charts a distinction must be made between the design and production phases (**Figure 5.2-4**). The first step is to determine the random sample testing function within the QCC during the design phase. The statement capabilities of the feature value as well as the effort required to determine the function have to be taken into account. By setting the target values and the detection probabilities, control and warning limits are calculated. Furthermore, the determination of timing and testing scope of random samples is important in relation to the significance of the control chart.

Random sample test result

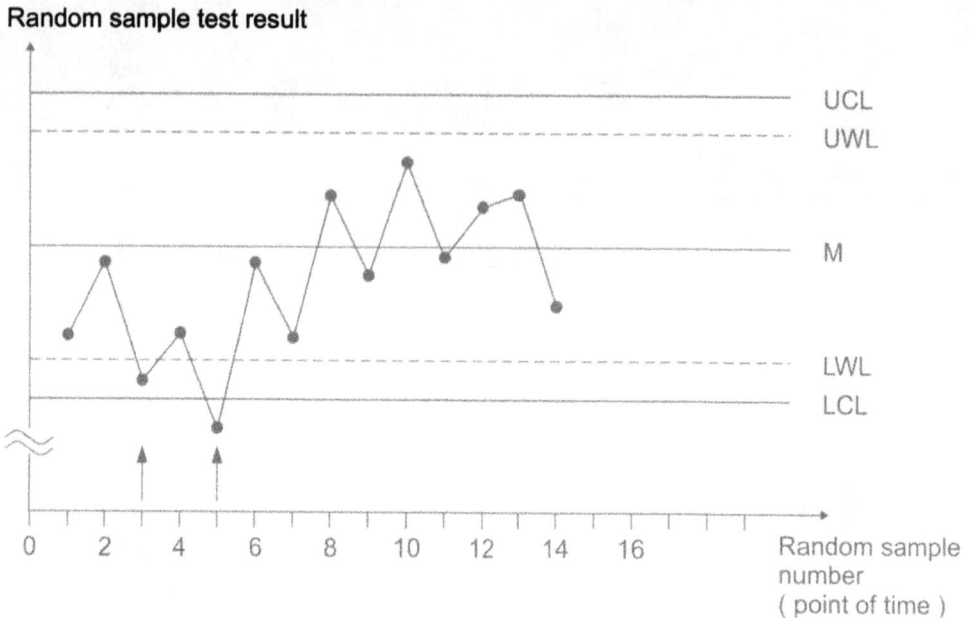

Figure 5.2-3:   Structure of a two-sided Shewart quality control chart (Source: [Rin 95])

During the production-phase of a QCC, random sample tests are taken at given points in time from the population and tested with the appropriate measuring aids. The subsequent consolidation of the test results is usually carried out with the aid of computers. After filing the values, the test results can be read off directly by the end user.

When using a control chart, which in addition to the necessary control limits also shows warning limits one of three possibilities will be indicated every time a random sample test result is logged. [Mit 93] **(Figure 5.2-5)**.

The reaction of the machine operator to the respective results must be consistent to guarantee a smooth process. Normally, the following decision process is used:

The value can be within the warning limits. It is acceptable that the process will run without interference. No corrective measure needs to be taken.

Should the random sample test result lie in between the warning and control limits, it can be assumed that a disruption may occur. The immediate requirement is to take and analyse another random sample test of the same statistic. If the second result lies within the warning limits, it is assumed that no disruption has occurred and no corrective measures are needed. However, should the second result lie outside the warning limits, it implies that some disruption has occurred and the user has to interrupt the process using corrective measures.

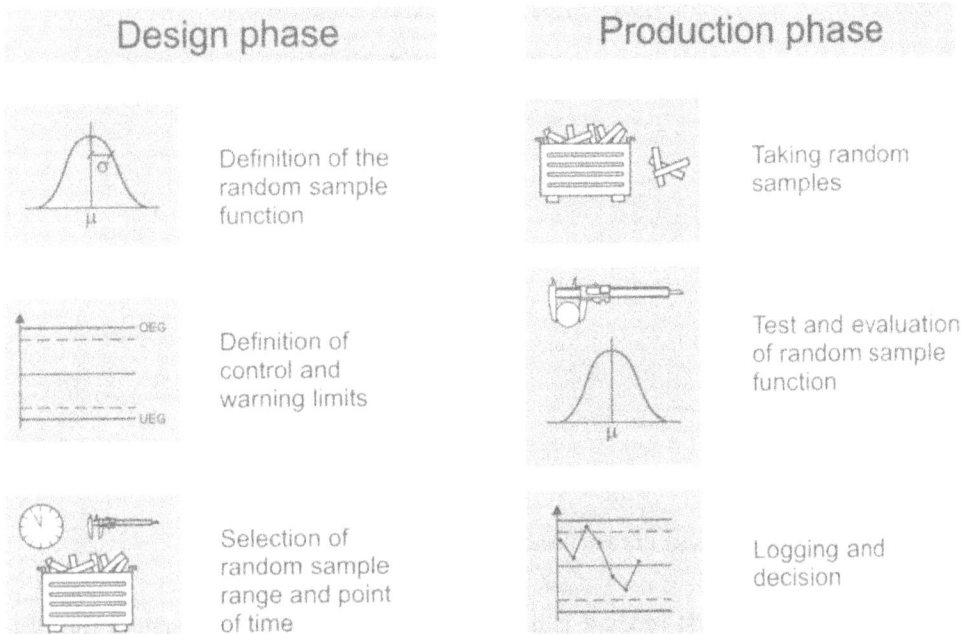

Figure 5.2-4: Phases when using quality control charts

A result outside the control limits clearly shows a deviation from the ideal process course. The process must be interrupted immediately, followed by an analysis of the deviation and correction of the process.

The control chart method can be applied to monitor counting as well as measuring. Within this section, however, only the test related to measurement commonly used for the surveillance of production processes is described. Five different random sample testing functions are used to consolidate the data from a random sample **(Figure 5.2-6)** [Pf 96].

The arithmetic mean of the individual tests is seen as a suitable approximation to determine the expectation value of the normal distribution if sufficient random sample tests have been carried out. The mean is utilised when using control charts as a descriptive distribution form for the population. This enables the surveillance of the production process as a whole. Alongside this, the standard deviation is an approximation of the distribution parameter σ, the width of the distribution and therefore the dispersion of the process.

Random sample test result

UCL
UWL

M

LWL
LCL

Random sample test result within the warning limits
- process under statistical control
- no action needed
- process continues

Random sample test result between warning and control limits
- disruption possible
- further random samples taken
  a) result within the warning limits
     ⟹ the process continues
  b) result outside the warning limits
     ⟹ the process must be corrected

Random sample test result outside of control limits
- process not under statistical control
- corrective intervention needed
- 100% test of the previously produced units

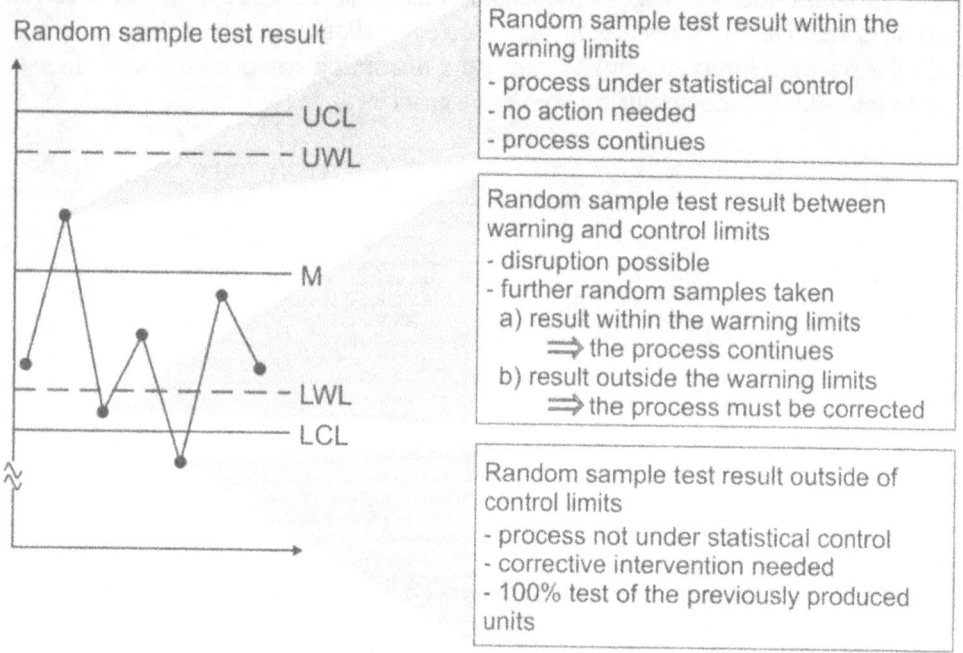

Figure 5.2-5:   Significance of control and warn limits

The random sample test median is a further feature value for evaluating the production process. This is derived from the mean of all individual samples. The process dispersion can furthermore be described by the random sample test statistic, which results from the difference between the smallest and largest value of the individual samples.

The surveillance of the production process and the dispersion is aided by recording data on an original data card. The results of the test are plotted on the time axis. The process course can be read off from the width and position of the resultant bar graph.

5.2.3 Statistical Background

The simple graphical form of individual random sample results leads to the assumption that the quality control chart is only an aid to visualise results with randomly chosen control and warning limits. In reality, this practise is unfortunately further reinforced as control charts are often used without any theoretical background knowledge. However, it must be stressed that the definition of boundaries is based on fundamental statistic considerations. Disregarding them inevitably leads to the wrong decision by the machine operator.

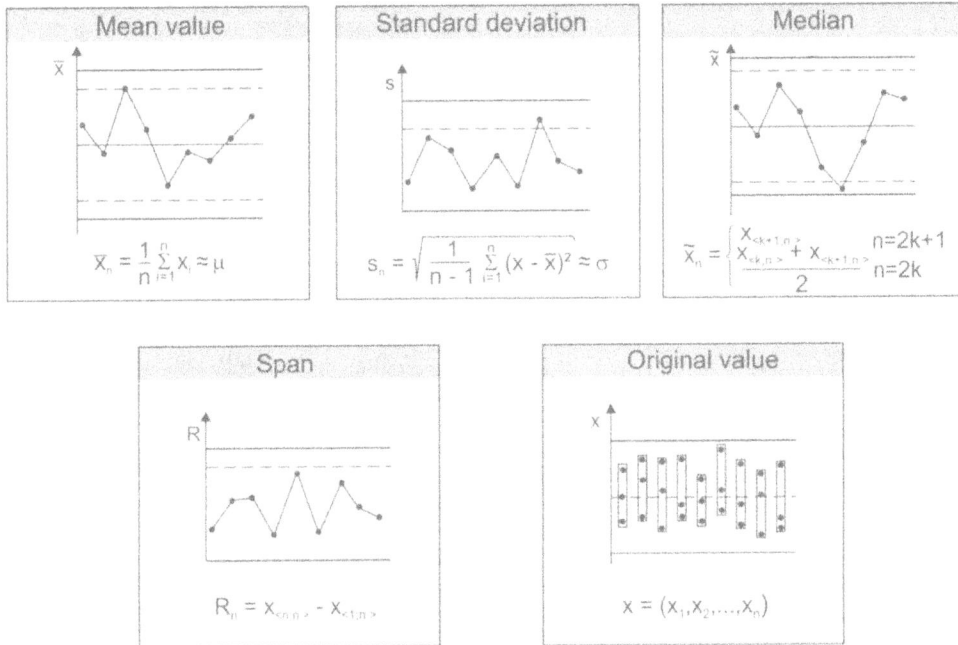

Figure 5.2-6:   Example for frequently used random sample functions

Using a quality control chart is always based on carrying out a parameter test **(Figure 5.2-7)** [Her 96]. Such a statistical test gives the opportunity to establish whether a random sample test result, which has been described in the random sample test function, matches a previously formulated null hypothesis H_0 (section 5.1.3.2).

With regard to the application of the quality control chart, the null hypothesis is always formulated as a correct process course, i.e. a process under statistical control. This implies an alternative hypothesis H_1, which is assumed for a disturbed process course. Taking the random sample test as well as recording its function on the control chart is therefore also testing the compatibility with the null hypothesis. From the result, a decision is deducted on the process performance.

A statistical parameter test is always based on the random sample test result and therefore on a random event. It cannot verify the hypothesis, i.e. prove its validity beyond doubt. An erroneous decision is probable when comparing the diminished reliability of the statistical test with a proof. It is basically possible to arrive at two types of incorrect conclusions when using the quality control chart. (Section 5.1.3.2).

Approach		Meaningfulness

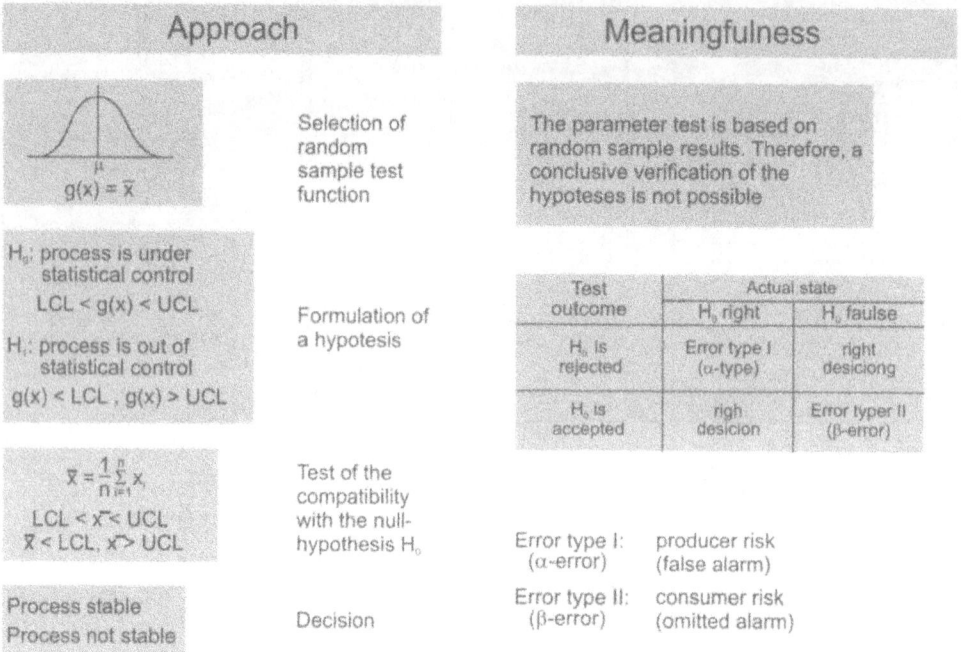

Selection of random sample test function

The parameter test is based on random sample results. Therefore, a conclusive verification of the hypoteses is not possible

$g(x) = \bar{x}$

H₀: process is under statistical control

LCL < g(x) < UCL

Formulation of a hypotesis

H₁: process is out of statistical control

g(x) < LCL , g(x) > UCL

Test outcome	Actual state	
	H₀ right	H₀ faulse
H₀ is rejected	Error type I (α-type)	right desiciong
H₀ is accepted	righ desicion	Error typer II (β-error)

$\bar{x} = \frac{1}{n}\sum_{i=1}^{n} x_i$

LCL < x̄ < UCL

x̄ < LCL, x̄ > UCL

Test of the compatibility with the null-hypothesis H₀

Error type I: producer risk
(α-error) (false alarm)

Process stable

Process not stable

Decision

Error type II: consumer risk
(β-error) (omitted alarm)

Figure 5.2-7: Theoretical basics for the parameter test

Firstly, the hypothesis, although correct, can be rejected based on the evaluation of the random sample testing. The erroneous rejection of a correct decision is seen as a type I error. As the probability of occurrence is generally identified by α, it is also called a α-error. During production surveillance this error amounts to interference in a fault-free production process (false alarm). This error characterises a production risk as it causes an unnecessary increase in workload.

The other possible wrong decision is to reject a true hypothesis. This is called a type II error or β-error. The result would be that a disruption would not be recognised (omitted alarm) and is therefore a consumer risk .

This can be illustrated with the following example. **Figure 5.2-8** shows a production process, which only functions in one of two possible conditions. The quality features, being produced in the desired conditions normally have an expectation value of μ=40,00 mm and a standard deviation of σ=0,01 mm. The undesired condition produces μ=40,05 mm with the same process dispersion.

The two hypotheses

      H₀: μ=40,00 mm (desired process)

      H₁: μ=40,05 mm (undesired process)

can be deducted for the parameter test. These are tested with a random sample function of the arithmetic mean $\overline{X}$. If H_0 is true then $\overline{X}_n \sim N(40,00;0,0001/n)$, otherwise $\overline{X}_n \sim N(40,05;0,0001/n)$.

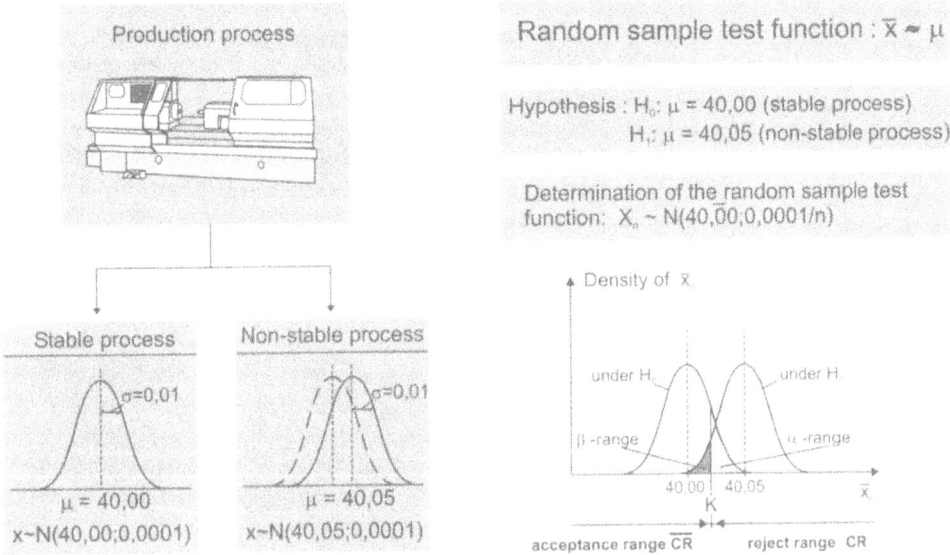

Production process

Stable process

$\sigma=0,01$

$\mu = 40,00$
$x \sim N(40,00;0,0001)$

Non-stable process

$\sigma=0,01$

$\mu = 40,05$
$x \sim N(40,05;0,0001)$

Random sample test function : $\overline{x} \sim \mu$

Hypothesis : H_0: $\mu = 40,00$ (stable process)
H_1: $\mu = 40,05$ (non-stable process)

Determination of the random sample test function: $X_n \sim N(40,00;0,0001/n)$

Density of $\overline{x}$

under H_0
under H_1

β-range
α-range

40,00 40,05
K

acceptance range $\overline{CR}$ reject range CR

Figure 5.2-8: Example of a parameter test

If the distribution of the quality features is plotted in a common graph, it becomes clear that all values of $\overline{x}$ are always possible within both hypotheses, due to the overlapping of the functions, albeit with different probabilities. To formulate the test, the variance range of the test statistic must be divided into an acceptance range $\overline{CR}$ and a rejection range CR for the null hypothesis. A figure K must be defined between the expected values of the distribution to divide the two ranges. By defining K, the probabilities of type I errors and type II errors can be determined.

As can be seen from the diagram, it is not possible to minimise both error probabilities simultaneously. A reduction in α-errors always leads to an increase in β-errors.

In accordance with a general convention, the control limits are determined when the quality control charts are designed such that the probability of α-errors is 1% (**Figure 5.2-9**). For two-sided quality control charts the production risk for exceeding or undercutting the control limits are therefore $\alpha/2=0.5\%$. The α-error probability for exceeding the warning limits is 5%.

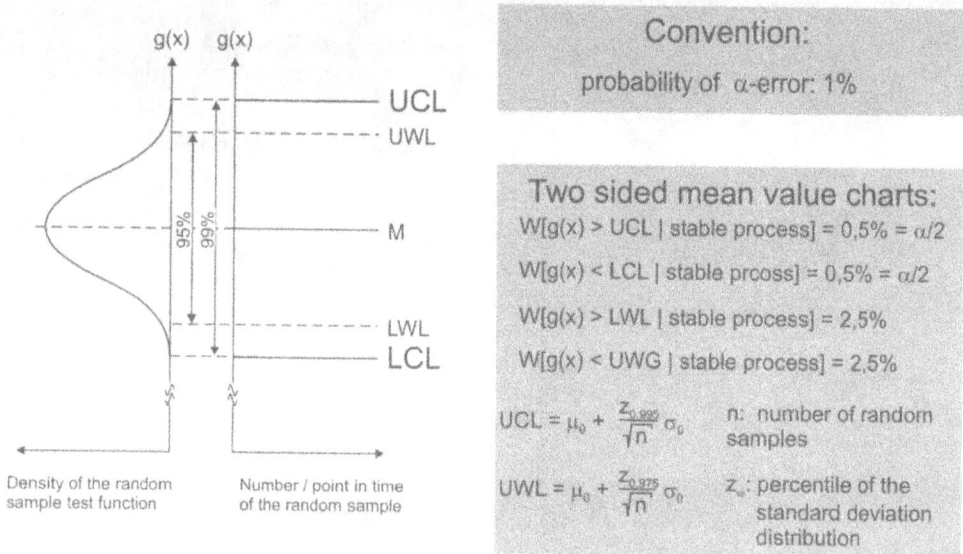

Figure 5.2-9:   Definition of control and warning limits

Taking these conventions into consideration, the limits for each random sample function can be accurately defined. For the arithmetic mean for example, the percentile function of the standard deviation with a probability of $1-\alpha/2$, which can be found in the respective tables, as well as the random sample test statistic n has to be given. Furthermore, the target values of the expected values and standard deviation of the normal distribution describing the process are needed.

5.2.4 Practical use of the Control Chart Technique

Before using quality control charts, basic considerations must be made regarding the surveillance strategy. In particular, it must be clarified which process changes are not permitted and which sample function identify these in the most cost effective way.

There are two different forms of process disruptions **(Figure 5.2-10)**. Firstly, the position of the distribution describing the process can move from the centre of the tolerance field to the boundaries. Once the control limits have been exceeded, it will show that an inadmissibly high percentage of the quality feature is produced, which does not conform to specifications. This can also happen due to changes in the form of the distribution, which is described by dispersion parameters.

Determination of the random sample test function

Stage of process: $\bar{x}, \tilde{x}, x$

Process scattering: s, R, x

$\bar{x}, s$

Determination of the target values

Nominal value μ_s, σ_s	Empirical value μ_e, σ_e	Estimated value $\hat{\mu}_0, \hat{\sigma}_0$
- production regulation	of earlier comparable	of an undisturbed
- norm	processes	first run
- law		

μ_0, σ_0

Design

Monitoring of the state of process

UT   UEG   OEG   OT

μ_0

Correction

Monitoring of the process scattering

UT   EG   OT

σ_0

μ_0

Interruption

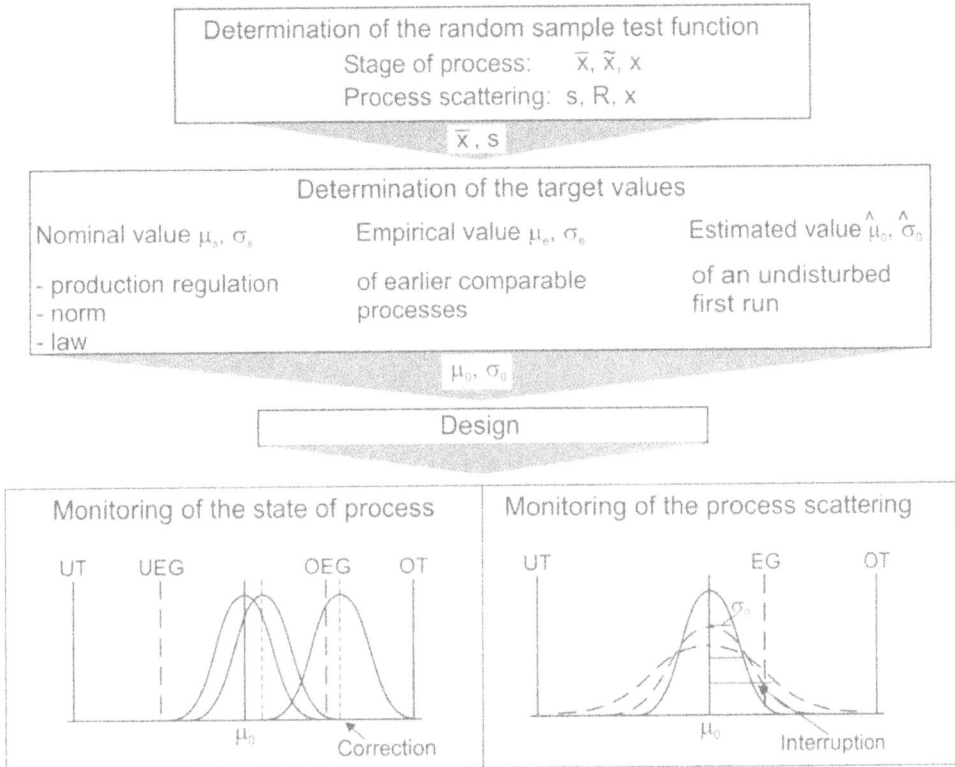

Figure 5.2-10: Simultaneous monitoring of process position and deviation

Monitoring the production process with only one random sample function is therefore not sufficient to arrive at a conclusive decision. A concurrent monitoring of the production status and dispersion must also be carried out. The position can be evaluated by recording the median or the original measured values as well as the means from random samples. A dispersion evaluation is made possible by using a standard deviation-, range- or original value chart.

After determining suitable random samples functions, the targeted feature values must be determined to define the control and warning limits. In an ideal case, this data is taken from the target values (μ_s, σ_s) derived from the manufacturing instructions, norms or regulations. If such regulations do not exist, empirical values (μ_e, σ_e) from previous, similar processes are often used. If this is not possible, meaningful estimated values must be determined. Particularly in mass production, extensive process test runs are carried out for this purpose. After producing up to 100 parts in an uninterrupted production process, an individual test follows as well as the determination of the necessary distribution parameters. After designing the control chart, the simultaneous surveillance of the production process is carried out.

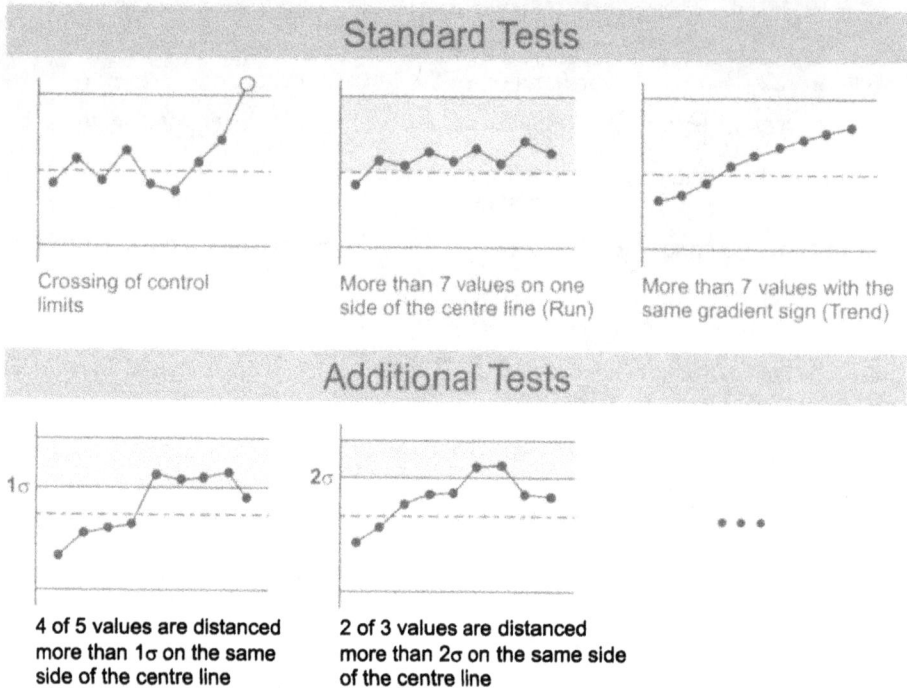

Figure 5.2-11: Test method for non-random processes on control charts

In general, the machine operator can correct a deviation from the ideal process, e.g. by changing a tool. The reasons for changes in the process dispersion are, however, usually much more complex. The only reaction available is the immediate interruption of the process as well as a thorough analysis of the original operating mechanisms.

Apart from exceeding the warning and control limits, there are a number of other test criteria, for identifying non-random events with control charts. (**Figure 5.2-11**).

A fault may be indicated if more than seven consecutive values lie on one side of the centre line (run) or are shown in a rising or falling sequence (trend). The probability that such performance is actually due to a process fault, and not just a random result, is more than 99%.

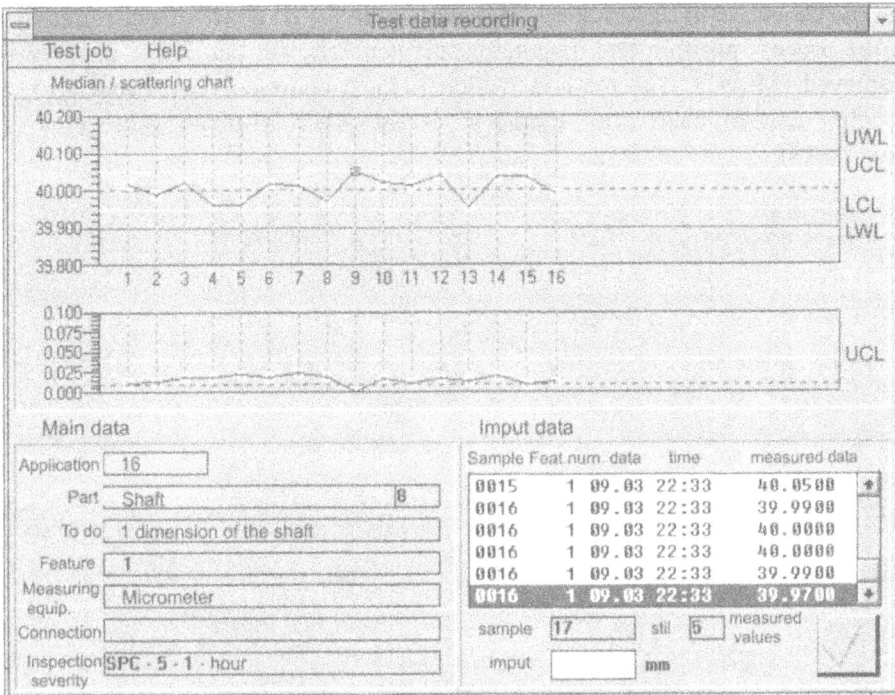

Figure 5.2-12: Example of a computerised quality control chart

In addition to these standard criteria, it can be tested whether unnaturally high amounts of consecutive random sample results are close to the centre line or the limits. Apart from the checks described in here, there are more than 30 further test methods. For reasons of clarity, the user is advised to remain constrained to the methods described herein, which are generally used in practice.

In order to reduce the workload when setting up control charts, they are nearly always operated with the help of computers. **Figure 5.2-12** shows a sample graphic for the simultaneous surveillance of mean and standard deviation.

5.2.5 Newer Types of Quality Control charts

The traditional quality control charts, which were described in the previous section, are based on a principle developed by W.A. Shewart. The decision whether a process needs to be corrected is solely made on the result of the most recent random sample. For some time now other processes of production monitoring with the help of random sample exist, which include in their results the outcomes of earlier tests. The advantage of this approach is seen in the more sensitive reaction of the sample functions to disturbances. Already in the initial phase of undesired drifting in the process position and dispersion, the necessity for corrective intervention is displayed [Mit 93]. The method, which is most commonly used and also called "quality control chart with memory" is the KUSUM-mean value chart (**Figure 5.2-13**).

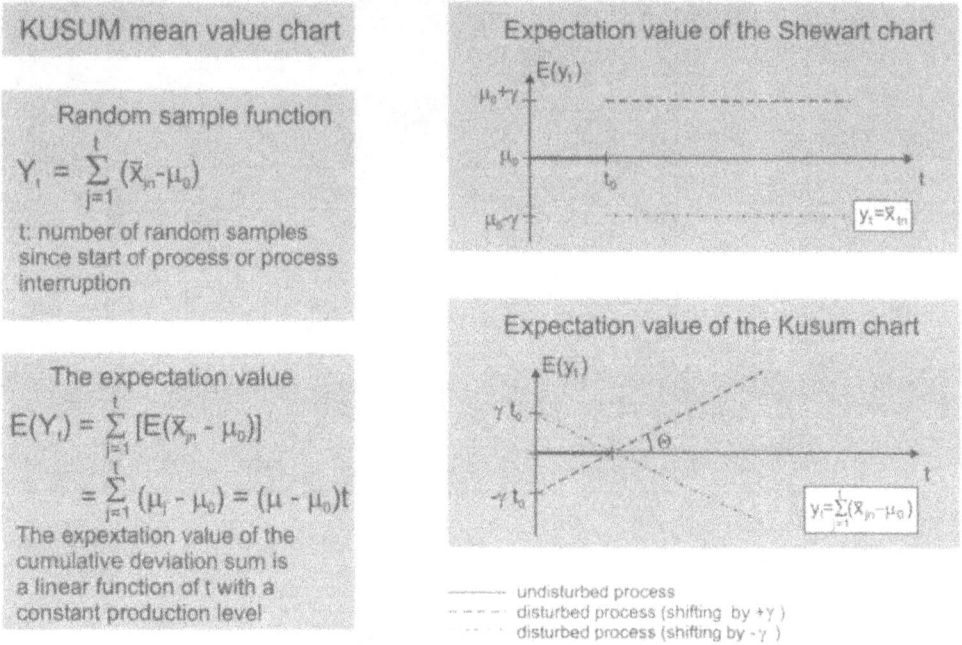

KUSUM mean value chart

Random sample function

$$Y_t = \sum_{j=1}^{t} (\bar{X}_{jn} - \mu_0)$$

t: number of random samples since start of process or process interruption

The expectation value

$$E(Y_t) = \sum_{j=1}^{t} [E(\bar{X}_{jn} - \mu_0)]$$

$$= \sum_{j=1}^{t} (\mu_j - \mu_0) = (\mu - \mu_0)t$$

The expextation value of the cumulative deviation sum is a linear function of t with a constant production level

Expectation value of the Shewart chart

Expectation value of the Kusum chart

--------- undisturbed process
- - - - disturbed process (shifting by +γ)
·· ·· ·· disturbed process (shifting by -γ)

Figure 5.2-13: Quality control chart with memory (according to [Rin 95])

In this the deviations between every j-th mean value $\bar{x}_{jn}$ and the target value μ_0 is added over all the random samples j from the start of production or the last process intervention and is summarised as the random sample function Y_t. For the expectation value of the test statistic defined in such a way, follows:

$$E(Y_t) = (\mu - \mu_0) \cdot t \tag{5.2-1}$$

The expectation value of the cumulative deviation sum therefore forms a linear function of t, for constant production levels, in which the respective line shows a gradient μ-μ_0. If the random sample function is shown over the random sample number t, the non-disrupted process shows an oscillating process around the t-axis. However, even small disturbances can lead to the test statistic moving away from the zero line in the middle of a level shift of the upward gradient μ-μ_0. Contrary to Shewart control charts, which indicate secure axis boundaries, the curve path of a

KUSUM mean value cards must therefore be judged on the basis of its upward gradient. Two different methods exist for this procedure. [Rin 95].

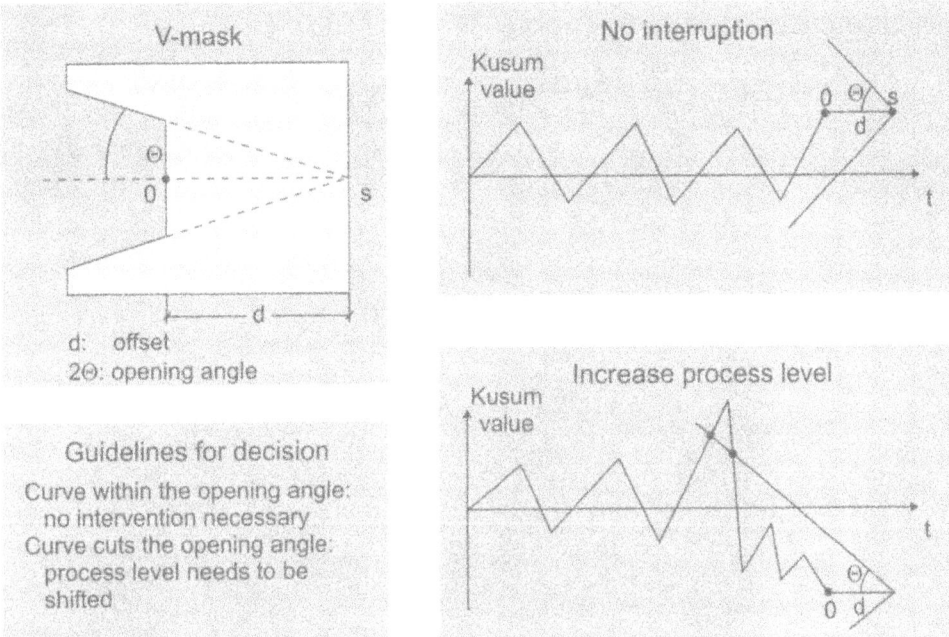

d: offset
2Θ: opening angle

Guidelines for decision
Curve within the opening angle:
no intervention necessary
Curve cuts the opening angle:
process level needs to be
shifted

Figure 5.2-14: Graphical assessment of KUSUM-mean value cards

The graphical analysis with the so-called V-mask was developed by von Barnard towards the end of the 1950s **(Figure 5.2-14)**. The test is aided by a template, which geometric features define the distance d as well as the opening angle 2Θ. The values d and Θ are determined with the aid of the maximum admissible process level shifts and analogue to the warning and control limits of the Shewart card with the specification of the statistical boundary conditions.

If the distance course of the KUSUM control chart up to the point of time t is to be examined, then point 0 with the v-mask is set to coincide with the appropriate value of the sample function Y_t. If the entire KUSUM graph, and thus the point sequence connected up to the point of time by distance courses within the opening angle is processed, it can be assumed with high probability that the manufacturing position did not deviate too strongly from the desired value μ_0. A correcting intervention is not necessary.

If the curve path intersects one of the two legs of the opening angle, it is assumed that the process course shifted inadmissibly at one point in time and must be corrected. Depending on whether the upper or the lower leg is intersected, the process level must be increased or lowered.

Test parameter	Test variables	Use of control chart
H: limit of decision interval K: positive constante	$s_i^+ = \sum_{j=1}^{t} (\bar{x}_{jn} - \mu_0 - K)$ $s_i^- = \sum_{j=1}^{t} (\bar{x}_{jn} - \mu_0 + K)$	- not guidance for: $s_i^+ \leqslant 0$ i.e. $s_i^- \geqslant 0$ - start of the guidance for: $s_i^+ > 0$ i.e. $s_i^- < 0$ - end of the guidance for: $s_i^+ \leqslant 0$ i.e. $s_i^- \geqslant 0$ - intervention for: $s_i^+ > H$ i.e. $s_i^- < -H$

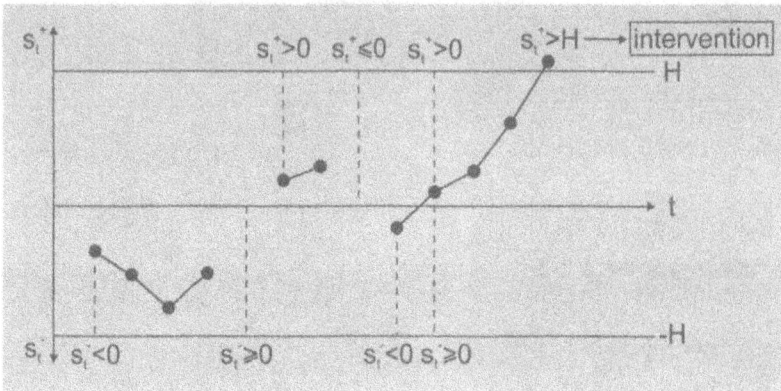

Figure 5.2-15: Assessment of KUSUM mean value card according to the EIS-method

As the description of this testing method shows, the analysis of KUSUM mean value charts with a v-mask becomes extremely complicated as well as non-descriptive. An acceptance of this method by the machine operator can thus hardly be attained, even though it is possible to substantially simplify the test by using computers. For this reason, Page developed the analysis with the so-called Decision Interval Pattern (Entscheidungsintervallschema - EIS) in the 1950s **(Figure 5.2-15)**.

KUSUM charts, which are used on the basis of EIS, also have two positive parameters, analogue to the analysis with the V-mask: value H, which determines the limits of the EI, and a positive constant K. Both parameters are calculated after specification of statistical assumptions. This does not calculate the conventional random sample function, but two test variables, which are derived from it, S_t^+ and S_t^-, which are formed from Y_t by adding or subtracting from K.

If

$$\bar{x}_{jn} - \mu_0 - K \leq 0 \text{ and} \qquad (5.2\text{-}2)$$

$$\bar{x}_{jn} - \mu_0 + K \geq 0 \qquad (5.2\text{-}3)$$

are met at the same time, no values are recorded in the respective chart.

If

$$\bar{x}_{jn} - \mu_0 - K > 0 \qquad (5.2\text{-}4)$$

the sum S_t^+ is formed, increasing or decreasing, depending on the added summands.

If the value sinks below zero again or reaches the upper boundary H, then the accumulation ends. When exceeding the control limit, the process level must be lowered through intervention. Accordingly the test variable S_t^- is formed as soon as

$$\bar{x}_{jn} - \mu_0 + K < 0 \qquad (5.2\text{-}5)$$

results for the test. The formation of the test variables ends here, as soon as a positive value is met or the negative limit is exceeded. The exceeding of -H indicates the necessity for a process level increase [Rin 95].

Although the described pattern seems quite complicated, an automated handling of such charts is possible without problems. The special advantage for the operator lies in the descriptiveness and similarity in comparison with traditional control charts.

5.2.6 Boundary Conditions for the Use of Control Charts

As the previous sections have shown, the process-accompanying guidance of quality control charts is quite cost intensive. This expenditure is justified by the possibilities of the process evaluation. The meaningfulness of the control charts is only given if some basic requirements are met, regarding the SPC suitability of the observed feature (**Figure 5.2-16**).

Firstly, the correlation of the quality feature with the remaining features of the component is to be examined. It must be clarified to what extent a necessary intervention would influence the parameter of the remaining values. An example for this is the feed of the turning tool with the simultaneous handling of several shaft shoulders. If this influence is given, suitable guidance features can be determined and regulated. A further substantial criterium is the ability to measure the feature. Above all, it must be ensured that the measurement procedure used has as few inaccuracies as possible. As a consequence of using inaccurate measuring instruments, the sample findings are interpreted incorrectly, so that a reliable result is not given. This increases the risks for both the producer and consumer unnecessarily. Great importance is therefore attached to the selection of suitable measuring instruments when carrying out SPC.

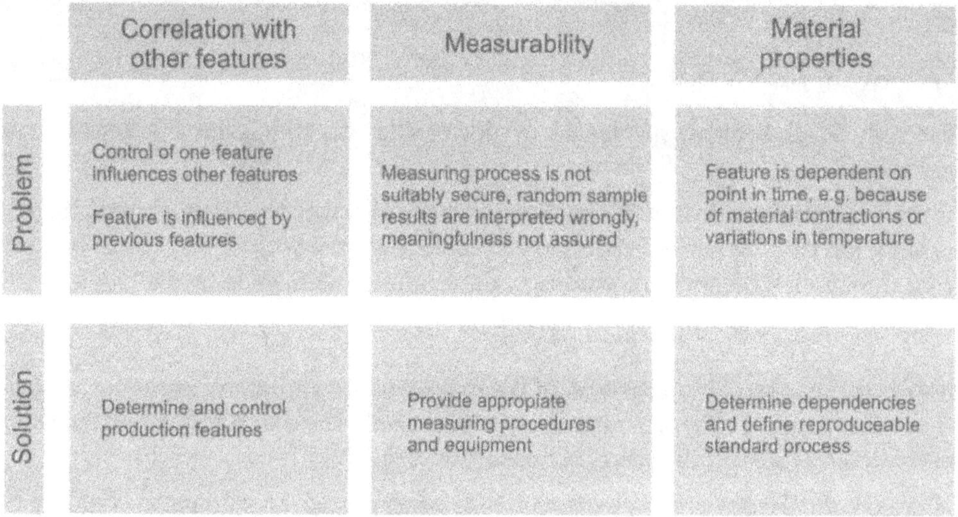

Correlation with other features	Measurability	Material properties
Problem Control of one feature influences other features Feature is influenced by previous features	Measuring process is not suitably secure, random sample results are interpreted wrongly, meaningfulness not assured	Feature is dependent on point in time, e.g. because of material contractions or variations in temperature
Solution Determine and control production features	Provide appropiate measuring procedures and equipment	Determine dependencies and define reproduceable standard process

Figure 5.2-16: Criteria for SPC-suitability of a feature

Finally, the material properties of the examined components must be considered. A time dependence of the measured variable is frequently observed due to variations in temperature or material contractions. Particularly with closely tolerated quality criteria, this can lead to a false evaluation of the course of the process. These dependencies must be determined before the process starts and considered during monitoring.

5.3   Capability of Production Processes

Due to the varying influences in the field of production, more or less strongly marked deviations from target requirements occur during the manufacturing of the quality feature. During the course of the processes, this effect leads to a dispersion of feature parameters within or outside of the tolerance range. This is described with statistical distributions. Position and form of the distribution is very important with regard to the produced number of substandard parts.

Figure 5.3 - 1 shows the possible parameters of distributions, which describe the process. The upper left section firstly shows a process, for which all manufactured features are within the tolerance range and dispersed evenly in both directions around the nominal value. This is known as a capable process. The upper right section shows a process of identical form, which has been shifted by a fixed amount from the nominal value. This change leads to the production of a high proportion of substandard parts. The section at the bottom left hand corner shows a trial process, in which the maximum value corresponds with the target value, but for which the dispersion is substantially broader. A high reject rate is produced again. The lower right hand section finally shows a superposition of the described effects.

Capability Indices

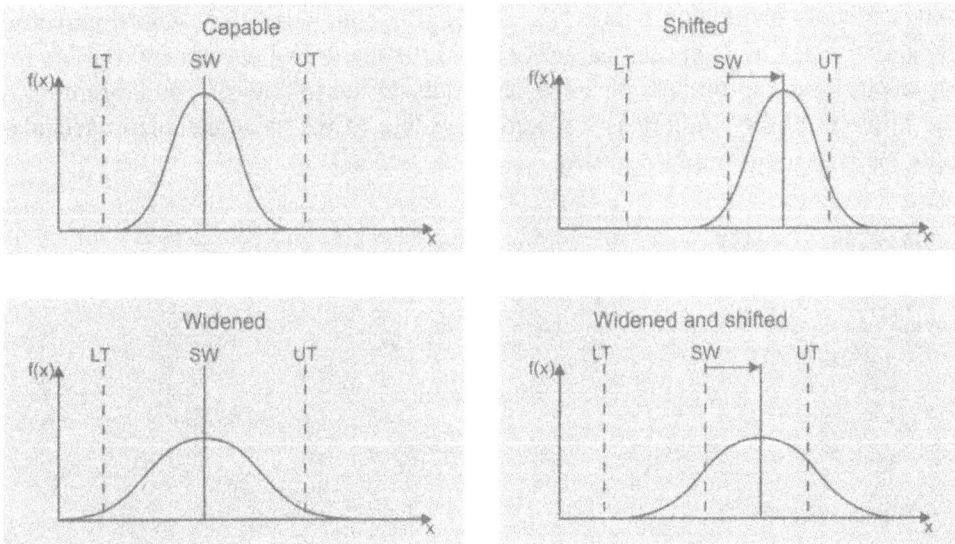

Figure 5.3-1:   Capability of processes

For reasons of clarity it is essential to define features, which enable a unique statement about the independent capability of specific geometrical parameters. In practice, this occurs by defining capability indices for the manufacturing equipment and processes **(Figure 5.3 - 2)** [Bos 91], [Chr 91], [Die 95]. A differentiation is made between potential and actual capability indices. Potential capability indices evaluate only the dispersion, since an ideally centred process is implicitly assumed. The actual capability indices (characterized by an index K) evaluate process position and dispersion. The process position can usually be corrected by small process interventions, for example the use of a tool with a different drilling diameter. However, the decrease of the process dispersion requires more extensive measures, which could go as far as changing the processing machinery.

To determine the capability indices the referred width of the tolerance interval (UT-LT) is applied onto the so-called natural tolerance 6s in an assumed normal distribution. The value s corresponds to the estimated value for the parameter $\sigma-$, the standard deviation, as determined by a random sample. The area of natural tolerance contains approx. 99.73% of the total area of the probability density function. A potential ability index of 1, for which the tolerance width corresponds to the natural tolerance 6s, permits the statement that 99.73% of the quality features are manufactured with a parameter conforming specification. The feature range, within which 99.73% of the feature values are to be expected, is

called process dispersion width. From this it is concluded that, with a potential capability index of 1, in each case 0.135% of the feature values are appropriate for an ideally centered process on either side outside the tolerance. The definition of the different ability indices is effected according to the same calculation regulation, but with a different set of data.

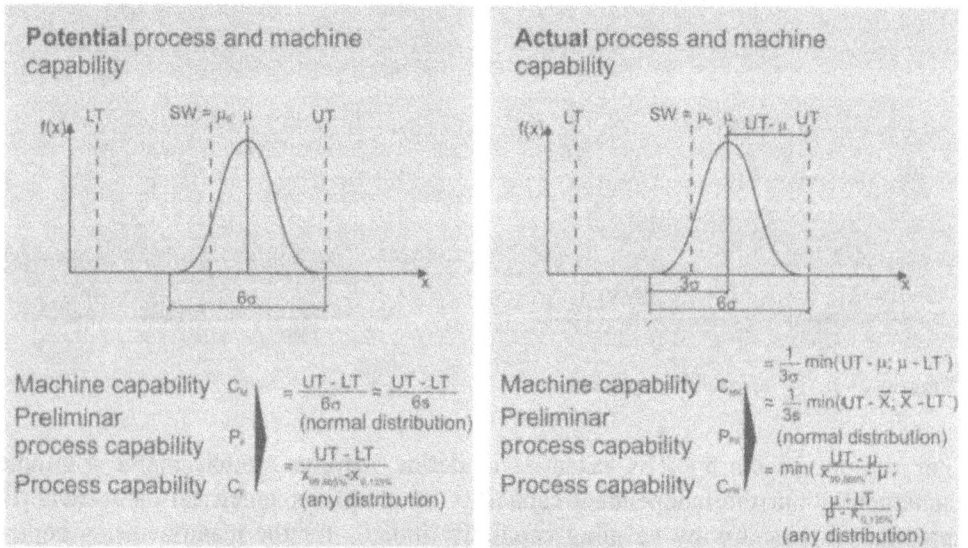

Figure 5.3-2: Capability Indices: Feature values to describe machine and process capability

The machine capability indices C_M and C_{MK} are calculated with data from a short-term study. It is based on a random sample to determine the dispersion components of the machinery. A typical application for the machine capability index is the acceptance of a machine tool by the manufacturer. The provisional process capability P_P and P_{PK} is determined in the context of an advance investigation. The basis is a minimum scope of 100 parts or a scope appropriate to the process. The actual process capability C_P and C_{PK}, is determined during the series run by taking out samples from the process.

Process and Machine Qualification

When planning production processes, DIN EN ISO 9001 requires these to run under controlled conditions. In order to fulfil this requirement in the entire process chain, suppliers frequently coordinate elements of the quality management system with their customers i.e. disclose quality data and maintain quality standards. This includes quality data in the form of capability proof for the processes and machines. The described capability indices are used with the general methodology for the qualification of a process and its manufacturing equipment (machine tool) according to VDA 4.1. The quintessential point is a multi-level methodology, as it described in **Figure 5.3 - 3**.

Some procedures are company specific and subject to variations (e.g. [Bos 91], [Chr 91], [For 91a], [Mer 91], [Ope 96], [Psa 91], [Sie 93], and [Vol 95]). The basis for machine procurement and machine selection is the machine capability index. The machine acceptance is based on the machine capability at the manufacturer. Before the start of a production series, all known systematic influences determined through an advance inspection are to be switched off and the distribution model of the examined process to be determined. Subsequently, the provisional process capability is determined. On this basis the end user can accept the machine. Accompanying the process, the (long-term -) capability is determined. The effectiveness of all process-determining factors (humans, machine, material, method and environment) needs to be guaranteed.

The customer usually defines minimum requirements for the potential and actual process capability, which partially impact on many C_P- and C_{PK}-values. If C_P-values are greater than or equal to 1, a capable process is indicated. Customer requirements have always kept as secondary conditions. Values, which strongly deviate or fall below the customer specific limit value (minimum of 1) C_{PK}-values, require a centring of the process.

Process analysis before start of series production time	**Process analysis once series production has started**

| Short term capability-test | Preliminar process capability test | start of series | Long term process capability test |

Application to be assessed:
* new applications at manufacturing
* new applications after instalation for process
* before series production

Application to be assessed:
* current series production

Machine	**Process**
Production equipment	Human Machine Material Methode Milieu	Continuous quality improvement

Short term capability	**Preliminary process capability**	**Long term process capability**
minimum of 50 parts or suitable number for process.	minimum of 100 parts or suitable number for process. At least 20 random samples are needed for use in control chart.	suitably long period of time to ensure that all parameters are tested. Suggested time: 20 days with normal production conditions.
C_m, C_{mk}	P_p, P_{pk}	C_p, C_{pk}

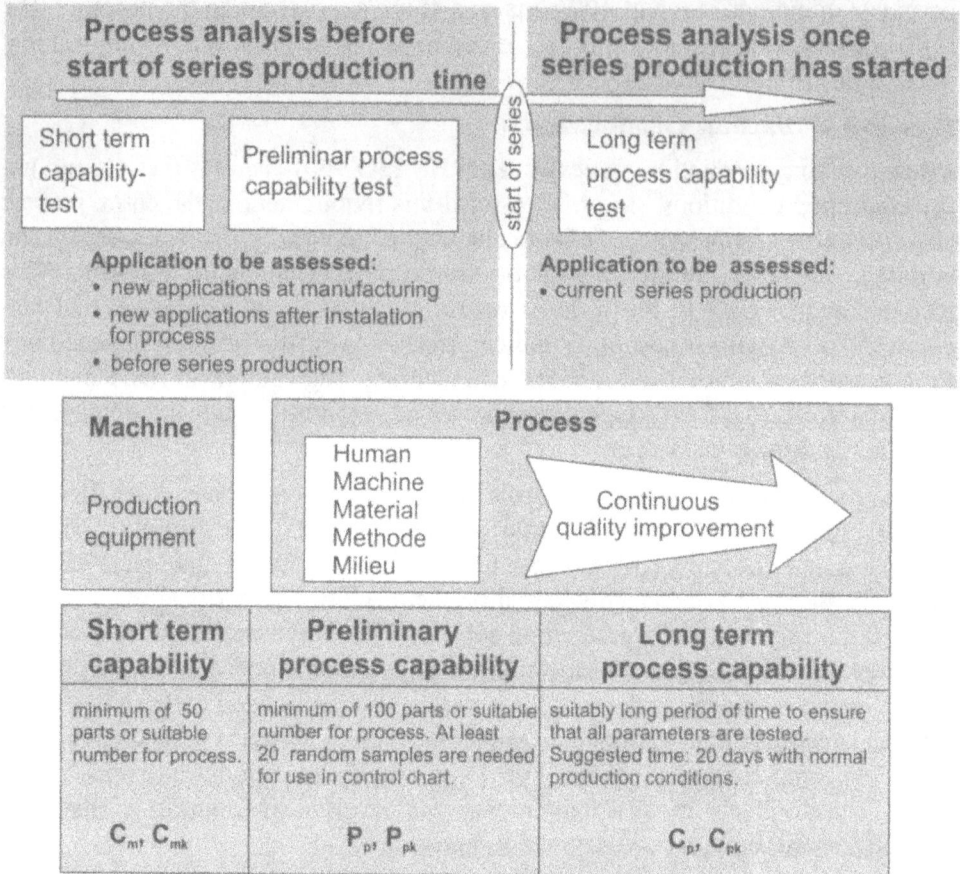

Figure 5.3-3:   Method for evaluating process capability according to VDA 4.1

Process Model

In order to arrive at the correct evaluation of a process, a suitable model is required to accurately describe statistical behaviour. To describe the model, mechanically manufactured sections are measured, as long as a capable measuring procedure and enough random samples are in place. With the help of the statistical procedures presented in Section 5.1 (probability network, χ^2-Test) the measurement data can be used to test the mathematical model, based on the process given as a hypothesis.

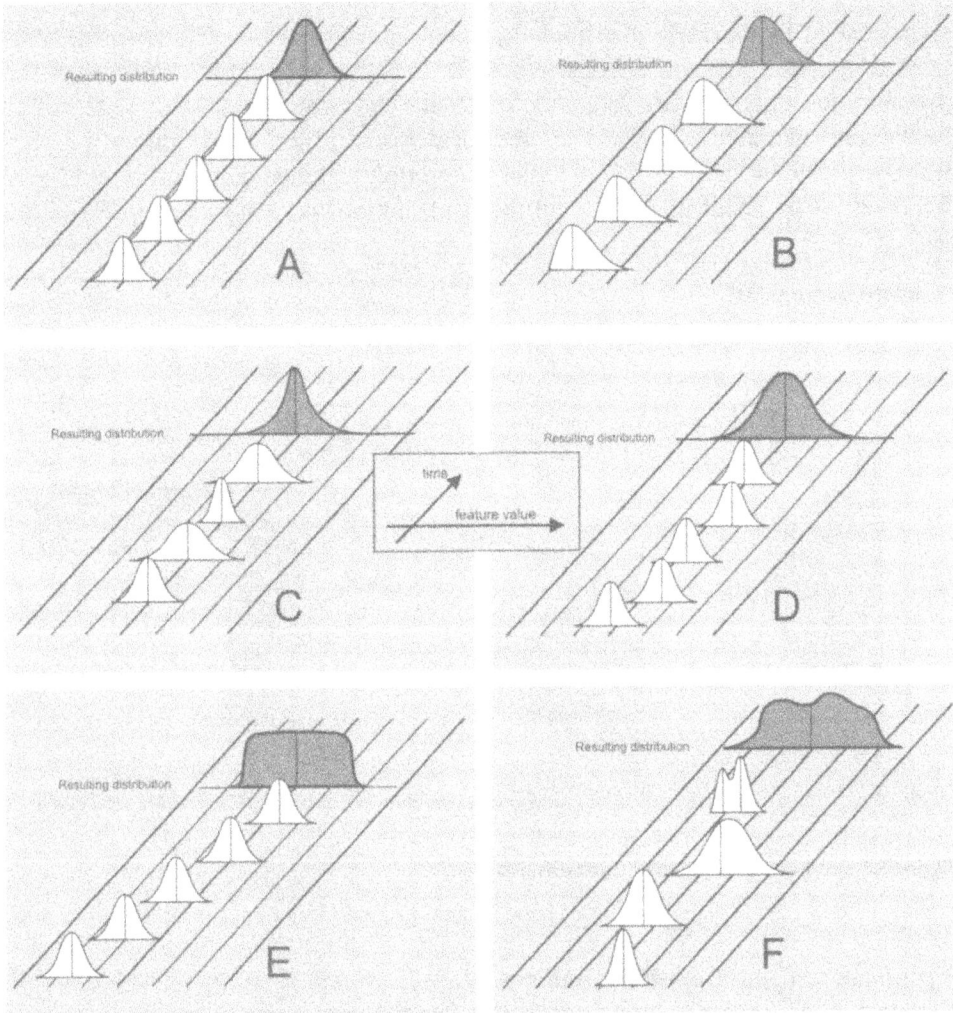

Figure 5.3-4:   Frequently occurring process models

On this basis, the selected model can be given an estimated value for the width of the process dispersion and the process position can be indicated. Different process models are specified in **Figure 5.3 - 4**. Process model A is the traditional Shewart-process; model B is the representation for one-sided limited features (physical or by calculation). Process model C shows non-normal distribution through deviation in dispersion, and models D and E through deviation in position. The combination of deviating position and dispersion, as well as deviating sample distributions leads to process model F.

In the case of the normally distributed process, the calculation of the capability indices simplifies as shown in **Figure 5.3 - 2**: It is formed from the quotient of the minimum distance between tolerance limits and the expectation value with half of the natural tolerance 3s. However, the results of a global investigation of 1000 processes resulted in a car manufacturer presenting a mixed distribution in 95% of the cases. The acceptance of a normal distribution was only met in 2% of the cases [Die98].

Distance dimension

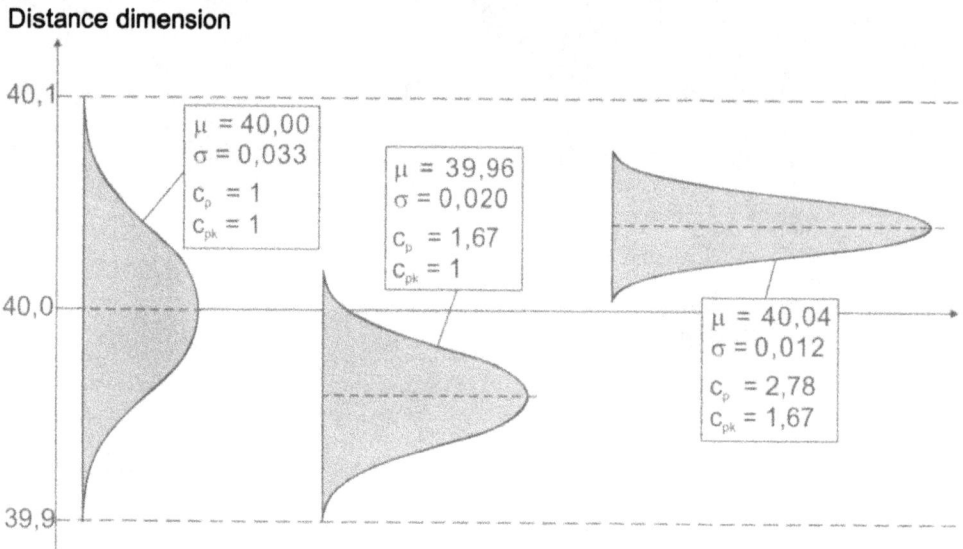

Figure 5.3-5:   Significance of Capability Values

Significance of the Capability Indices
Figure 5.3 - 5 shows the position of the distribution in the tolerance field of a normally distributed quality criterion for different values of the process capability indices C_P and C_{PK}. A value of one for both indices, the entire tolerance range is used up by the process dispersion width (6s). This is the borderline case for a capable process. A minimal shift of the process level already causes an increased production of substandard parts and must be corrected accordingly. For a relatively high c_p -index of 1.67 a perfect process course is not inevitably guaranteed, since the distribution can shift the boundaries of the tolerance interval. A securely controlled process only takes on the distribution represented on the right for the respective case. Status ability indices of c_p=2.77 or c_{pk}=1.67 result. In this case, a further process improvement can be achieved by the centring of the process, i.e. by a shift of the process position, so that the process capability indices take on the value c_p=c_{pk}=2.77.

ppm
3000
2700
2500
2000
1500
1000
500
0
176,89
1   1,05   1,1   1,15   1,2   1,25
C_p

ppm
200
176,89
150
100
50
6,80
0
1,25   1,3   1,35   1,4   1,45   1,5
C_p

ppm
7
6,80
6
5
4
3
2
0,15
1
0
1,5   1,55   1,6   1,65   1,7   1,75
C_p

ppm
0,16
0,15
0,14
0,12
0,1
0,08
0,06
0,04
0,02
0,0019
0
1,75   1,8   1,85   1,9   1,95   2
C_p

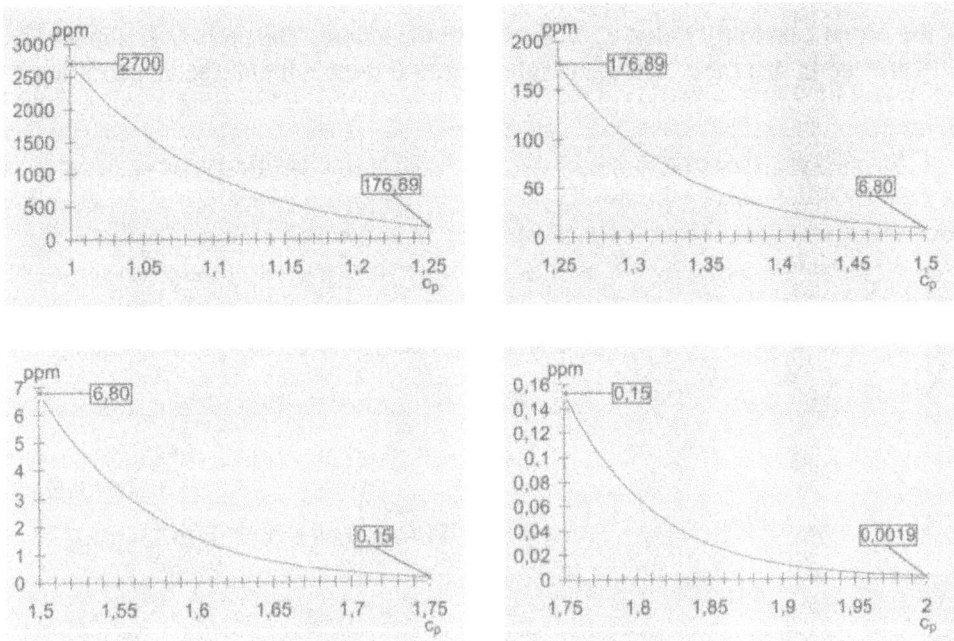

Figure 5.3-6: Connection between capability values and ppm-rates

The simply defined and thus descriptive process capability indices often do not clarify the extreme demands. As a feature value of 1 symbolises a capable process, the customer requires certain assurances, which are expressed in c_p- and c_{pk}-values from 1.33 to 2. The manufacturers frequently underestimate the significance of these specifications. The maximum reject ratios derived from this, which are generally indicated in the unit ppm (parts per million) obtain a better impression about the quality **(Figure 5.3 - 6)**.

Certified ppm-rates drop exponentially as a function of the C_p-index. A c_p-index of 1 permits a production of approx. 2700 reject units. This is usually realisable for the regarded quality feature of a well laid out process. A c_p-value of 1.5 is extremely demanding with a maximum of 6 reject units. For a value of 2 practically no reject becomes certified over the entire product life span. These circumstances clarify that the producing enterprises are nowadays exposed to an extreme pressure. Particularly the automobile industry determines the demands and causes huge expenditure for preventative measures as well as with regard to the monitoring.

Due to demands made in practice (it is not possible to ensure an ideal centring) processes are only regarded as capable when starting from a potential capability index above $c_p=1.33$ **(Figure 5.3 - 7)**. In addition, it is called a controlled process if the actual capability index c_{pk} is a comparable value. The process is almost ideally centred in this case. The same subdividing is done with incapable processes.

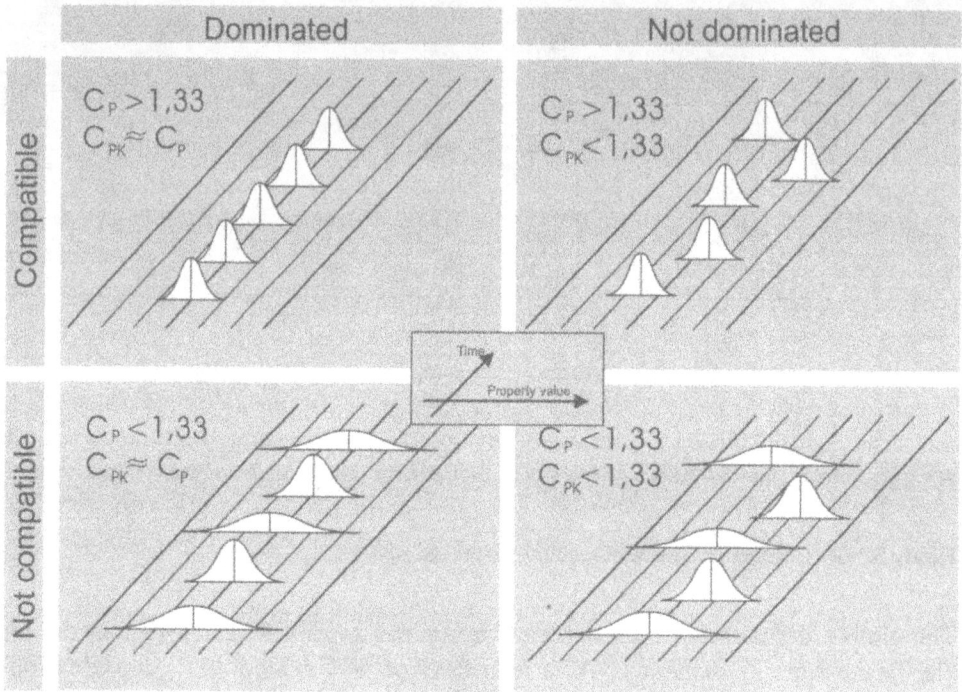

Figure 5.3-7: Capable and controlled Processes

Further Reading

[Ang 92]	Anghel, C.; Hausberger, H.; Streintz, W.: Unsymmetriegrößen erster und zweiter Art richtig auswerten. In: Qualität und Zuverlässigkeit 37 (1992) 12, S. 755-758 und 38 (1993) 1, S. 37-40
[Bam 96]	Bamberg, G.; Baur, F.: Statistik. München, Wien: R. Oldenbourg Verlag, 1996
[Die 95]	Dietrich, E.; Schulze, A.: Statistische Verfahren zur Maschinen- und Prozeßqualifikation. München, Wien: Carl Hanser Verlag, 1995
[Die 98]	Dietrich, E.; Schulze, A.: Richtlinien zur Beurteilung von Meßsystemen und Prozessen, Abnahme von Fertigungseinrichtungen, Carl Hanser Verlag, September 1998.
[Dut 84]	Dutschke, W.; Illig, W.: Statistische Auswertemethoden. In: Warnecke, H.-J.; Fertigungsmeßtechnik. Berlin, Heidelberg, New York, Tokyo: Springer Verlag, 1984
[Ehr 86]	Ehrenberg, A.: Statistik oder der Umgang mit Daten (Titel der Orginalausgabe: A Primer in Data Reduction). Weinheim: VCH Verlagsgesellschaft, 1986

[Gim 91]   Gimpel, G.: Qualitätsgerechte Optimierung von Fertigungsprozessen. Dissertation RWTH Aachen, Düsseldorf: VDI-Verlag, 1991

[Har 95]   Hartung, J.; Elpelt, B.; Klösener, K.-H.: Statistik, Lehr- und Handbuch der angewandten Statistik. München, Wien: R. Oldenbourg Verlag, 1995

[Her 96]   Hering, E.; Triemel, J.; Blank, H. P.: Qualitätsmanagement für Ingenieure. Düsseldorf: VDI-Verlag, 1996

[Kai 99]   Kaiser, B.; Nowack, H.: Nur scheinbar instabil. QZ 44 (1999) 6, S. 761-765.

[Kir 94]   Kirschling, G.: Qualitätsregelkarten. In: Masing, W.: Handbuch Qualitätsmanagement. München, Wien: Carl Hanser-Verlag, 1994

[Mit 93]   Mittag, H.-J.: Qualitätsregelkarten. München, Wien: Carl Hanser Verlag, 1993

[Pf 96]   Pfeifer, T.: Qualitätsmanagement. München, Wien: Carl Hanser Verlag, 1996

[Rin 95]   Rinne, H.; Mittag, H.-J.: Statistische Methoden der Qualitätssicherung. München, Wien: Carl Hanser Verlag, 1995

[Slö 94]   Schlötel, E.: Auswertungsverfahren. In: Masing W.: Handbuch Qualitätsmanagement. München, Wien: Carl Hanser Verlag, 1994

[Sta 70]   Stange, K.: Angewandte Statistik Teil 1 Eindimensionale Probleme. Berlin, Heidelberg, New York: Springer Verlag, 1970

Norms and Regulations

DGQ 172b   DGQ 172b: Anleitung und Beispiele zum Auswerteblatt für (annähernd) normal verteilte Werte. Berlin: Beuth Verlag, 1976

DIN 53804   DIN 53804: Statistische Auswertungen (Teil 1, 2, 3, 13). Köln, Berlin: Beuth Verlag, 1981 bis 1990

DIN 55350   DIN 55350: Begriffe zu Qualitätsmanagement und Statistik (Teil 11, 33). Köln, Berlin: Beuth Verlag, 1993 bis 1995

ISO 2859   ISO 2859: Annahmestichprobenprüfung anhand der Anzahl fehlerhafter Einheiten oder Fehler / Attributprüfung (Teil 0, 1, 2). Köln, Berlin: Beuth Verlag, 1989 bis 1993

ISO 5479   ISO 5479: Tests auf Normalverteilung. Köln, Berlin: Beuth Verlag, 1982

QS-9000   Chrysler Corp.; Ford Motor Comp.; General Motors Corp.: Quality System Requirements QS-9000, 1994

VDA 4 Teil 1   VDA 4 Teil 1: Sicherung der Qualität vor Serienansatz - Partnerschaftliche Zusammenarbeit, Abläufe, Methoden. Frankfurt: Verband der Automobilindustrie e.V. (VDA), 1996

VDMA 8669   VDMA - Verband Deutscher Maschinen- und Anlagenbau e.V. - VDMA 8669: Fähigkeitsuntersuchung zur Abnahme spanender Werkzeugmaschinen. Beuth-Verlag GmbH, Berlin, 1995.

6 Management of Inspection Equipment

In recent years, many companies have been certified for their quality capability according to guidelines and norms on the other hand. This has occurred against a background that only high quality products and processes can secure and strengthen the market position of a company in the long-term. On the other hand, many customers now ask their suppliers for a certificate of assured quality management [Die 97], [Gei 97], [Rin 95].

To ensure the economical production of high quality products, the quality of the production process must be monitored and continuously optimised. The starting point for these actions is the acquisition of quality features with inspection equipment.

According to DIN EN ISO 9000ff, inspection equipment monitoring lies within the "quality element inspection equipment" part of the company's quality management system. QS-9000 requires for the testing of measurement systems: "Adequate statistical tests for the assessment of measurement systems and inspection equipment must be carried out".

Furthermore, due to the globalisation of markets, it is necessary that test and measurement results can be compared within a company as well as world-wide. The term "traceability" (traceable to national and international standards) gains further meaning within this context [Tra 96].

The central task of inspection equipment management is to ensure the accuracy, reliability and availability of all measuring and inspection equipment of a company at any point in time [Dut 96], [DIN ISO 10012-1]. For this, inspection equipment is seen as a reference, which measures the quality of the products. To find the actual parameter value of a workpiece with regard to economic aspects, the faultless state of the inspection equipment used at the time of the test must be guaranteed [Pf 96]. The necessity of inspection equipment monitoring is given by the situation that faulty inspection equipment or inspection processes, with only vague empirical information, will lead to the wrong decisions when assessing the actual product or production process quality.

With regard to product quality, on the one hand, a positive test result can be given although a rejected part has been tested. On the other hand there is the danger for the manufacturer to declare good parts as rejects or carry out unnecessary adjustments [Pf 96], [Rin 95]. The wrong decisions, which were taken on this basis, have been caused by faulty quality data and its consolidated feature values and will consequently lead to higher production costs.

As the quality of products is directly influenced by the quality of the production processes, it is important to monitor these processes and adjust them when necessary [Pf 97], [Die 95], [Rin 95] (section 5.2). In order to be able to draw conclusions as to the quality of the production process, the suitability of the used measuring and inspection equipment for this test must be assured. Here, the measured values are only observed values of a quality feature, which has been produced in a production process. The produced quality characteristic is covered with the systematic and statistical influences of a measurement process during the measurement. It is therefore necessary to recognise and quantify systematic and random influences on the test under actual test conditions [Pf 96], [Bos 95].

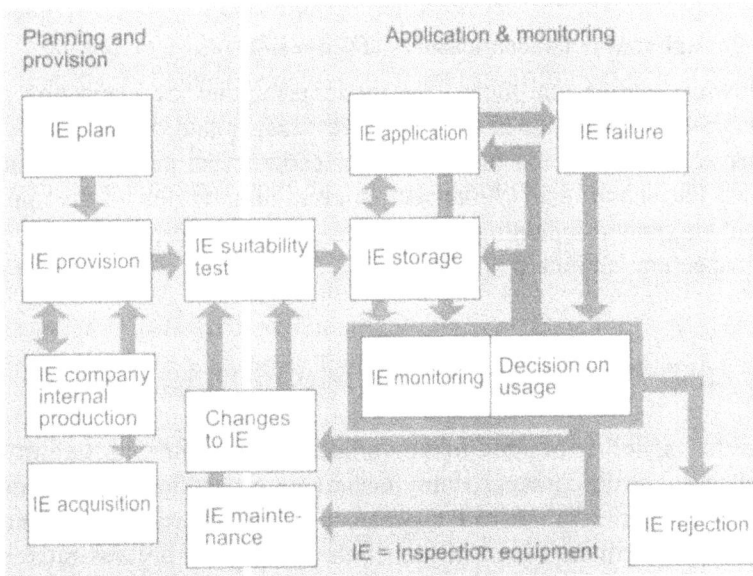

Figure 6-1: Integration of inspection equipment monitoring into inspection equipment management

The tasks which are part of inspection equipment management can be divided into three areas:

- Inspection equipment monitoring
- Inspection equipment planning and acquisition
- Inspection equipment management

Inspection equipment planning and acquisition includes planning the usage, characteristics, requirements, specifications and the field of application of the inspection equipment as part of production planning and acquisition or manufacturing [Pf 96]. The suitability test after acquisition checks whether the equipment meets all requirements (contract specifications, drawings, norms, regulations). When the equipment is cleared for use, its data is stored and the equipment is put in storage. While in use in the production process or stored, the equipment is monitored in pre-defined intervals. Unobjectionable inspection equipment is cleared for further usage. Objectionable inspection equipment or that which has failed during an application, is tested in a so-called "usage decision". This decides if inspection equipment can be

- used for certain tasks,
- used for other or similar test tasks in a changed form, or

repaired through maintenance measures (**Figure 6-1**).

The following section introduces the basic principles of inspection equipment monitoring and the techniques used, their advantages and disadvantages. It is limited to inspection equipment monitoring for geometrical measurement and testing techniques. The second part of the section gives an overview of the application of inspection equipment planning, acquisition and management within company-internal inspection equipment management.

6.1   Inspection Equipment Monitoring

To assess the capability of a production process, its position and stability and to be able to intervene in the process if any deviations occur, the accuracy and stability of the measuring process must be guaranteed. Consequentially, the characteristics of inspection equipment which influence the measuring process must be known. The task of inspection equipment monitoring is to test and guarantee the changing characteristics in periodic intervals. The characteristics of inspection equipment can be described as follows [Die 95], [Bos 95], (**Figure 6.1-1**).

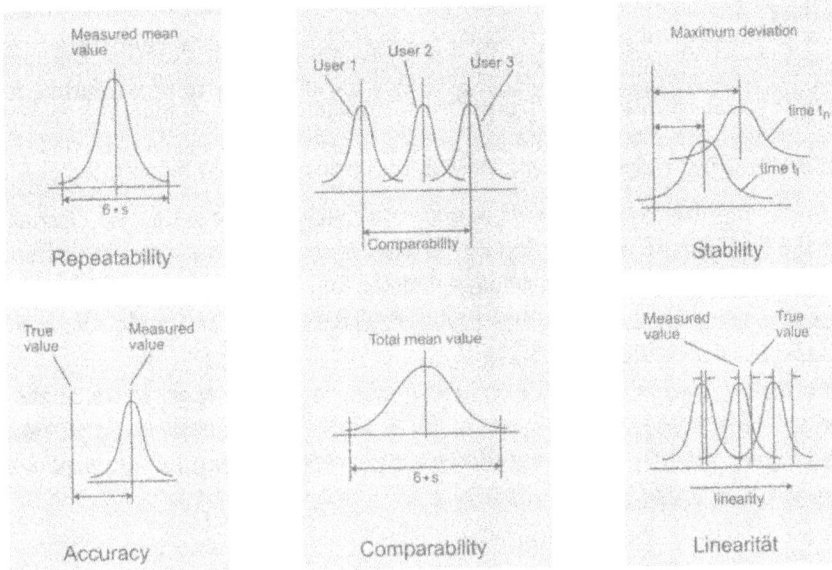

Figure 6.1-1:   Characteristics of inspection equipment

Accuracy
The deviation between a mean value of a measuring series for repeated measurements of the same feature and the actual value of the feature is called accuracy.

Repeatability
Repeatability describes how precisely the measured value is repeated. A dimension for the repeatability is the standard deviation or the span width of a measuring series.

Comprensibility
In order to quantify the influence of an external condition, such as operator, test site or applied inspection equipment, comprensibility can be used.

Stability
The behaviour in time of the inspection equipment is called stability. For this, measuring series are carried out in defined intervals and the differences of the statistical feature values are compared.

Linearity

The characteristic that a line with a known gradient can describe the measuring deviation with increased measuring values as an initial approximation is called linearity.

When monitoring inspection equipment, two principles can be adopted:

- *Monitoring with regard to the inspection equipment* looks at the characteristics of the equipment without regard for its capabilities to solve a specific inspection task. These tests are usually carried out in accordance with equipment-specific norms and regulations under ideal conditions and under the supervision of trained staff (Section 6.1.2).

- In contrast, *monitoring with regard to the inspection task*, looks at the usefulness of the inspection equipment for a single specifically defined monitoring task under specific testing conditions, e.g. within the production area with varying temperatures (Section 6.1.3.).

6.1.1 Traceability

Tracing back and monitoring of measuring equipment (general inspection equipment) is done by comparison with a standard, which represents the value of the dimension, which is assumed to be correct and linked by an uninterrupted chain of such reference standards to the national standard . Reference and working standards are such standards. National standards to describe SI-units are provided in the Federal Republic of Germany by the Physikalisch-Technische-Bundesanstalt PTB, which also enables the link to the international standards [Pf 96], [DKD 92].

Calibration Chain

As reference standards for inspection equipment calibration are not often directly linked to the PTB, so-called calibration chains often have one or more intermediate steps built in to trace back to the measuring equipment. It must be taken into account that measuring equipment and standards have an uncertainty attached to them (**Figure 6.1-2**). When calibrating a vernier calliper the shown measurement value is compared with the known value of the gauge block [VDI/VDE 2617]. All standards which are used for the calibration or monitoring of inspection equipment must be traced back when they are being used. The uncertainty of a dimension of the working standards used (alignment ring and gauge block) lies at 1/100 of the uncertainty of the traced back inspection equipment (vernier calliper). The working standards used in the company can be calibrated with a suitable test process either internally or at a DKD-calibration laboratory. The inspection equipment is traced to the national standard by an uninterrupted chain of calibration processes.

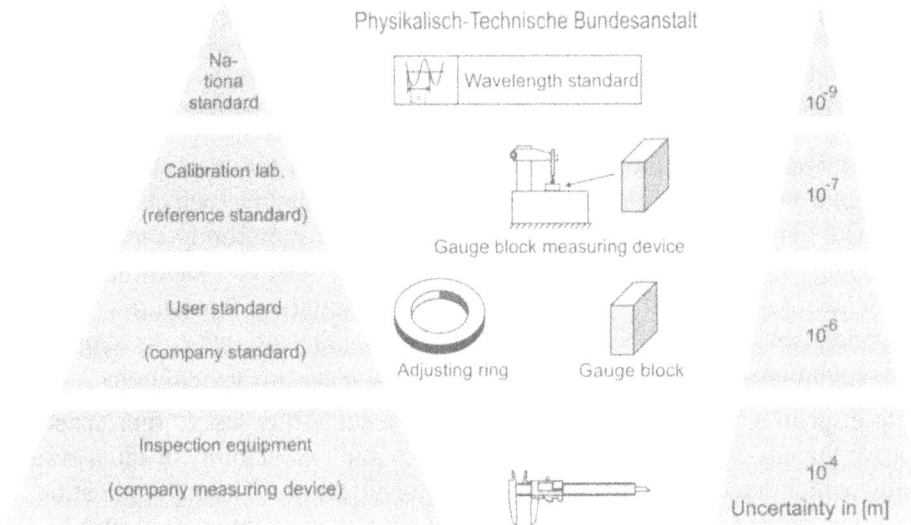

Physikalisch-Technische Bundesanstalt

National standard

Wavelength standard

10^{-9}

Calibration lab. (reference standard)

Gauge block measuring device

10^{-7}

User standard (company standard)

Adjusting ring

Gauge block

10^{-6}

Inspection equipment (company measuring device)

10^{-4}

Uncertainty in [m]

Figure 6.1-2: Feedback of a vernier calliper

Calibration Service

The calibration chain in Germany is completed by the Deutscher Kalibrierdienst DKD, the industry or other institutions (research institutions, TÜV), which offer calibration as a service. It is the DKD's task to secure the linkage of the measurement and inspection institutions for industrial metrology to national standards or standard measurement applications [DKD 92].

In recent years, a new service area has been established in the Federal Republic of Germany, which offers calibration services for inspection equipment for small and medium sized companies or for special inspection equipment. These are measuring laboratories, which have been established especially for measurement and calibration tasks. The manufacturers of inspection and measuring equipment have also fulfilled these monitoring tasks. The service providers also offer inspection equipment management and the automatic notification of monitoring dates for inspection equipment. The calibration providers of the DKD are certified by the PTB and in a PTB newsletter information is publicised which measurands and processes have been certified [DKD 92]. The DKD calibration certificates handed out by accredited calibration laboratories are proof of the traceability of calibrated measuring equipment and dimensions to national standards, as required by DIN EN ISO 9000ff and DIN EN 45 001.

6.1.2 Inspection Equipment-Specific Monitoring

Equipment-specific monitoring describes the testing of the inspection equipment with regard to its device characteristics. Inspection equipment which is often used in industry is specially regulated with regard to its monitoring process. Worth mentioning are those regulations with regard to manual measuring devices [VDI/VDE/DGQ 2618] and coordinate measuring machines (CMM) [VDI/VDE 2617], [DIN EN ISO 10360-2]. The monitoring, which is subjected to these regulations and norms, is usually carried out under ideal conditions.

These inspections are often carried out during a suitability test or final inspection to prove the characteristics, especially accuracy and repeatability of the measuring results, which have been guaranteed by the manufacturer. During application, the inspection equipment is monitored periodically to guarantee its usability. The manufacturer, or regulations and norms define the conditions which have to be maintained, e.g. temperature.

Manual Inspection Equipment

To monitor inspection equipment which is regularly used in a company, a joint committee of the VDI/VDE Gesellschaft Meß- und Automatisierungstechnik (GMA) and of the Deutschen Gesellschaft für Qualität e.V. (DGQ) have written the regulation VDI/VDE/DGQ 2618, "Testing instructions for inspection equipment monitoring ". This regulation consists of 27 pages and has an introduction as well as inspection instructions. This facilitates a standardised assessment of new or used inspection equipment. The aim of this regulation is to give manufacturers and users of inspection equipment a joint basis for their monitoring efforts.

The testing instructions can be used as a test and work schedule for the initial inspection and for the periodic monitoring of the inspection equipment in use. The content and structure is based on the requirements of the company's procedures. The test instructions are recommendations, which can be varied within a company or even between the customer and manufacturer. Usually, the test instructions for the monitoring of inspection equipment describe the following points (**Figure 6.1-3**):

- preparatory work steps,
- features to be tested and permissible deviations,
- usable working or inspection equipment for the monitoring task

- as well as evaluation, decision and documentation criteria.

The regulations include testing instructions for gauging and measuring inspection equipment as well as material measures.

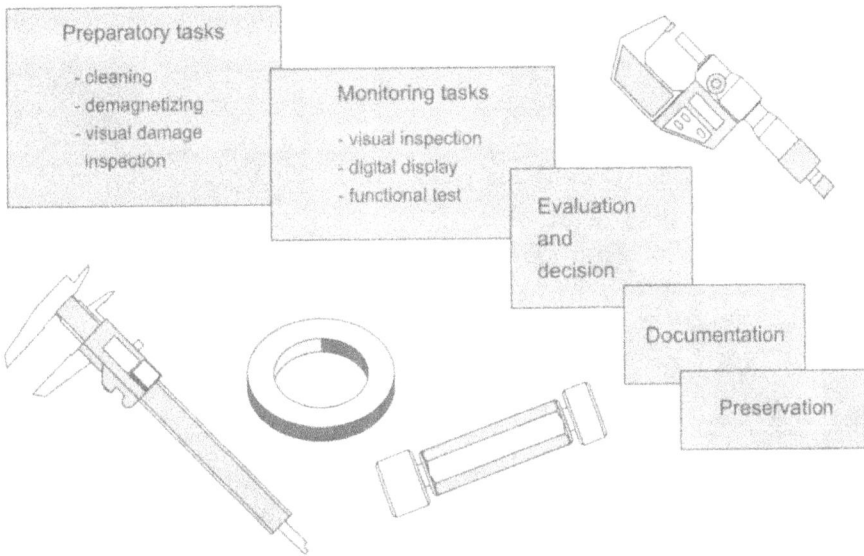

Figure 6.1-3: Principal course of inspection equipment monitoring according to VDI/VDE/DGQ 2618

A complete list of all available test instructions can be found on page 1 of the regulation VDI/VDE/DGQ 2618:

- Gauging inspection equipment   e.g. plug gauges, snap gauges, cone gauges
- Measuring inspection equipment   e.g. vernier callipers, dial gauges, micrometer gauges
- Material measures   e.g. gauge blocks, alignment rings

Permissible deviations, which are indicated in the regulations, are orientated toward norms, norm drafts and scientific publications of defined details for commercial inspection equipment. The permissible deviations quoted in the norms relate to new inspection equipment, i.e. the values, which are still to be determined have to take the wear and tear of the equipment into consideration. Not all equipment has the relevant specifications.

Normally, the monitoring process is less detailed than the entry or final inspection, which tests the characteristics or the guaranteed features of inspection equipment.

Coordinate Measuring Machines

To monitor coordinate measuring machines, regulation DIN EN ISO 10360-2 was published, which is based on regulation VDI/VDE 2617. They provide testing instructions to the operator, which not only helps to monitor characteristics of the machine but also facilitates the inspection. The regulations define feature values, which allow a quantitative comparison of various coordinate measuring machines.

Figure 6.1-4:   Ball plate in a universal coordinate measuring machine

It is of special interest, which measurement uncertainty must be taken into account when carrying out measurements. The value of the measurement uncertainty, usually called u, is dependent on the measuring task and is described as a length-dependent measurement uncertainty:

$$u_i = A_i + K_i \cdot L < B_i \tag{6.1-1}$$

A_i, B_i and K_i are constant specific measuring tasks, L is the measured length. The index i can take on the values $i = 1,2,3$ and shows if the length measurement is one, two or three-dimensional. The measurement deviations, which occur when measuring with coordinate measuring machines, can be defined with various measurement processes [Pf 92], [Tra 96], [Neu 93], [DIN EN ISO 10360-2]. Processes and test pieces for this are e.g. laser interferometers as well as standards, which have been calibrated with low measurement uncertainty, such as a ball plate or levelled gauge block (**Figure 6.1-4**).

A ball plate has a geometrically long-term stable base, to which highly precise contact elements (ceramic spheres) are attached in defined distances. The positions of the ceramic spheres to one another are calibrated by a governmental calibration service and are the reference values for the tests, which will be carried out with the ball plate. The measurement values shown on the coordinate measuring machine during monitoring must not exceed the calibrated values by more than the values specified by the manufacturer. The measured deviations can be shown with the aid of graphs for length measurement uncertainty.

6.1.3 Inspection Task-Specific Monitoring

Due to the varied influence parameters on the monitoring process, monitoring with regard to inspection equipment shows some deficits. These are mainly due to test conditions being different from actual operating conditions. Influences which are particularly critical are e.g. the testing staff, the temperature of the workpiece or of the equipment as well as dirt on the workpiece.

Furthermore, the inspection task is often not directly comparable with those inspection tasks during inspection equipment monitoring. This is especially true for coordinate metrology. The monitoring processes, which are based on the above-mentioned regulations, only prove the capability of a coordinate measuring machine to measure lengths (distances) within a space. It cannot form a predication about the accuracy of complex measuring tasks e.g. during gear manufacturing.

Apart from accuracy, which is described quantitatively through the measurement uncertainty, further characteristics of the inspection equipment , e.g. repeat precision and the comparability, must be described and monitored under actual conditions [Die 91], [Ang 97]. A differentiation must be made analogue to an SPC between the testing of inspection equipment, which decides on the principal suitability of a measuring procedure and the continuous monitoring in the processing environment. Monitoring with regard to the inspection task is at present regulated by company-internal procedures of larger companies in the automotive or automotive supply industry or by quality management companies [Die 95], [Bos 95], [For 91], [Die 97].

Capability Indices
As the test can be seen as a process, it invites the establishment of test process capability (long-term stability of variance and position of the test process) analogue to the definition of the capability of a production process. Therefore, the capability indices c_g and c_{gk} are calculated periodically and documented in a control chart.

The definition of the capability indices c_g and c_{gk} varies depending on the user [For 91], [Bos 95]. The tolerance width $OT\text{-}UT$ of a feature or the process variance $s_{process}$ with the variance of the measuring device $s_{measuring\ device}$ are put into relation with one another and valued with a factor [Pf 96], [Die 95], [Die 91]. To reach a measured value, a calibrated feature of a master workpiece or alignment standard is measured 50 times with a measuring device (**Figure 6.1-5**).

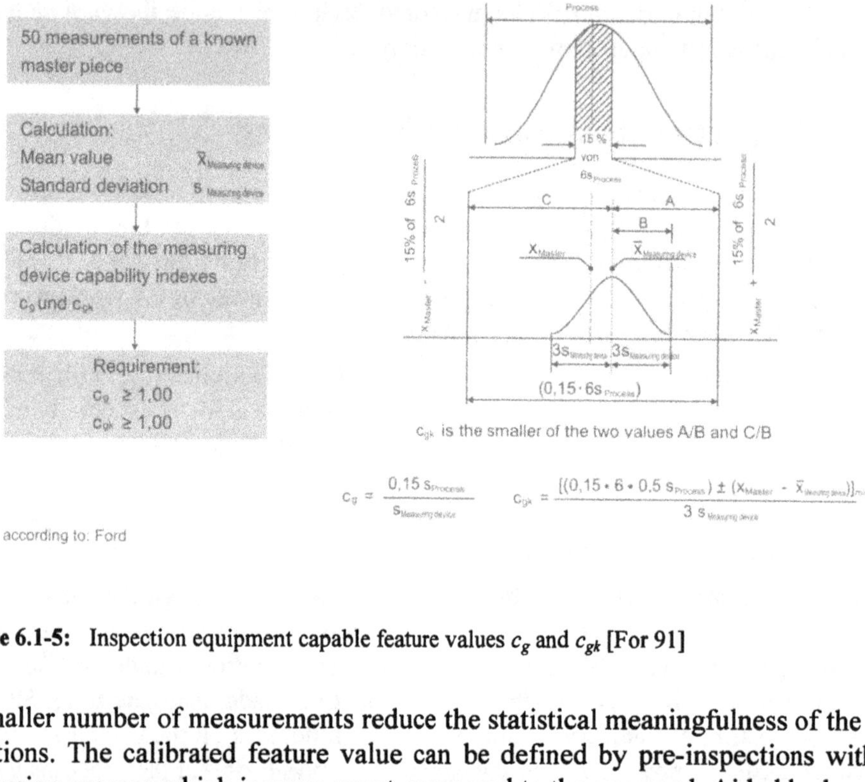

Figure 6.1-5: Inspection equipment capable feature values c_g and c_{gk} [For 91]

A smaller number of measurements reduce the statistical meaningfulness of the inspections. The calibrated feature value can be defined by pre-inspections with a measuring process which is more exact compared to the one used. Aided by known formulas, the arithmetical mean value $\bar{x}_{measuring\ device}$ and the standard deviation $s_{measuring\ device}$ are gained after carrying out the series of measurement s. To calculate the inspection equipment capability indices c_g and c_{gk}, the standard deviation of the production process $s_{process}$ is needed in addition to the nominal values $\bar{x}_{measuring\ device}$ and $s_{measuring\ device}$ [For 91]. Instead of a standard deviation of a production process, the tolerance band width of a feature can be used [Bos 95]. The difference of the two similar procedures lies in the fact that the first process relates to the production process and the second one to the feature tolerance, without taking the production process into account.

Depending on the company (or regulation), different factors are used for the multiplication with the relation of tolerance field width $T = OT\text{-}UT$ (or standard deviation $s_{process}$) and test mean variance $s_{measuring\ device}$ to find the capability index c_g. Furthermore, companies place different minimum requirements on capability indices, e.g. c_g or $c_{gk} \geq 1.0$ or 1.33.

R&R Study

Apart from studies of the inspection equipment capability indices c_g and c_{gk}, there has been another study in recent years to establish the repeatability and reproducibility of a measurement process, the *R&R study* [For 91], [Bos 95], (**Figure 6.1-6**). This procedure makes it feasible to assess how well a measuring method finds differences between products [Gim 94], [Whe 86], [For 91]. For the R&R study, 3 operators test 10 parts with 3 repetitions under actual conditions. A reduction of the test to 5 parts, 3 testing staff and 2 repetitions is permissible in exceptional cases.

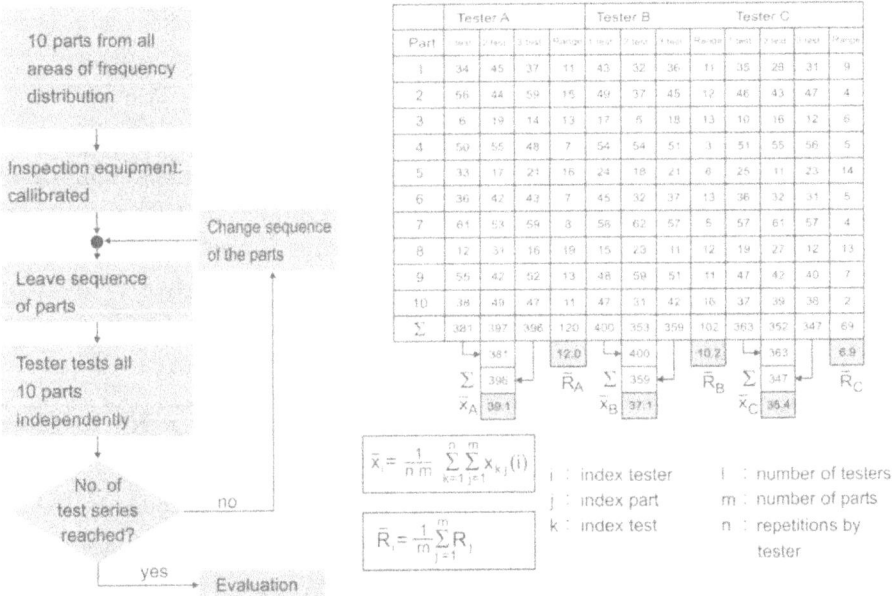

	Tester A				Tester B				Tester C			
Part	1 test	2 test	3 test	Range	1 test	2 test	3 test	Range	1 test	2 test	3 test	Range
1	34	45	37	11	43	32	36	11	35	28	31	9
2	56	44	59	15	49	37	45	12	46	43	47	4
3	6	19	14	13	17	5	18	13	10	16	12	6
4	50	55	48	7	54	54	51	3	51	55	56	5
5	33	17	21	16	24	18	21	6	25	11	23	14
6	36	42	43	7	45	32	37	13	36	32	31	5
7	61	53	59	8	58	62	57	5	57	61	57	4
8	12	31	16	19	15	23	11	12	19	27	12	13
9	55	42	52	13	48	59	51	11	47	43	40	7
10	38	49	47	11	47	31	42	16	37	39	38	2
Σ	381	397	396	120	400	353	359	102	363	352	347	64

$$\overline{58} \qquad \overline{12.0} \qquad \overline{400} \qquad \overline{10.2} \qquad \overline{363} \qquad \overline{6.9}$$

$$\Sigma\ 398 \rightarrow \overline{R}_A \qquad \Sigma\ 359 \rightarrow \overline{R}_B \qquad \Sigma\ 347 \rightarrow \overline{R}_C$$
$$\overline{x}_A\ 39.1 \qquad \overline{x}_B\ 37.1 \qquad \overline{x}_C\ 35.4$$

Flowchart (left column):

- 10 parts from all areas of frequency distribution
- Inspection equipment: calibrated
- Leave sequence of parts
- Tester tests all 10 parts independently
- No. of test series reached? — no → Change sequence of the parts
- yes → Evaluation

$$\overline{x} = \frac{1}{n\,m} \sum_{k=1}^{n} \sum_{j=1}^{m} x_{k,j}(i)$$

$$\overline{R} = \frac{1}{m} \sum_{j=1}^{m} R_j$$

i : index tester   l : number of testers
j : index part   m : number of parts
k : index test   n : repetitions by tester

©98

Figure 6.1-6: R&R study

To show the influence of the testing staff on the measuring result, all tests are carried out with the same measuring equipment. Variable influences can be due to different testing facilities or the usage of inspection equipment at a different position. It must be guaranteed that only one influence is changed and all others stay the same [Die 97].

To explain the R&R study (**Figure 6.1-6**), a wide variance range of the measured values was chosen. The results shown by the measuring device the result read off by the testing staff is listed in a table. After the first and second trial series, the sequence of the parts is changed. The separate results, derived from the measurement series, are compiled to a mean span $\overline{R}_i$ and a mean deviation $\overline{x}_i$ of the measurements of one tester:

$$R_j = x_{j,max} - x_{j,min} \tag{6.1-2}$$

$$\overline{R}_i = \frac{1}{m} \cdot \sum_{j=1}^{m} R_j \tag{6.1-3}$$

$$\overline{x}_i = \frac{1}{n \cdot m} \sum_{j=1}^{n} \sum_{k=1}^{m} x_{k,j} \tag{6.1-4}$$

These denote:

i	: index tester	l	: tester
j	: index part	m	: parts
k	: index trial	n	: repetitions per tester

By arithmetically averaging the mean span $\overline{R}_i$ and defining the maximum difference of the mean values $\overline{x}_i$ of the separate testers, a further consolidation of the test data is achieved. As the number of measuring series (3 repetitions) or of testers is limited, both tester-specific feature values must be corrected with an additional factor. This way repeatability and comparability (*RP* and *CP*) of the tested inspection process can be calculated with the following formula:

$$WP = K_1 \cdot \overline{\overline{R}} \qquad\qquad \text{with } \overline{\overline{R}} = \frac{1}{l} \cdot \sum_{i=1}^{l} \overline{R}_i \tag{6.1-5}$$

$$VP = K_2 \cdot x_{Diff} \qquad\qquad \text{with } x_{Diff} = \text{max. difference } \overline{x}_i \tag{6.1-6}$$

At a confidence level of 99.00% or 99.73%, the following correction factors K_1 and K_2 can be found in literature [Die 95], [Die 97] (**Table 6.1-1**)

Table 6.1-1   Correction factors for an R&R study

Confidence level	99.00 %	99.73 %
K_1 (number of repetitions)		
2	4.56	5.32

3		3.05	3.54
K$_2$	(number of testers)		
2		3.65	4.28
3		2.70	3.14

Repeatability RP incorporates the variance of the testing results through the random influences of the measuring processes and the tester, while comparison precision CP clarifies its influence on the position of the measurement result. The repeat precision RP and the comparison precision CP can be combined to the total scattering S_m:

$$S_m = \sqrt{WP^2 + VP^2} \qquad\qquad (6.1\text{-}7)$$

The total/overall variance of a measuring process can be put into proportion with six times the range of process scattering $6s_{process}$ or, if this is unknown with the tolerance band width T

$$S_{m\%} = S_m \cdot 100 \cdot T \qquad\qquad (6.1\text{-}8)$$

Assessment of the measurement process capability takes place in three classes:

$\quad$ 0 %$\quad$ < $S_{m\%}$ < 20 % usable

$\quad$ 20 %$\quad$ < $S_{m\%}$ < 30 %$\qquad$ conditionally usable

$\quad$ 30 %$\quad$ < $S_{m\%}$$\qquad\qquad$ not usable

Practical Example
With the procedures of the inspection task –specific monitoring-, the usability of a multi-point measuring device can be proven on a master workpiece (**Figure 6.1-7**). The multi-point measuring device can be used to simultaneously collect multiple diameters of a shaft with coupled inductive length measuring devices.

While the separate length measuring devices can be monitored in accordance with the regulation VDI/VDE/DGQ 2618 page 26 for device-specific inspection equipment monitoring, the compatibility of the separate measuring devices in the overall inspection advice is the most important matter. The suitability of the inspection device can be proven with the aid of capability indices c_g and c_{gk}. The master shaft, with its measurements having been taken on a more precise measuring device, is measured 50 times and the above described evaluation is carried out. This procedure can be carried out in the production environment. The difference between calibrated value and measured mean value can be used to correct the multi-point measuring device. This method is also appropriate for the suitability test as well as for periodic monitoring and the use of control charts for the multi-point measuring device.

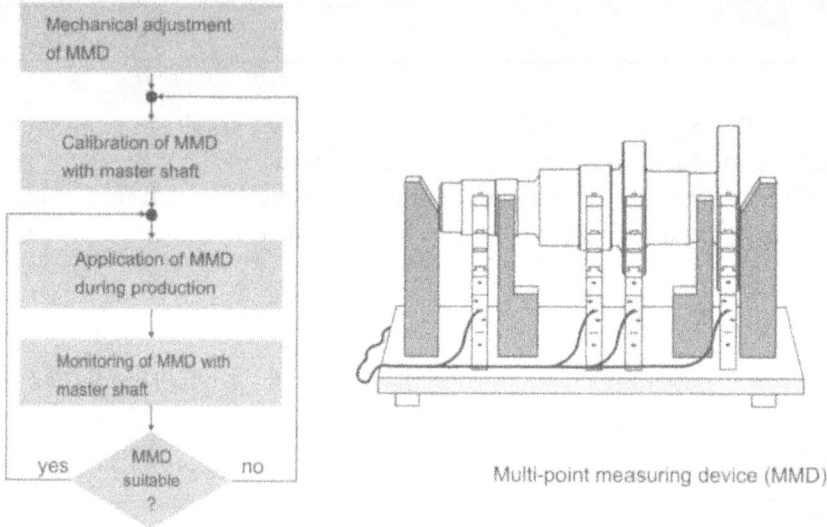

Figure 6.1-7:   Set-up and calibration of the multi-point measuring device

6.1.4 Dynamic Inspection Equipment Monitoring

Systems for the inspection equipment monitoring often work with static periods T_{stat} to define a monitoring time frame. If the inspection equipment in a company is individually identifiable, it can also be put under a dynamic monitoring strategy, which takes into account the method and period of their usage [Pf 96], [Dut 96]. The existing rationalisation potential is economically viable if the number of the inspection devices in a company is large and the logistical effort of the system in making the inspection equipment monitoring more dynamic is small. To reduce monitoring costs with dynamic inspection equipment monitoring, the application conditions of inspection equipment must be collected and evaluated.

The following application conditions can be used to calculate the next monitoring date:

- period of use and
- application environment of inspection equipment

The application environment of the inspection equipment is taken into account by a so-called environmental factor, which characterises a cost centre specific utilisation and can be altered when experience suggests doing so. Should no criteria exist for a suitable calibration interval, data must be collected with regard to the long-term stability of the inspection equipment. In such a case, the monitoring tests must be repeated in short time intervals.

Even though inspection equipment monitoring has been made more dynamic, the separate inspection equipment must be monitored at a pre-determined safety inspection date.

6.2   Inspection Equipment Planning and Provision

Various types of inspection equipment can be used for the quality tests, as they cover different testing functions. The central function of inspection equipment planning is to select the correct devices and procure the required testing devices at the right time (procure or produce). During the provision phase the suitability tests are carried out and test instructions for the periodic monitoring are drawn up (**Figure 6.2-1**):

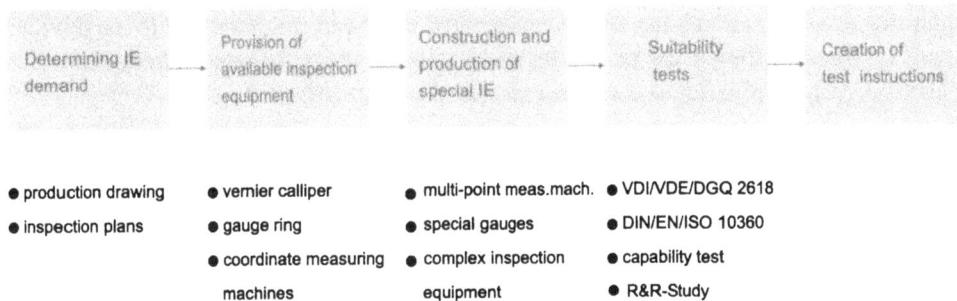

Determining IE demand	→	Provision of available inspection equipment	→	Construction and production of special IE	→	Suitability tests	→	Creation of test instructions
● production drawing ● inspection plans		● vernier calliper ● gauge ring ● coordinate measuring machines		● multi-point meas.mach. ● special gauges ● complex inspection equipment		● VDI/VDE/DGQ 2618 ● DIN/EN/ISO 10360 ● capability test ● R&R-Study		

Figure 6.2-1:   Inspection equipment planning and provision

Inspection Equipment Planning
The inspection equipment planning is aimed at finding the optimal inspection device for each inspection feature [Dut 96]. When evaluating the inspection equipment demand, all requirements for inspection equipment are collected in contract specifications or in a requirements list. The requirements are derived from production drawings or test plans, which include quality features and also from the surrounding conditions in which inspection equipment is used. As various test devices can often fulfil test functions, it is also necessary to look at the costs in addition to the measurement uncertainty. The testing costs should principally be only as high

as necessary for the test function at hand [Pf 96]. The costs incurred during the usage of a test device include:

- acquisition costs,
- running costs,
- maintenance costs, and
- repair costs.

Metrological as well as economic viewpoints will lead to a decision during the test-planning phase as to which test device is optimal for the inspection task. Due to the complexity of the boundary conditions when choosing a inspecting device, automatic selection with the aid of a CAQ-system is difficult [Dut 96]. Systems which provide suggestions for one or more test devices can be used in these circumstances. An expert will still make the manual decision as to which test device should be used.

Acquisition

If the required test devices are not in stock at a company, they can either be chosen from inspection device suppliers' catalogues or, if a special device is needed or if various quality features should be tested, special inspection devices are manufactured. The company must decide whether it can manufacture these devices itself or whether they must be produced by another manufacturer. This decision is based on technological and economical boundary conditions.

Supply

Once the suitability inspection has shown that a test device fulfils the characteristics guaranteed by the manufacturer or that it is suitable to test for a certain quality feature, it can be incorporated into the production process. The methods normally used for the inspection equipment or inspection function-related inspection equipment monitoring have already been described (Section 6.1).

In this phase, suitable testing instructions must be provided on a basis of national and international norms and company regulations.

6.3 Inspection Equipment Management

Inspection equipment management includes all administrative tasks which are required for the management of inspection equipment [Pf 96], [Dut 96].

When introducing systematic inspection equipment management, all on-hand equipment, which will be needed in the near future, is collected in a catalogue. All inspection equipment must be analysed, identified, described by its characteristic features and, if necessary, classified. To describe all relevant inspection equipment features, the required description criteria must be defined before the actual description is carried out. The description criteria known from literature must be completed by further aspects, which describe and sort inspection equipment. The first step, when building up an inspection equipment management system and developing a descriptive model of the test devices for a company, is the clear identification of the test devices. This is the only possibility for carrying out inspection equipment monitoring efficiently and in accordance with the company's requirements. Apart from the clear identification of a test device, the calibration status must be unambiguous at any point in time. For this, a suitable notification tag must show the next monitoring date (e.g. similar to an MOT sticker).

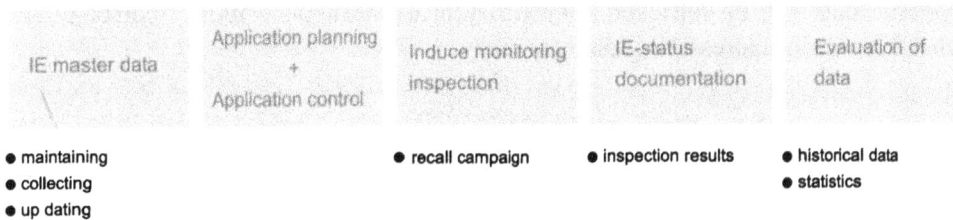

IE master data	Application planning + Application control	Induce monitoring inspection	IE-status documentation	Evaluation of data
● maintaining ● collecting ● up dating		● recall campaign	● inspection results	● historical data ● statistics

Figure 6.3-1: Tasks of inspection equipment management

The tasks of inspection equipment management can be listed as follows (**Figure 6.3-1**):

- Documentation of the master data

  After collecting inspection equipment-specific master data such as type of inspection equipment, supply source, acquisition cost, it is maintained and updated during the life span of inspection equipment.

- Logistical tasks

  The logistical tasks include utilisation scheduling and control of the inspection equipment within the company, i.e. planning when the equipment is going to be used; for which task; at which site. Particularly the initiation of monitoring inspections and calibration due dates, as well as the associated recall campaigns are part of the logistical tasks.

- Documentation of historical data

  To be able to prove the suitability of inspection equipment at a later stage it is necessary to document the results of inspection equipment monitoring. By evaluating the test results, such as feature values, using statistics and control charts, the suitability of inspection equipment can be assessed over the long term.

Computer-Supported Inspection Equipment Management

For most companies, the administrative effort of functional inspection equipment management is only justifiable with a suitable IT system. The advantages of computer-supported management result from the monotonous and repetitive tasks, such as management of databases, collection and saving of data, management and distribution tasks as well as date monitoring. Suitable software and database systems for these tasks and the evaluation of inspection equipment-specific information are already available on the market from various companies or are integrated into the CAQ system. Through suitable interfaces and data formats, inspection equipment-specific data can be retrieved from different departments within a company and also from various sites of a global company.

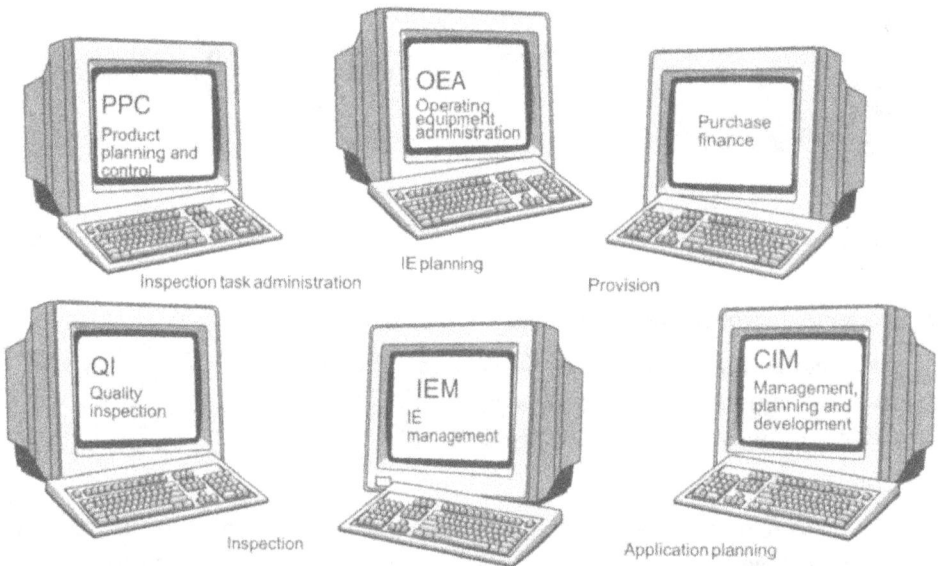

Figure 6.3-2:   Integration of computer-supported inspection equipment management

With the support of computers, the data management becomes more transparent and is more rationally realisable. Through selective functions, the aimed retrieval of information is possible. A database's functionality offers interesting possibilities as it allows the structured saving of data as well as focused data retrieval. Due to the IT-supported status control it is possible to check the current stock status at any time and, with appropriate networking, from any site in the company. In this way, a higher availability of the test devices and a more transparent data management keeping is reached and recall and reminder lists are easily and quickly drawn up.

Due to the automation of procedures and reduction of errors, a higher degree of reliability of inspection equipment management is reached as well as a reduction of costs due to higher efficiency from the optimum utilisation of the company's inspection equipment. Current stock lists of all in-house inspection equipment are easily and quickly produced with the aid of the system for monitoring inspection equipment and providing inspection equipment plans. Particularly when looking at the linking of other computer-supported components (networking, intranet, internet) of a company, the advantage of the computer-supported inspection equipment monitoring becomes clear.

An example of communication between different areas of a company is the production planning and control PPS, which exchanges information with the quality assurance system CAQ and the inspection equipment management PMV. Parallel to the production order, the CAQ-system draws up a test order. During the test, a calibrated test device is used and information about its current position and degree of usage is given to inspection equipment management system.

Data Structures
The data structure of a managed test device is dependent on the functions which must be fulfilled and which information is needed. In general, it should be orientated towards the requirements of a company and the task areas of inspection equipment management. A data model for inspection equipment management can be structured as follows:

- master data
- logistical data
- historical data

Master data includes, e.g. identifying data, which describes the inspection equipment with regard to organisational concerns. This includes name, type and identification of the inspection equipment. Furthermore, all information with regard to the usage of the inspection equipment is summarised. This includes geometrical sizes and information about the measuring range of a measuring device and also details about the preparation and usage cost per measurement with this inspection equipment. The classification of inspection equipment and applications can be shown here, as it provides a rational inspection equipment choice. Furthermore, the required information includes technical data with regard to the monitoring pro-

process and the description of the tested features as well as their tolerance limits or reference to the respective test specifications and regulations.

Logistical or movement data only incorporates current and short-lived data concerning specific devices. It provides information about the current position, the responsible tester and the degree of utilisation of the inspection equipment. This data permits optimal planning and efficient usage of on-hand inspection equipment. Dynamic inspection equipment monitoring is made possible by the logistical data, by summarising the level of utilisation in accordance with given algorithms and when reaching a defined overall usage a test or a re-calibration is ordered.

The historical data includes all data, which must be documented over a longer period of time, such as life cycle of inspection equipment. Parts of the logistical data are saved over a long period. This can be used at a later point in time to check which inspection equipment was used at which site and at what point in time. Apart from the lasting documentation of logistical data, results of monitoring tests are also documented and statistically evaluated.

Further reading

[Ang 97]	Anghel, C.: Meßgerätefähigkeit im Prozeß, in : Qualität und Zuverlässigkeit 42 (1997) 4, Carl Hanser Verlag, München
[Bos 95]	N.N.: Robert Bosch GmbH: Schriftenreihe Qualitätssicherung in der Bosch-Gruppe Nr. 10, Technische Statistik, Fähigkeit von Meßeinrichtungen, Stuttgart 1995
[Die 91]	Dietrich, E.; Schlosser, D.; Schulze, A.: Fähige Meßverfahren - Die Basis der statistischen Prozeßlenkung, in : Qualität und Zuverlässigkeit 36 (1991) 3 Carl Hanser Verlag, München
[Die 95]	Dietrich, E.; Schulze, A.: Statistische Verfahren zur Maschinen und Prozeßqualifikation, Carl Hanser Verlag, München 1995
[Die 97]	Dietrich, E.; Schulze, A.: Die Beurteilung von Meßsystemen in : Qualität und Zuverlässigkeit 42 (1997) 2, Carl Hanser Verlag, München
[DKD 92]	N.N.: Deutscher Kalibrierdienst, Ziele, rechtliche Grundlagen Akkreditierungsverfahren und -kriterien, Organisationsstruktur und Publikationen des Deutschen Kalibrierdienst (DKD), Braunschweig: PTB 1992
[Dut 96]	Dutschke, W.: Fertigungsmeßtechnik, 3., vollst. überarb. und erw. Aufl., Teubner Verlag, Stuttgart 1996
[For 91]	N.N.: Ford Motor Co.: EU 1880A und B. Fähigkeit von Meßsystemen und Meßmitteln, Köln 1991
[Gei 97]	Geiger, W.: Was bringt die neue ISO 9000-Familie?, in: Qualität und Zuverlässigkeit 42 (1997) 8, Carl Hanser Verlag, München
[Gim 94]	Gimpel, B.: Analyse von Meßprozessen, Arbeitsunterlagen der GfQS-Gesellschaft für Qualitätssicherung mbH, Aachen 1994

[Neu 93]	Neumann, H.J. (Hrsg.): Koordinatenmeßtechnik, (Kontakt und Studium, Bd. 426) Expert-Verlag Ehningen 1993
[Pf 92]	Pfeifer, T. (Hrsg.): Koordinatenmeßtechnik für die Qualitätssicherung VDI Verlag 1992
[Pf 96]	Pfeifer, T.: Qualitätsmanagement: Strategien, Methoden, Techniken, 2. vollst. überarb. und erw. Aufl., Carl Hanser Verlag, München 1996
[Pf 97]	Pfeifer, T.: Ohne Meßtechnik geht nichts, in : Qualität und Zuverlässigkeit 42 (1997) 9, Carl Hanser Verlag, München
[Rin 95]	Rinne H.; Mittag H.-J.: Statistische Methoden der Qualitätssicherung, 3. überarbeitete Auflage, Carl Hanser Verlag, München 1995
[Tra 96]	Trapet,E.; Wäldele,F.: Rückführbarkeit der Meßergebnisse von Koordinatenmeßgeräten, VDI-Berichte Nr. 1258, VDI Verlag GmbH, Düsseldorf 1996
[Whe 86]	Wheeler, D.J; Chambers, D.S.: Understanding Statistical Process Control, Statistical Process Control, Inc.Knoxville, Tennessee 1986

Norms and guidelines

DIN EN ISO 9000ff	N.N.: DIN EN ISO 9000ff: Normen zum Qualitätsmanagement und zur Qualitätssicherung/QM-Darlegung, Beuth Verlag, Berlin 1992 bis 1994
DIN EN ISO 10360-2	N.N.: DIN EN ISO 10360-2: Coordinate metrology Part 2 Performance assessment of coordinate measuring machines, Reference number ISO 10360-2 :1994
DIN EN 45001	N.N.: DIN EN 45001 ff: Normen zum Betreiben, Beurteilen und Akkreditieren von Prüflaboratorien, Beuth Verlag, Berlin 1990 bis 1995
DIN ISO 10012-1	N.N.: DIN ISO 10012-1: Forderungen an die Qualitätssicherung für Meßmittel - Bestätigungssystem für Meßmittel, Beuth Verlag, Berlin 1992
QS-9000	N.N.: QS-9000 Chrysler Corp.; Ford Motor Comp.; General Motors Corp.: Quality System Requirements QS-9000, 1994
VDI/VDE/DGQ 2618	N.N.: VDI/VDE/DGQ 2618, Blatt 1-27, Prüfanweisungen zur Prüfmittelüberwachung, Beuth Verlag, Berlin 1991
VDI/VDE 2617	N.N.: VDI/VDE 2617 Blatt 1-6: Genauigkeit von Koordinatenmeßgeräten - Kenngrößen und deren Prüfung, Beuth-Verlag, Berlin und

7 Index

www.ingramcontent.com/pod-product-compliance
Lightning Source LLC
Chambersburg PA
CBHW081525190326
41458CB00015B/5456